Advances in Intelligent and Soft Computing

161

Editor-in-Chief

Prof. Janusz Kacprzyk
Systems Research Institute
Polish Academy of Sciences
ul. Newelska 6
01-447 Warsaw
Poland
E-mail: kacprzyk@ibspan.waw.pl

T0143047

For further volumes:
http://www.springer.com/series/4240

Jia Luo (Ed.)

Soft Computing in Information Communication Technology

Volume 2

 Springer

Editor
Jia Luo
National Kinmen Institute of Technology
University Road
Jinning ownship
Kinmen
Taiwan

ISSN 1867-5662   e-ISSN 1867-5670
ISBN 978-3-642-29451-8   e-ISBN 978-3-642-29452-5
DOI 10.1007/978-3-642-29452-5
Springer Heidelberg New York Dordrecht London

Library of Congress Control Number: 2012934955

Printed on acid-free paper

Springer is part of Springer Science+Business Media (www.springer.com)

Preface

The main theme of SCICT2012 is set on soft computing in information communication technology, which covers most aspect of computing, information and technology. 2012 International Conference on Soft Computing in Information Communication Technology is to be held in Hong Kong, April 17–18.

Soft computing is a term applied to a field within computer science which is characterized by the use of inexact solutions to computationally-hard tasks such as the solution of NP-complete problems, for which an exact solution cannot be derived in polynomial time.

Soft Computing became a formal Computer Science area of study in early 1990's. Earlier computational approaches could model and precisely analyze only relatively simple systems. More complex systems arising in biology, medicine, the humanities, management sciences, and similar fields often remained intractable to conventional mathematical and analytical methods. That said, it should be pointed out that simplicity and complexity of systems are relative, and many conventional mathematical models have been both challenging and very productive. Soft computing deals with imprecision, uncertainty, partial truth, and approximation to achieve practicability, robustness and low solution cost.

Generally speaking, soft computing techniques resemble biological processes more closely than traditional techniques, which are largely based on formal logical systems, such as sentential logic and predicate logic, or rely heavily on computer-aided numerical analysis (as in finite element analysis). Soft computing techniques are intended to complement each other.

Unlike hard computing schemes, which strive for exactness and full truth, soft computing techniques exploit the given tolerance of imprecision, partial truth, and uncertainty for a particular problem. Another common contrast comes from the observation that inductive reasoning plays a larger role in soft computing than in hard computing.

The conference receives 480 manuscripts from the participants. After review process, 149 papers are finally accepted for publication in the proceedings.

The Editors would like to thank the members of the Organizing Committee for their efforts and successful preparation of this event.

We would like to thank all colleagues who have devoted much time and effort to organize this meeting. The efforts of the authors of the manuscripts prepared for the proceedings are also gratefully acknowledged.

Jia Luo
Publication Chair

Contents

Discusses Display Space Information Dissemination to Appeal to the Point

Yuhua Li, Lili Xu, and Futing Wang

Institute of Art & Fashion, Tianjin Polytechnic University, Tianjin, China
{Liyuhua0633,Xulili8688}@163.com

Abstract. Display activities is a kind of dissemination of information activity, it contributed to the people and things dialogue between process, embodies a relationship between disseminator and receiver, the fundamental purpose is to make the populace most effective to accept the related information in the limited space and time. And appeal to point for designers can generally been understood as through certain media to target audience gambits and pass on in order to achieve the desired response. This article from the display space information transmission of several appeal point expounds display space as the design process is the most direct medium of communication that play the role of dissemination of information.

Keywords: space of display, information dissemination, display for information, appeal Point.

1 Introduction

People in daily life will come in contact with all sorts of display activities, such as information communication, promotion, for the purpose of education, etc, it needs in a certain period and space will message content presented to the public. Display is the activity which the public participation, the public while accepts the information the also feedback information, in display activities, the information demonstration's medium are varied, demonstrated that the space should be is most direct and the direct-viewing one way, it lets the exhibit article and the visitor talks directly, in fact explains the display space design is a kind of to convey information as the main function of activities. With the continuous development of commercialization, exhibiting space design gradually, in boomed display design process, either to display exhibits to audiences, convey exhibits information is the main goal.

Along with the development of the display activities and significance of outspread, display activities have become an important information dissemination means, dissemination of information and communication also evolve into the protagonist of display design, the display design of the core target and fundamental principle became to realize the effective spreading of people and things "dialogue". Therefore, it revealed that a space design should transform from past of the exhibition, displaying, storage, people and safety, etc, as the main basis of space design transformation until now to serving the spread of information effectively implemented as main basis, namely with certain display design gimmick, definite object relation, with the author, exhibits and visitors be conducive to formed between information dissemination and interpretation of coordinating relations.

Act as the media to reveal a space, often is neither a two-dimensional graphic image, also is not a simple 3d box, but by a series of yaps, different formative individual space and space of all kinds of information carrier common combination becomes the complex space, the space is filled with flow and information flow conversion. Therefore, display the information transmission system and works to reveal a space composition is intertwined, and jointly shows multivariate composite state. Meanwhile, in display information transmission system, different information units, different information carrier will inevitably lead to display information presents complex works of multiple state.

2   Display Space Information Dissemination to Appeal to the Point

The appeal to point for designers can generally been understanded as through certain media to target audience gambits and pass on in order to achieve the desired response. The display space with the unclear appeal point can't either to convey clear information, therefore cannot be successful display space. Display space is information transmission of public space, and transmitted information is show stylist to dimensional function and understanding, stylist to reveal a space put himself through the experience of beauty, of space design concept and the understanding to some kind of oneself, understanding the suitable way of life transmitted to visitors and public. Display space message is varied, the content is the spread of rich and colorful, the designer must understand himself then to express and transmit information demands point, accordingly designed display space to get the best performance, the maximum effect.

A.   Display Information Transmission

The exhibit article is demonstrated space the lead simultaneously also demonstrated in the information is the most important sources of information, space exhibits the spread of primary expression and content, its to reveal a space themselves attached information both constitutes the spread of exhibition space content. First, exhibits through own shape, color, texture, such as design of their own information directly across to the audience, such as through a silent language with audience to formation dissemination of information and communication.. Secondly, exhibits itself attached information, often cannot express complete meaning, usually also need other media or device to help it statements, their exhibits attached information into words, pictures, electronic explanations, multimedia and other forms, convey the information to the audience, make the audience can get information more easily through these media information.

The Contemporary Arts Center was designed by Zaha Hadid, which is located in a busy neighborhood corner. Architects in display space design, use high and long white metope or fresh tablecloth, the artwork salience, making the audience in a relatively independent pure environment art appreciation and analysis, easy to obtain space display information.

Cologne art galleries and concert hall have different display activities in accordance with the requirements, exhibits different its decorate means different also, some sites relatively open exhibits directly put on the ground, some use island type is decorated, some is hanged on wall, although arrangement forms different but exhibits position and very conspicuous, give a person with full and prominent display space to appreciate and exchange.

B.   Dimensional Function of Transmission

Any nature of space are made of various functional space composition, due to the particularity of exhibition space, it is through space to the audience the expression means, in which the information transfer information more phyletic, it is required to reveal a space to meet the needs of all kinds of display function, at the same time it also requires the expressional means to reveal a space more can reflect its function must be needed. Display function space refers to realize the basic exhibition venues, by information space, public space and auxiliary space composition. In carries on the show space design of space function subarea is priority and smooth satisfactory to complete the whole process of display basic guarantee. Functional partition is to display activities of various functions and they the relationship between spatial analyses, make the space division meeting the functional requirement. As to the exhibition booth design at independent, they need to whole space function analysis, the proper configuration such as exhibits display area, negotiate area, commodity or enterprise image exhibits etc function area, in order to achieve the rational use of exhibition space and space combination of natural harmonious.

C.   Spatial Significance Transmission

To display space character, has the different type division, the different function meaning, exhibiting space these characteristics to the audience expressed space meaning, such as museums exhibition space, exhibition space, sells the display space create atmosphere is different, give a person the feeling is different, these different is the spatial information attached itself different meaning.

Spanish architect JoseRafaelMoneo designs the Merida merida national Roman Art Museum, with extremely concise pure form and material express profound historical sense and strong narrative. The museum selected location in a Roman Empire time's historical site, there are many limiting conditions to the design of the monuments, the request and the environment are coordinated, Rafael in the design and architectural style do not against roma, but adopt strict imitation of refined form and local materials to design, both guardian historical integrity and continuity, and keep the building of independence and perfect. In exhibit space design, using a series of parallel equi-distant wall will space partition into several galleries, each other through different sizes of the GongQuan connectivity, among them along the line perpendicular to the wall of the visitors a series of scale direction GongQuan form a truly great perspective scene, from the function and form, this a similar church of classic space is undoubtedly the most valuable part of building, is the most inspiring space clip.

D.   Space Beauty Convey

For the construction is concerned, space not only has direct practical value, but also has the important aesthetic value, it not only gives building takes "live" purposes, causes its "function", but also give the levites architecture "beauty" construction to its "watch" interest. Display space of "beauty" performance they particularly prominent, the audience enjoyed exhibits, space surrounding environment on the exhibits foil and on human psychological influence has very big effect to show appreciation of the space, also put forward some corresponding requirements. Space beauty is a formal beauty, a beauty lies in the display space itself in, also lies in the people to the form of understanding and knowledge of the modelling of space, proportion scale, materials, simple sense, light environment, the color and so on, by appropriately handle obtain perfect image results, this image can give a person with a certain mental feeling, and display activities closely linked to show, deep into the dissemination of information process. At the same time display space itself is an art, it is a kind of beautiful expression, these information and exhibits the information at the same time passed to the audience, all of the audience impact. The audience in the receiving display space transmitted the aesthetic information process, aesthetic standards for his itself also produced significant influence.

3   Conclusion

We know, the information of ordered degree and organized the higher level, the effective information is greater, the easier it is accepted by the audience. So, in the design process, the designer must be good at screening each nformation units within each inherent order, using proper permutation and combination, realize disorder to orderly about transforming information, realizes the disorderly information to the order information transformation, collection of information unit to scientific combination, sorting, make its closer to leading information requirements of order, and to promote the viewer to the effective information acquisition and use, to realize the effective information transmission. In addition, orderly information also depends on the display space on the various material carrier can exist and was the recipient of information, the receiver of contact only by understanding also combination of the space of of all kinds carrier achieved.

Display space is not only hold all kinds of information carrier of containers, and its different division and combination, also stipulates the viewer to walk therein, and affects the line of space in the viewer to various information process of cognition and understanding order, which requires the display space combination and arrangement order must be is to explain the agreement and information order, namely display information structure and display of spatial structure of the logic is unified. Only under such space, the viewer understand, accept the order information may and information of the inherent order, and obtain a lot in approaching the effective information. Therefore, only display space structure and display information structure can meet orderly and unification, can improve display information of effective spreading.

With social progress and economic development, show exhibition space design also has increased, and spread increasingly rich content, display form, tend to be more diversified and the display space design requirements are increasingly high. Display space is actually information exhibition space, use propagation mode as a display space design method just fits the display space characteristic. In display space design process, the person, content and the spatial relationship is very close, exhibiting space design process as a type of communication process is the basis of investigation, this paper aims to implement the effective information transmission, to achieve the anticipated display purposes.

References

1. Arnheim, R.: Art and visual perception. China Social Sciences Press, Beijing (1984)
2. Dollo, L.: Individual culture and popular culture. Shanghai People's Press, Shanghai (1987)
3. Dai, Y.: Dissemination study theory and application. Lanzhou University Press, Lanzhou (1998)
4. Peng, J.: Art study introduction. Beijing University Publishing press, Beijing (1999)
5. HanBin: Display design learning. HeiLongjiang Art Press, Harbin (1996)
6. ZhuXi: Exhibiting space design. Shanghai People's Art Press, Shanghai (2006)
7. Heping, Z.: Space. Southeast University Press, Nanjing (2006)
8. Bryan, L.: Space language. China Architecture Building Press, Beijing (2003)
9. Sanoff, H.: Visual research methods in desing. Van Nostrand Reinhold, New York (1991)
10. Harrigan, J.E.: Human factors research: methods and applications for architects and interior designers. Elsevier, Amsterdam (1987)

Brief Discussion of "Concept Illustration Art" of Information Age

Futing Wang

Institute of Art & Fashion, Tianjin Polytechnic University, Tianjin, China
Ftw55623@163.com

Abstract. This paper is based on the evolution of traditional illustration to modern illustration. With the brief review of the development of art, it stated the changes between the development of human society information and the way, the speed of information dissemination; The areas of modern illustration are mainly used in printing media from the past to TV, the Internet and other electronic media quickly, thus formed more beyond the traditional "illustration" concept; Illustration will continue to seek and enlarge self-development space with the development of human information age, which will continue playing an important role in modern society information dissemination.

Keywords: illustration, traditional concept illustrations, Modern concept illustrations, illustration form, information dissemination.

1 Brief Review of the Development of Illustration Art

Illustration generally refers to the pictures that inserted in the text to help explain the contents, to explain the significance of the text and enhance the fun of reading and visual enjoyment. Yu Fenggao mentioned in the book 《Illustration of Cultural History》 "the most important thing of illustration is to make clear explanation of script and presswork, meanwhile it has the function of decoration and beautify" [1] Recalling the long history of development of illustration, although the basic functions of illustration do not fundamentally change, with the development of printing technology, especially the rapid development of modern science and technology and the human face of the enormous changes, illustrations have been more extensive development, especially the relationship with the text is far more flexible.

Illustration has a long history, its origin can be traced back to primitive man's cave rock paintings, but the real illustration appears after the books. China's "book" straightforward indicated the close relationship between diagrams and books. Books are the common name of pictures and works in ancient times. "Left picture right book", "Left picture right history", "picture fills the words, words fill the picture", illustrations complement the book is a fine tradition. The earliest illustrations also appeared in Europe in religious writings, such as the famous hand-painted illustrations in European Middle Ages between missionaries; Illustrations in the books are not only stories about the religious, but also birds, flowers, figures and other decorative artwork. Worth remembering that the early history of Chinese and foreign illustrations

J. Luo (Ed.): Soft Computing in Information Communication Technology, AISC 161, pp. 7–10.
springerlink.com

both relevant woodcut engraving, it reflects the relationship between illustration and printing at that time; After the German metal carpenter Gutenberg invented letterpress printing, printing books have become convenient and inexpensive. Printers in Europe are more likely to use woodcut as illustration pages, together with metal type, this gradually replaced the previous labor and time-consuming production of the manuscript production industry.

Seen from the illustration history, in the early of low literacy rates, the illiteracy and semi-literate people are in common needs of illustration. Therefore, the traditional concept of illustration is to assist explanation of diagrams and maps of literature and text. It is used for plates of books and printed material. There will be no words and texts without illustrations, they are subsidiary and have no independent character of painting, so the history of the great artists often make illustrations in spare time or as extra income. In addition, illustrations need books as carrier, contact the readers by printing, not the original form, this high-volume print production and print copy are closely linked. It develops with engraving and mask-making technology, copy prints also derive several species and engraving techniques because of the needs of copy illustrations. The first dissemination form of illustration is books and newspaper, thus the relationship between illustration and print, books and newspaper as the carrier make traditional illustration "small". Until late in the 19th century, with the development of lithography's large scale, illustrations have become "large" and appear in the street as advertisement dissemination carrier in the true sense.

In the 20th century, with the development of education of culture, people began to improve their reading competence. People are easy to understand the text and propose more fastidious request of illustration arts. Meantime, modern printing with the promotion enables more and more people to read, while illustration artists provide text rich imagination.

After the 1950s, while "new performance consciousness" rising, the illustration broke through its position of attaching text. While it plays its own role in "explaining text illustration", changing and developing all the time, becoming illustration artists find a special way to convey thoughts and self-discovery process, thus the independence of illustration begin to be established. So the performance of the interesting illustrations, personal expression ideas have become the goal that illustration artists are pursuing. After 1970s, with the photographic plate technology continues and printing technology has become more beautiful, illustration has undergone fundamental changes. Illustrators who directed and painting by illustrate, created a variety of forms of comic books, poetry and painting of the narrative, children's books, informative books and records, and science fiction art painting books. The most important feature of this kind of painting is illustrators who became self-expression of the original people in the out of the "literary vassal," the shackles of the location, they opened up a free, self-perception of the new world of direct communication. It is based on the above changes, so there will be some "traditional illustration" and "contemporary illustration," differences, which illustration is no longer a vassal of literature or writing articles, it has become a kind of visual imagination to convey thoughts through the unique type of visual communication.

2   The Development Trend of Modern Illustration

In 1980s, printed media greatly increased, the pattern of life continues to expand, which makes illustrators' expression space expand to daily life and environment. This makes the rapid increase in the number of illustrators in developed countries, their work extends to drawing energy from the books of various design categories, such as product packaging, shirt design, public advertising, brand image, product image, toys, cartoons, showcase, etc., and the environment decoration, wall murals and other public arts. It can be said that illustrations exist in many areas. The brighter side of illustration art in developed countries makes the industry developing, and increases the people who work there. Illustration studios are everywhere; every year there are several versions of the "Illustration Yearbook" album that available to introduce hundreds of thousands of dollars of artwork and artist studios of individuals, for enterprises and departments under the style choose to use.

With the development of human information society, the speed and manner of information dissemination is changing. The areas of modern illustration are mainly used in the print media from the past and quickly to the TV, the Internet and other electronic media fields of application extensions. In the development of modern illustration, technological development, computer graphics means, new materials development and category painting rich, making the technique of expression and visual style of illustration very numerous and varied: modern style of illustration can be either inch of room or very large "giant"; either be flat, or it can be three-dimensional or three-dimensional half. It is a creative advertising posters in the graphic or photographic image; it is also out of the game to show pictures of various scenes and characters; it is a web page that is static or swim, or pictures you want. It can be said that almost all the visual means can be used by the form of illustration; illustration has become a new testing ground for instruments and performance practices. Thus, modern artwork is no longer just the traditional sense of the literature and text for illustrations, broadly speaking, modern illustrations are illustrations for modern life, the information explosion of the intelligence community. Illustration has become an object to the public as to the form of visual communication as the main graphic design.

Widely used in the illustration design today, modern illustration as an important carrier of cultural transmission for the culture, makes important contribution to culture transmission and carrying, also promotes the development of arts and culture. As illustration has a huge market in Japan and South Korea. In Japan, animation is the illustration of an important branch of industry, commercial animation developed quite mature, and already has a very large-scale operation of the team, and illustrations greatly promoted the development of animation industry. While in South Korea, the rising of games provides a large market for illustrations, especially digital illustration.

Looking back on the track of the development of illustration art recent years, we can conclude the following general characteristics: First, the illustrations will continue to be in the text in the form of pictures and words as added illustration, also generally in the general magazines, books, textbooks, etc. Second, the illustrations will be inserted into the media image of its visual, together with words of the same function, which are used in many poster advertising, product packaging and other print media and electronic media and multimedia commonly. Third, illustrations have nothing to do with words, it conveys vision ideas directly, becoming an independent significance and charm of the visual modeling language, to form the "illustration" more than the concept. Although we can not predict the future illustrations, it can be expected, illustration will continue to seek and enlarge self-development space with the development of human information age, which will continue play an important role in modern society information dissemination.

References

1. Yu, F.: Illustrations of cultural history, P1. New Star Press, Beijing (2005)
2. Cao, F., Joe, S.: Perceptual Schema. Jiangsu Art Press, Nanjing (2008)
3. Sheng, R., Wang, J.: Illustration Art. Hefei University Press, Hefei (2006)
4. Yang, H.: Illustration Art and Culture. Tsinghua University Press, Beijing (2008)
5. Wang, S.: Illustration history of the United States. China Youth Press, Beijing (2002)

Nature·Humanity·Soul Porcelain·Ganzi

Kuikui Zhao and Jicheng Xie

Institute of Art & Fashion, Tianjin Polytechnic University, Tianjin, China
vchinesekui@163.com, zhangwuren@126.com

Abstract. This essay aimed at exhibiting the national culture and nature environment of Tibetan region as well as the kindness, honesty and devoutness of the Tibetan people with the inspiration from carved stones and the porcelain arts as the carrier. The aesthetic conception of modern porcelain arts has been changing as the application of ethnical culture in modern porcelain arts has become more and more attractive for viewers, Porcelain·Ganzi, with the characteristics from combining the porcelain and MANI cairns, showed a particular kind of regional culture of Tibetan to us through reappearing the MANI cairns with porcelains, which can bring viewers affection related with nature and intense aesthetic experience, which aim at carrying the Tibetan culture forward and invest with the innovations by trying to provide interaction and communication between different emotions.

Keywords: MANI cairns, modern porcelain arts, material conversion, Tibetan culture.

1   Introduction

When I was in my journey to Tibetan and Sichuan religion, MANI cairns could be found almost anywhere, they are composed of carved stones and flints, I was very excited when the MANI cairns occurred to my eye sight for the first time, and it is this particular kind of carved stones that widely distributed in the Tibetan regions. Stones, with its quality of never rotted, has been the carrier of humane social culture along the mankind history, and has also firmly witnessed the founding and developing of the long history civilization, for example, grotto fresco of the Paleolithic Period; megalith building of the Neolithic Age (the typical representative is the Stonehenge in the south of England); icon in stone of Han dynasty; modern stele arts and so on. And because the aesthetic experience I had on my journey was so intensely moving that it motivated me to gather the materials as many as I could, during this processes, I had understood the MANI cairns much more and moved by the culture they were imparted with. All these intimations of the spirits of the Tibetan culture make me addicted with them, and I was going to express the emotion that it roused in my heart by remodeling the MANI cairns.

J. Luo (Ed.): Soft Computing in Information Communication Technology, AISC 161, pp. 11–19.
springerlink.com   © Springer-Verlag Berlin Heidelberg 2012

2   The Material Conversion in Porcelain·Ganzi

A.   The Origins of MANI Cairns

Some one says that MANI cairns tracing the origins of Tibetan and Mongolian cairns. There are lots of versions of opinion on where did the MANI cairns come from, however, legends are just legends, where did it come from on earth and who arrange them in organization cannot be learned completely. It is also impossible to make a widest guess at the quantities of MANI cairns with so huge scale; it was uncountable in terms of its distribution. This mystery has motivated my curiosity upon the MANI cairns more intensely.

B.   Cognition of Formula and Content of the MANI Cairns

Some of the MANI cairns are in big scale, and they were arranged in a shape of trapezoid with the big ones complements the foundation and small ones at above, there are also some sorts of MANI cairns are in small scale, they are in the shape of cone which without stairs in it, there are even walls which were consisted of MANI cairns, and MANI cairns have invested the Qinghai-Tibet Plateau with a hint of mysteries and spectaculars, and the stones which complemented the MANI cairns are integrative or deformities or overlaps as well as many other formulas, and they have gradually merged in the universal environment with its harmonious color scheme with its surrounding environments, and the humane emotions was in those stones after they were carved by folk sculptures. Sculptures carved those stones using different techniques, which are flexible and innovative. Some common techniques are line cutting, garland cutting, intaglio cutting, bass-relief cutting as well as high relief cutting and so on. And it is obviously clear that the sculptures had elaborate skills, they drew the composition according to the shape of the stone, choosing the knife according to the texture of the stone, and they created the classic masterpiece of Tibetan folk arts with their elaborate skills, and used different techniques of exaggeration, generalization and simplification.(Fig.1~ Fig.4)

MANI cairns' function of directing has been remained till nowadays, they were created to bless and protect the outside pedestrians with peace, health, lucky and safety. Then the places where there are existences of MANI cairns have become the arena where the Tibetan peoples hold the ritual activities on every festival day, while trying to prevent the devils by adding coprolites to MANI cairns. This kind of ritual was first the spiritual sustenance of the Tibetans who lived in poverty; they conveyed their beautiful hope for their spectacular future by embodying the natural stones and make them sacred.

So grand and imposing shock, such funky folk flavor characteristics, and unique hard rock, heavy, to convey the snow on the Tibetan Buddhist religious beliefs and longing for a better life for the future. It is Sichuan-Tibet region of the natural and cultural landscape that what I want to express in **PORCELAIN·GANZI.**

Fig. 1.   Fig. 2.

Fig. 3.   Fig. 4.

C.   The Choosing of Porcelain Material

The MANI cairns have remained because they were carved in stones, so in my work, I had chosen the porcelain as the material. We all know that porcelain is nearly related to stones in many respects, both of them are rigid, be carved, never be decayed, and both of them can be the witness of the developing of social civilization with the information carved on them. Porcelain, with the qualities of easy-shaped, good plasticity, pretty texture, can be carved with a similar sense of texture with stones. Otherwise, porcelain has a rich and varied color glaze, nearly to mud in some characters, which can be invested with humane culture. Porcelain art is a combination of mud and fire; they have merged to each other in porcelain products. Clay comes from nature, which has been the preference of mankind for a long time, and clay will become porcelains after taken out from the porcelain spark plug, then they will be imparted with the spiritual touching of humans. All these properties and characteristics have boosted me to choose the porcelain as the material of my final works.

D.   Overview of My Works

My works aimed at exhibiting the national culture and nature environment of Tibetan region as well as the kindness, honesty and devoutness of the Tibetan people with the inspiration from carved stones and the porcelain arts as the carrier, because of their natural similarity to each other, I have been trying to exhibit the unique aspect of Tibetan culture through the material conversion from stones to porcelain, and the porcelain material itself has a natural affinity and physical properties of historical witness that can best serve the effects of the finished products, I have reappeared the MANI cairns in materials of porcelain according to my conception of this particular variety of Tibetan arts, which I have been fascinated with, the form, the beauty. As you can see in my work, the reappearing is not simply coping; it is a kind of innovation based on the basic formula of the MANI cairns after I had understood this Tibetan art, my work was created to provide a feeling of unique and pristine of Tibetan culture by the elaborate intrinsic rhythm, scale and harmonious proportion in the composition of my work, and the content has changed along with the conversion of the materials, and the original spirit has also changed. And the heritage and innovation of Tibetan culture was wished a sound future in my work while the Tibetan culture were being applied and exhibited in my work of Porcelain·Ganzi.

3   The Conversion of Form and Content in Porcelain·Ganzi

A.   Form Conversions

a)   The formation

PORCELAIN·GANZI have used concrete presentation techniques," Figurative meaning is carried by the spirit of the idealized realm of the surreal "Parts of the works used the natural stone prototype directly, some others used turning model to manufacture prototype material and add a new interpretation of syntax, this techniques have made my works in the vision to create a natural scenery, Tibetan humanities and trying to make people produce madding crowd, in xanadu feeling. Tibetan cairn of external image depicts is the ladder of seek Tibetan memory, the pursuit of spirit realm, is my pursuit of the life in the comprehension of the Tibetan.

The items invest the stones with in real sense with various kinds of formation complete and incomplete, chiseled, and fruity; then forming echo with rich and colorful nature. The stone or flake have different size with each other, this is one characteristic of Tibetan cairn in reality. Rocks and flake in different size form contrast and echo increase the rich and harmonious of ceramic language.

For the using of colors, the major schemes of color were the main colors that formed after the porcelain being put into the fire; Match again another kind of porcelain clay which color is white for ornament. Not only unity, but also not lacked variety. A few rare stone were glazing, but still on the basis of keep the whole situation in big tonal.

Entries overall show is the accumulation of individual. Tibetan cairn accumulation has various ways. From afar it looks like ladder, circular pile and wall. We can accord specific environment and to decide one way, thus make works better in expressing ideas into the environment, and at the same time the separation between the environment and it will not be generated.

b)   Textural effect of Porcelain·Ganzi

The texture that appears after firing the glaze and mud of porcelain materials is one of the important factors which invest the modern porcelain arts intense aesthetic function. "Skin texture affect of the material of clay is not simply a piece of surface, but the coat of composition consisting." Suitable and fantastic textural effect can benefit the expressing of the spirit inside the works to the viewers.

Porcelain clay, with the quality of easy shaped and dimension keeping, is easier in creating of the skin texture. My works have used plaster re-prints, stamping and sculpture methods for texture production. Gypsum re-prints texture are using natural's real stone texture, stamping and sculpture is by imitation stone texture. When making texture, you should take into consideration of the aesthetic feeling of rhythm, like the continuation, overlap, cross, density, etc. The humanistic trace texture on the works give people a warmth experience.

Without glair, the pottery and porcelain show original silent and pure after be fired. With the use of glaze increased the richness of the works' texture. Through the gloat's chemical changes, the texture can form the effect of smooth, party-colored, Ice crack etc. In addition, glaze can make natural soil formed luster effect and matt effect after coalmining. Highly reflective stone is not true with nature stone. And it is also one of the means to leave the traces of human.

Porcelain clay and ceramic soils can give off the feeling of quiet and innocence, these two kinds of chemical composition of mud, formed feelings after firing. Pottery gives a person with coarse, bold, and unconstrained feelings. While porcelain makes human beings feel exquisite, white, smooth and hard. After firing the two kinds of clay materials form a bright contrast. In a single stone material selection, some use pottery clay, some use porcelain clay, some choose two kinds of mud. Mixing mud makes a small area of the white porcelain clay in the dark, pottery clay freely appeared, and forms the visual characteristics of sedimentary rock. "The curt combination" mainly wants to leave the traces of human on the lifelike stones.

Texture effects were happened to formed after firing is the difference of ceramic materials from other materials. This cannot control textural effect can make people happy, also can make people lost.

c)   Decoration methods and items

As we can see, the modern contour of blue and white as well as carving were applied to decorate the items of my collection, the contour of blue and white, as a method of decoration, has a very long history. And has kept his youth through hundreds of years when most of people are fond of the contour of blue and white. This sort of contour has always been prized of its quality of quiet and elegancy. And this particular kind of contour can extrudes an intense sentiment of purity and liveliness, which can impart observer with a feeling of coziness.(Fig.5) And those affections are just what I am going to convey in my MANI carvings as a result of I was moved by the meanings of MANI carved stones.

MANI carved stones are also applied with the decoration of carvings, those carvings include line cuttings and embossments. There are many kinds of line cuttings that can be deep or shallow, wide or fine. Some parts of those cuttings have combined with the celadon. It was the differences in the depths of cutting lines that extrudes different hues between blue and white, and it was the combination of those colors that present the features of quiet and elegancy of MANI carved stones.(Fig.6)

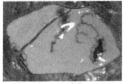   

Fig. 5.   **Fig. 6.**

The fonts, which were applied in the decoration, were deformed Chinese characters which can be imagined with the Tibetan characters, Tibetan language is a sort of language which is considerably mutual and excellent. And the using of Chinese characters, in decorations, which are similar with the Tibetan character has become more and more common. Hoh Xil, for example, is applied with the Tibetan like Chinese characters in the published posture in 2004. (Fig.7).The using of Chinese characters, in decorations, can serve the works with the lingering charm of Tibetan. There is 6-word-mottos being carved in the MANI carved stones, for my works, however, I did not cut the Tibetan characters, the reasons included two factors, the first factor is that there is such difficulty in recognizing the Tibetan character, the second factor is that I was afraid of that it may be conflicted with the mottos of Buddhism. The characters I carved are as follows: endure virtue real rite goodness beauty, which mean tolerance, virtue, sincerity, courtesy, kindness and nice in sequence. as well as the sentences in Classic of the Virtue of the Tao of LAOZI : The top class of virtue is like water, which benefits ten thousand objects without any demands for return, that means he highest good is like that of water. The goodness of water is that it benefits the ten thousand creatures. (Fig.8). Sincerity, kindness, modesty and charity without demanding appreciation are what I was feeling when I was with the Tibetan peoples. And those honorable qualities are precious deposits of modern civilization that we must put in heart forever.

Parts of the pictures depicted in my works majorly reflected the scene of the daily life of Tibetan peoples and some beautiful landscapes in Xizang; others are sacred churnings that the Tibetan put their faith in. (Fig.9~Fig.11); all the carvings were aimed at memorizing the wishes, of Tibetan peoples, of the daily and mental life. This benefited the aesthetic performance of my carvings. However, the cutting could not be too complicated because of the limits of line cuttings and embossments, the contours were carved in a generalized and simplified way.

Fig. 7.

Fig. 8.   **Fig. 9.**

Fig. 10.   **Fig. 11.**

B.   Content Conversion

a)   Humanity concerning

The mental pursuit of clay of contemporary peoples is just like the quality pursuit of jade of humans in Song dynasty. Clay has been considered as the spirit shelter of contemporary people because of its particular individualities, the wishes of returning back to nature motivate contemporary designers to work hard in the area of man-made environments, in order to impart folks a sense of returning back to nature with the construction of the primitive concept of the Qinghai-Tibet Plateau feelings, **PORCELAIN · GANZI** reappeared the MANI carved stones in Sichuan and Xizang regions with the new carrier of porcelain, it also extrudes humanity concerning of returning to nature through the concrete decoration which can be related with the imaginations of the natural and humanistic scenery in Qinghai-Tibet Plateau.

It is fairly clear that the porcelain has a long history and a rich cultural heritage. It has always been with us over the periods. It has also witnessed the development of human civilization. As nowadays the culture has been a combination of western culture and eastern culture, porcelain arts, rooted in this kind of culture, will benefit modern arts with quality and intension by the promiscuous characteristic of porcelain. For porcelains can then not only carry on the conventional culture and fuel it with new proved decoration of foreign culture, then a new cultural conception will be created. **PORCELAIN · GANZI** aimed at reflecting the features of Tibetans and illustrating the way that the Tibetans lived through their daily life, "this goal can lead to the spirits reaction between the MANI carved stones and its observers that extruded the concerning of humanity in my works to satisfy plural aesthetic."

b)   To satisfy plural aesthetic

It is differences in the visual experiences, sentiments as well as the costumes that lead to the different kinds of aesthetics, humans' pursuit of entities has been on the way to a field of plural conception and perspectives. So the pursuit of regional culture, humanistic affection as well as individualities has become more and more popular as a trend in design industry.

The items of **PORCELAIN · GANZI** have got specifically contour through carving realistically; there are factors in **PORCELAIN · GANZI**, such as the using of colors, creating of skin textures, extruding of different materials, the way organization and order of constructions complementing the spaces, decorations of contours of blue and white, techniques of carving, all these factors can serve observers' plural conception of aesthetic very well.

It is a good idea to create new design style for modern life with the applications of domestic culture, which can provides the design not only national features but also regional characteristics, the modern pottery arts, itself, have the natural attributes of returning to the nature. It can reflect its unique characteristics by mixing with the Tibetan culture and spirits with its magnificent momentum, **MANI** carved stones sculptures are mainly consist of flagstones, the contours carved on it were Sichuan-Tibetan scenes of people's daily lives, the natural landscapes, building materials and some auspicious patterns, which revealed Sichuan- Tibetan people's preferences, customs, values and aesthetic psychologies, The modern porcelain crafts can not only visually show the charming and splendid Tibetan culture and natural style of landscapes, but also can impart observers with a spiritual movement in the deeply souls.

4   Conclusion

Clay, the raw material of porcelain, with the traits of being close to nature, being harmony with the environment and holding communication with souls, that formed the historical context, so that people can really feel close to nature, simplicity, unsophisticated, people can experience more humanistic care. Porcelains' cultural attributes make it easy to combine with the culture, carry on the culture and make culture better. Modern porcelain, which is a bridge between Chinese traditional arts form and the concepts of modern arts, become the carrier of my works. We find a meeting point between the regional and national aesthetic needs of the audience in Tibetan areas, and between the spiritual needs and cultural needs through the conversion of the material, which has made people's spirit be consistent with the Sichuan-Tibet's nature and the spirit realm, so that we can achieve the ultimate concerning of humanity.

Porcelain · Ganzi depicts the application of the natural and humanistic landscape of **MANI** carved stones. Instead of staying on the simple copy, I have been trying to understand the inside meanings of it, refine its forms and the spirits, and finally I attained a magnificent feeling of pristine Tibetan culture. By melting my own feeling into Tibetan culture, I showed the Sichuan-Tibet magnificent landscape and unique local culture though my works. Moreover, the ways we treat Tibetan culture not only affect the destiny of Tibetan culture, but also have an influence on the use and development of Tibetan culture in contemporary art.

Acknowledgment. This paper PORCELAIN · GANZI was finished under the guidance of Teacher Zhang Bin. During the course, he, a teacher with unique insights on the professional knowledge and great concentration of teaching for years, gave me the patient advice and help, which urges me to do the article. To engage in education seriously, to think about study insightfully, to work diligently, teacher Zhang's words and deeds are so admirable that benefit me all through my life. Finally, I'm obliged to teacher Zhang for his encouragement and care for my study in recent years; here I want to express my respect and appreciation!

References

1. Bai, M.: Overview of the World Modern Ceramics. Jiangxi. Arts press, Nanchang (1999)
2. Liu, B.: Modern Landscape Design. Donghua univerxity press, Nanjing (1999)
3. Tseden, L.: Tracing the origins of Tibetan and Mongolian cairns. Academic Journal of Tibet Universit 21(1), 66–68 (1997)
4. Kang, G.: The grand and chaste folks culture of Tiberan MSNI cairns, pp. 21–26 (January 2003)

An Analysis of Cost Optimization in Promoting Activity-Based Costing Management

Xiaoqiang Li

School of Business, Tianjin Polytechinc University, Tianjin, 300387, China
lixiaoqiang1997@126.com

Abstract. The rapid development and application of modern science and technology cause the market environment to change from gradually to abruptly, at the same time, the global financial crisis has not receded, all the nations actively seek to protect the environment and keep the sustainable economic development. In this new business environment, traditional cost management is not enough to achieve the strategic goal while Activity-based Costing Management can, by analyzing the value chain, distinguish value-added activities and non-value-added activities and measure the value-added costs and non value-added costs accurately to reduce the number of intermediate links that can lower the cost, and to provide strong impetus for the maximization of enterprise value along the value chains.

Keywords: Activity-based Costing Management, Activity, Value chain.

Activity-based Costing Management is a mode of management that transfers the starting point and core of cost management from "product" level to "activity" one. This is the fundamental improvement of traditional cost management because in Activity-based Costing Management, the enterprises pay more attention to a series of closely-related activities that can satisfy customers' needs and realize the strategic goal of their value maximization, and they think that products or services consume activity, and activity consumes resources in turn, and process of resource consumption is also the process of value accumulation, in other words, value transfers from an activity to the next one, and is accumulated in the final goods or services finally. From this process, it can be found that the activity chain of an enterprise is also its activity chain. Activity-based Costing Management requires that cost management must reach each activity of the enterprise, and eliminate the activities that cannot create value as far as possible, thus preventing the resources from wasting and reducing cost and regaining value for the realization of the goals of the enterprises' development.

1 The Motives Adopting Activity-Based Costing Management to Minimize Cost

1.1 The Objective Requirements of New Business Environment

1.1.1 To Adapt to the New Economic and Technological Environment and to Realize Enterprise Strategic Objectives

Since 1980s, science and technology has being accelerated toward the development of informatization, and has produced significant influence in commercial activities. First,

J. Luo (Ed.): Soft Computing in Information Communication Technology, AISC 161, pp. 21–26.
springerlink.com

in modern enterprises, the automatization and infomationization of production process makes product manufacturing system more complicated. Secondly, the traditional few-variety and mass production mode is transformed into multi-variety, individualized and small-batch production mode. Finally, in order to satisfy the market needs as well as the customers' increasing expectations about the quality, function of products, and in order to capture the markets, the enterprises need to develop new products continuously, produce the products with higher qualities and lower prices and improve their technology and competitiveness, resulting in the shorter product life cycle.

Facing the changing markets and the fierce and more flexible competition, the enterprises try every means to reduce costs and to expand market share, and to gain profit. The traditional cost management mode, which regards "product" as the core and starting point of management, and the variance analysis and control of the standard cost and actual cost as the key method, does not adjust itself to this new and dynamic and unstable business environment. In this case, it is necessary to break through the traditional cost concept as a guide and transfer the foundation of theory and methods of cost control from a product cost to an activity cost, and to establish a new whole-process and all-scale mode of cost management based on the activity cost.

1.1.2   To Overcome the Negative Influence of Financial Crisis on the Enterprises and to Accelerate the Management Mode Reform

It is the wise choice to face the crisis actively rather than passively. The global financial crisis broke out in 2008 has started to seriously affect the development of the world economy and has not subsided, so many enterprises want to get out of the crisis, must adopt Activity-based Costing Management mode to optimize cost and sustain their development.

Additionally, activity-based costing management in China is still under theoretical research while according to a survey undertaken by Krumwiede (Pan & Tong,2005:177), 49% of the enterprises in U.S.A. have adopted activity-based costing management in 1996. This difference of activity-based costing management between China and U.S.A. is due to the internal and external resistance for the enterprises' reform in addition to the operational cost of activity-based costing management. We know that, when the economy is under steady development, the enterprise's reform and innovation will meet with some big internal and external resistance, but the current financial crisis will force many enterprises to reform for survival, thus make them meet with less resistance than before and helpful to adopt activity-based costing management.

1.1.3   To Be Suitable for the Objective Requirements of Economic Sustainable Development and to Promote the Enterprises' Sustainable Development

The resource scarcity for human survival makes people more and more aware of economizing on energy and protecting environment. Ineffective management accounting system will greatly harm the optimization of product development and manufacturing process as well as marketing activities, resulting in a series of low efficiency and various recourses wasting, so is the traditional cost management, which will regard portfolio as the only factor of cost, excessively simplify the cause of cost, make it easy to distort the cost, and cannot reflect the enterprise' resources consumption in the production and management. For example, a majority part of overhead costs are often distributed to some products with high production or a simple manufacturing process while a minority part of overhead costs are often distributed to other products with low production or a complicated manufacturing process, and this possibly will mistake the loss products for the profitable products to expand their production and the profitable products for the loss product to stop producing them, eventually resulting in the low efficiency and resources wasting of the whole society, but all of the above defects can be overcome in the activity-based costing management.

1.2   The Enterprises' Fundamental Changes in Product Cost Structure

With the development of technology, especially the utilization of computer has changed manufacturing environments greatly: on the one hand, the enterprises adopt the multi-variety, individualized and small-batch production management mode to satisfy the gradually changing and diverse needs of the customers; on the other hand, the technical progress causes the organic composition of capital to unceasingly increase, and makes the cost ratio of direct materials and direct labor to product decease greatly while that of overhead costs to product costs decreases less and less. Furthermore, it is an unarguable fact that the cost composition of products has changed fundamentally and the indirect costs in the production process correlate less and less with the direct materials and direct labor. If the enterprises continue to adopt the direct labor cost, which occupy less proportion in the commodity costs, to allocate indirect cost, which takes up more proportion in the commodity costs, it will result in severe distortion of cost information in commodity and misguide the enterprises to take a wrong strategic decision.

1.3   The Limitations of Traditional Cost Management in Cost Optimization

1.3.1   Ustainable Development rn ost, in Traditional Cost Management Mode, the Enterprises Consider Costs in Isolation and Do Not Connect It with the Enterprises' Strategies

The correct formulation and effective implementation of the enterprises' strategies requires relevant cost information, but in the traditional cost management mode, the enterprises focus their attention on how to reduce the cost of products, and neglect the coordination with the enterprises' strategies. Therefore, the only thing they can do is to consider the cost in isolation because the cost information provided in the traditional cost management mode is unconvincing in terms of accuracy, correlation and timeliness.

1.3.2   In Traditional Cost Management Mode, the Enterprises Consider the Cost Factors One-Sidedly, and Pay No Attention to Value Chain

In traditional cost management mode, the enterprises focus their attention on portfolio, but in the new business environment, application of advanced science and technology as well as their strategic choice often makes the portfolio not the major factor of affecting the enterprises' costs but a series of interconnected activities in which the enterprises create valuable products and service, namely, the cost in terms of the value chain analysis, including production value chain and industry value chain. By contrast, in traditional cost management mode, the enterprises find it had to do so because they regard the portfolio as the only factor of the costs.

1.3.3 In Traditional Cost Management Mode, the Enterprises Ignore the Non-Value-Added Cost

The non-value-added cost refers to the cost expensed in the activities that do not increase customers' value. In traditional cost management mode, the enterprises pay attention to the actual cost expensed in the operation process and try every means and take every possible measures to control these costs but neglect the non-value-added cost existed in the actual cost because in the current situation the non-value-added cost is not the unnecessary cost but one that can be eliminated through continual improvement. For instance, the inspection of such activity cost is necessary to prevent the unqualified products from flowing into the market, but it does not increase customers' value, can be eliminated by increasing the level of production technology and processes, which is neglected in traditional cost management mode.

2   The Advantages of Activity-Based Costing Management in Cost Optimization

All the limitations of traditional cost management in cost optimization can be made up in activity-based costing management mode. The advantages of activity-based costing management can be shown in the following:

2.1 The Combination of Activity-Based Costing Management and the Enterprises' Strategies

In the activity-based costing management mode, the enterprises take the value chain analysis as the core of activity-based costing management, and concern the cost as a whole for the service of the enterprises' strategic target.

At the same time, because the cost information provided in traditional cost management mode is unconvincing in accuracy, correlation and timeliness and has influenced the enterprises' decision-making while in activity-based costing management the enterprises can, through the activity management, identity the root cause of the cost and guarantee the quality of cost information for the formulation and implementation of enterprises' strategies to meet the different needs of cost management for different strategies, thus improving the customers' value and increasing the enterprises' profit.

2.2 In Activity-Based Costing Management, the Enterprises Emphasizerprises'mer'tategoes of Cost Manangement Expensed in the Operation Process the Value Chain Analysis and Consider the Cost Overall

The enterprises conduct a series of activities to realize the production management process of its value target, so they can distinguish the value-added activities and non-value-added activities through a systematic and comprehensive analysis of value chain. According to value chain theory, value chain can create value and consume resources. Through further analysis of value chain, the enterprises promote the optimization and coordination of the activities of local enterprises and ensure the consistence of local and overall interests, which will further enable them to eliminate the non-value-added assignments to reduce the waste, resulting in the optimization and reorganization of value chain to reduce cost.

2.3   In Activity-Based Costing Management, the Enterprises Pay More Attention to the Value-Added Costs and the Non-Value-Added Costs, and Emphasize the Elimination of Non-Value-Added Costs

The value-added costs are the costs when the enterprises conduct value-added activities efficiently, which is the continuous improvement goal for the enterprises, is helpful to the accumulation of the enterprises' product value and customers' value. The non-value-added costs are the costs arising from low efficiency of the non-value-added activities and value-added activities, of which the non-value-added costs are produced by the non-value-added activities, and value-added activities with a low efficiency can also produce the value-added costs. In activity-based costing management, the enterprises focus their attention on the activities, which consume recourses, which enables the enterprises to improve the efficiency of value-added activities, to eliminate wasting and to identify and report the non-value-added costs, to put forward the goal, and eventually eliminate the non-value-added costs existed in the actual costs but neglected in traditional cost management mode through continuous improvement.

3   The Ways to Continuously Optimize the Cost in Activity-Based Costing Management

In activity-based costing management, the enterprises can continuously optimize cost because it enables the enterprises to transfer the cost management from a traditional static process to a dynamic one and to analyze the activity chain from the activity level to eliminate the non-value-added activities, thus making the value-added activities more effectively and reducing cost. And it enables the enterprises to improve their competitiveness and profitability continuously and to realize the enterprises' strategic objectives ultimately.

From the process of whole management, the enterprises can take measures to improve the activity management and optimize the cost.

3.1   Optimizing Activity Chain

Cost is driven by the activities, so the enterprises must be clear about the structure of activity chain and the organization of activity flow, whether the cohesion between activity flows is smooth or not, whether the non-value-added activity need to be eliminated, in this way the enterprises can improve circulation speed of the activities and increase the efficiency of value-added activities. It must be noted that the optimization of an individual activity does not equal that of the whole activity chain, so the enterprises must regard the activity chain and analyze it as a whole body of value creation for customers. Thus they can avoid the disadvantages of reducing the costs for the sake of reducing the costs, and it helps consider the profitability of the whole enterprise from the perspective of the enterprises' strategies, therefore, optimizing activity chain is a dynamic and complicated process. The enterprises can distinguish the value-added activity and non-value-added activity to eliminate the non-value-added activity and to minimize the cost for the non-value-added activities that cannot be eliminated at all. For value-added activity, the enterprises can improve the overall efficiency by reforming process, such as the business process reengineering.

3.2   Reducing Activity Cost Continuously

In activity-based costing management the enterprises focus on activity chain in the logistics to enhance their efficiency while in traditional cost management the enterprises focus on their cash flows to reduce the cost continuously. According to cost information provided in the activity-based costing management, the enterprises must focus on the high cost, whether they are added-value costs or non-added-value costs, control the main activities in value chain and   increase efficiency and reduce cost by forwarding relevant activity. Meanwhile, the enterprises can eliminate the idle resources in the cost and reallocate them to reduce the consumption of time and resources, so as to achieve cost reduction. In addition, the enterprises can make full use of economies of scale to improve the efficiency of value-added activity, namely, increasing the ratio of output to input, can also reduce the cost of product allocated.

References

1. Brealey, R.A., Myers, S.C., Allen, F.: Principles of Corporate Finance. China Machine Press, Beijing (2007)
2. China CPA association, (ed.): Financial Cost Management. China financial and economic publishing house, Beijing (2010)
3. Pan, F., Dong, W., et al.: Value-based Management Accounting: Cases Study. Tsinghua University Press, Beijing (2005)
4. Ronald, W.H.: Management: Creating Value in a Dynamic Environment, 5th edn. McGraw-Hill Companies, Inc. (2002)
5. Wang, X.: Activity Cost Calculation: Theory and Application Research. Northeastern university press, Shenyang (2001)
6. Xiang, L., Wu, X.: The Activity-based Costing Method in The Application of Environment Analysis in China. Accounting Communication (1) (2006)
7. Zhou, B.: Management Accounting. Nankai university press, Tianjin (2004)

Analysis on the Influence of Visual Modeling on Information Transmission in Advertising Design

Cheng Chang and Jicheng Xie

Institute of Art & Fashion, Tianjin Polytechnic University, Tianjin, China
changcheng412@yahoo.com.cn, zhangwuren@126.com

Abstract. Information transmission is main characteristics of advertisement, which relies on visual modeling to transmit information. In order to achieve the perfect advertising results, the designer always adopts modern design to transmit commodity information. This paper will reflect the transmission result by understanding and analyzing visual modeling in advertising design.

Keywords: information transmission, advertising design, visual modeling.

The advertisement took the information transmission is the most fastidious in the artistic performance the visual effect, the advertisement work and audience's visual touch is extremely direct. But if the advertisement visual language expression is not good, can become the information transmission the barrier, greatly will weaken the advertizing info transmission the effect. Therefore, the advertisement visual modeling has the special research value, it is playing the extremely vital role regarding the information transmission effect. The advertisement mainly is plays its role by the visual transmission, must display visual in the advertisement design the positive role to have to grasp the visual modeling, can cause audience's attention. The advertisement is not the simple visual question, but is relating the advertisement final transmission effect, but the visual modeling must display the advertisement the characteristic to be able to cause audience's attention. The advertisement vision modeling mainly is through the visual vivid element in the plane space the effective arrangement and the performance, thus disseminates the commodity information method well. Below we come under the concrete analysis in the advertisement visual modeling.

1 Visual Modeling Elements

A. Point

The point is the succinct modeling, is all shape origin. As the visible modeling, a accumulation and the disperser can cause the picture to produce the person or household who refuses to move and bargains for unreasonably high compensation when the land is requisitioned for a construction project, the accumulation spot have the centripetal force, is in the picture core. Took by the constitution visual modeling, may through the method which fans out from point to area design the modeling, also may the initial point constitution design the modeling.

B.   Line

The line modeling extremely rich expressive force, the line along with the direction, the shape different may have the different visual feeling. For example: The horizontal line has the tranquil feeling; The vertical line has the strength feeling; The inclined line has the impact feeling; The curve both has in the length change, and has in the modeling change, the cosmetic on strong in the straight line. On often directly displays the commodity characteristic in the advertisement design with the curve modeling.

C.   Plane

The plane has the associative perception and the visual tension in the visual modeling. The entity mask has the impulse and the expansion feeling; The hypothesized mask has the attraction and the contraction feeling. In the visual modeling has the surface which produces by the color shape, has surface which produces by the line constitution. For example: The advertisement designs Chinese Library and the writing, the image is likely the color constitution surface, the writing is the line constitution surface.

2   Visual Modeling Manifestation

A.   Product Visual Modeling

The product visual modeling is refers to the direct performance product shape and the characteristic modeling technique, really unfolds in the advertisement design the product gives the audiences. A commodity visual modeling needs to want to give the human and causes the human by the direct feeling to remember profoundly, needs to pay attention to the choice prominent commodity contour, the color, the structure, the material quality in the visual modeling and so on, pays attention to the prominent modeling in the performance the visual impulse and the power, causes to produce the audiences in the vision the intense impression, and strengthens the product the confidence level and the third dimension, achieves the advertisement the goal. As shown in Figure 1.

B.   Originality Visual Modeling

The originality visual modeling is refers to the basis advertisement subject and the commodity demand, displays the modeling effect after the creativity. The creativity modeling is different with the product modeling, it is not displays directly by product itself for the audiences, but is expresses the product demand by the subjective intention constitution perceptual image. The creativity visual modeling is carries on the creative thought with the visual image, seeks in the visual modeling the unique

Fig. 1. Product advertisement

Fig. 2. Originality advertisement

expressive force, effectively transmits the advertizing info vividly. As shown in Figure 2.

3   Visual Modeling Performance Characteristic

A.   Intuitionalism

The advertisement and audience's visual touch is extremely direct. The visual modeling has the positive cognition function regarding the setting up product image, even more is advantageous for the advertisement transmission. At the same time, the visual modeling took one kind of visual performance, itself has the credible feeling which seeing is believing, then the advertisement uses the populace media the transmission, to has been provided the audiences the extremely direct-viewing visual transmission effect.

B.   Concreteness

A real advertisement can have faithfully the audiences. In the advertisement design needs really accurately to reappear the commodity information, a clear concrete visual modeling may have the persuasive power development commodity characteristic. It may clear demonstrate specifically the commodity the contour characteristic, the use function and so on, and let audience realize, the judgment, with is communicated effectively by the direct method the   audiences.

C.   Accuracy

The visual modeling needs to transmit the advertizing info accurately, simply depends upon the pure image to reappear the commodity information by no means, but is must contain questions and so on modeling creativity and image localization. The visual modeling only then locates accurately, the creativity is clear about, can have the command to receive the effect which the audiences believe, thus achieves the advertisement transmission the effect.

4   Visual Modeling Should Display the Characteristics of Advertising

A.   Visual Modeling Must Display Usefulness Information

In the advertisement design, the visual modeling cannot be separated from the concrete commodity image, must contain the rich commodity information, these related commodity characteristic image and the information are precisely the consumer want the understanding to obtain have the practical value information. The people accept the information always and oneself are interested the content related, is willing to accept on own initiative.

B.   Visual Modeling Must Display the Irritant Information

Arnheim thought that the vision is completely one kind of positive activity. Is interested by the eye selection the thing is one kind of instinct activity, was the stimulant has caused the response which is made audience's attention. In daily life, the people contact the advertisement often are accidental or instantaneous, all accepts the advertizing info carelessly, thus, requests the advertisement picture to have the irritant characteristic specially, but the irritant advertisement is studied the proof is disseminates best the effect.

5   The Influence of Visual Modeling to Information Transmission

The advertisement transmission takes the method by the visual communication, by receives the visual modeling which the audiences easy to accept to take the transmission commodity information, demonstrated the commodity content the way, achieves the advertisement the transmission effect.

A. Visual Modeling Infectiousness

The visual modeling is in the advertisement dissemination most has the emotion factor the information dissemination way. Thus the modern advertisement design demand the numerous emotions, the visual modeling is cut into take the emotion form, causes to infect the numerous unconsciousness transforms as accepts. For example the CHINA MOBILE M-ZONE advertisement image localization is "My site, listen to me", makes to order the mobile communication brand for the young people market quantity body, provides the personalized service. Because the young people pursue the fashion, worship idol, therefore the modeling design on infects by the fashion element and the star effect, moves the audiences, builds the brand-new expense idea, in receives in the audiences to set up the bright brand image. An shown in Figure 3.

Fig. 3. M-ZONE advertisement

B. Visual Modeling Communication

The advertisement is an art which convinces, activates by the sentiment influence demand audience's mood, causes to find the audiences the appropriate own demand the good feeling. The visual modeling is based on and the individual analysis carries on regarding the commodity information assurance to the materiality. The visual modeling may demonstrate the commodity not merely the information content, but also may transmit certain special psychological feeling. The visual modeling took the advertizing info communication the form, is precisely transforms the advertisement subject content for the emotion manifestation moves the audiences, with is carried on audience's mind the communication, and guaranteed disseminates the information is received the audiences to accept.

The 21st century are an informationization time, but the advertisement is biggest, quickest, most widespread information vector, but the advertisement design also most draws close to populace's artistic performance, has the pivotal value in the dissemination domain. The visual modeling is constitutes the advertisement important element, the use vision art method molds and transmits the information, needs to express the advertisement subject accurately, and must have the good dissemination effect. To the visual modeling research into advertisement design in the main demand spot, in the advertisement visual modeling must digest the advertisement the subject to devise completely, and has the intense personalization and the distinctive quality, causes the subject and the design performance perfect union, thus attracts audience's attention, better display information dissemination effect.

References

1. Arnheim, R., (composer), Teng, S., Zhu, J., (trans.): Art and Visual Perception. Sichuan People's Publishing house, Chengdu (1998)
2. Bin, L.: Introduction to Communication Theory. Xinhua Publishing house, Beijing (2003)
3. Ping, L.: Advertisement vision design foundation. Xiamen University Publishing, Xiamen (2008)

Interpretation of the Symbol of Ancient Chinese Pottery

Xin Lin and Jicheng Xie

Institute of Art & Fashion, Tianjin Polytechnic University, Tianjin, China
linxin_880907@163.com, zhangwuren@126.com

Abstract. The invention and use of pottery have facilitated and consolidated the settlement of human life and expanded the meaning of the agricultural production to progress of human society. The pottery is not the language and text, but it is the physical legacy of a bygone era and the fossil of human life from the passing era in another sense. This article is trying to analyze the class and decorative patterns of the ancient Chinese pottery from the perspective of the semiotic cognitive view, in order to find his semiotic meaning.

Keywords: pottery, semiotics, signifier, signified.

Pottery appeared with the sedentary human life from the prehistoric time into the Neolithic. According to archaeological findings and scientific determination the earliest extant pieces of pottery date back to about 9000 ~ 10000 years ago. In the long course of historical development significant changes in terms of form, name meaning and function of the pottery have taken place. There changes not only reflect the development of socio-economic activities, but also reveal the cultural evolution during the time. This article attempts to use the knowledge of modern semiotics to re-examine and interpret the development of the pottery.

The representative of semiotics, the Swiss linguist Ferdinand de Saussure, has an important theoretical insight, "Language is a structured social system and value system, and it is also a historical convention". If people want to communicate with the language, they must abide by it. The "language" here is constituted by a number of symbolic elements. People get the value and meanings of each symbol by comparison with other symbols. We use the above insights in the reinterpretation of pottery in order to understand the pottery from a new perspective.

1   Class of Pottery and Semiotics

Potteries in the Neolithic are mainly daily necessities. According to different purposes of use the can be divided into drain devices, cooking devices, drinking vessels, tableware, container and some other things. The use of some potteries can be guessed and determined today with modern people's thought. But the use and purposes of some classes of pottery are still unknown. The naming of these classes of pottery is only based on their general distinctions. The scope of their use should not be limited to what we know.

J. Luo (Ed.): Soft Computing in Information Communication Technology, AISC 161, pp. 33–37.
springerlink.com   © Springer-Verlag Berlin Heidelberg 2012

In different areas and in different culture types, the details of the different potteries differ from each other quite well. The combinations of the pottery classes are also different. This reflects the different lifestyles of different community groups. In fact, the pottery with the same name may have different forms because of the different regions and different culture and there may be quite obvious differences. If we use Saussure´s semiotic theory to analyze it, the different form and use of the potteries with the same name are called "signified" and the sound and the image of the name are called "signifier". As a result, we find that the relationship between "signified" and "signifier" of the pottery is not inevitable. If we believe that their relationship is "conventional" – that means, their relationship is "established by custom" -, then the meaning of "signified" and "signifier" should be ruled by a group of people who lived in a particular historical period and at a region with particular geographical factors. But, this ruled relationship indicates just the human contact between the different regions. The same "signifier" has different correspondent "signified" with it. How can the same "signifier" be known and recognized by people living in different regions and be given different correspondent "signified" by these people in different regions? There may be a reason for this: the forming and understanding of "custom" shall be from a specific period on, when not many people were living together. But in this period the same "signifier" referred to the same "signified". People had common understanding of "signifier" and "signified" on this basis. After that people began to live dispersed and settle in different regions. People in these regions lived in a same, similar or different way and formed different "customs". Because of the common life experience and the formed common understanding of "signifier" before the dispersed living there were no changes on sounds and images even after the dispersed living. "Convention" should be understood based on this level. The meaning of the "signifier" here deviates from the definition in the concept of Saussure. Maybe the "signifier" of sign language should include not only its acoustic image, but also its auditory image. Both can be regarded as "signifier". If this works, the relationship between "signifier" and "signified" turns to the relationship between the word, which exits as a sound and physical shape, and the represented concept by the word. Thus, it is not one to one correspondence between "signifier" and "signified".

2   How to Name the Ancient Objects

1. The ancient objects are generally named by comparing the objects with the similar objects today, such as cylinders, cans, cups, bowls etc.. But the ancient and modern way of life are different, the name and use of the objects may not be exactly the same. This shows that "signifier" and "signified" in the language of semiotics are strictly temporal and spatial concepts. The reflect the language of the contemporary human concept, not the ancient language and intellect continued till today. For the same object there may be break in term of language and symbol during the language development. Even if modern people make a thorough study of the behavior of the ancient people, they are still not creator and maker of the ancient objects. We are talking about the language and symbol of an era or a period, not only the language or symbol for the object itself.

2. If a similar pottery was still in use in the Bronze Age, it will be named based on the recorded name in the inscriptions of the Bronze, such as Ding, Gui, He, Dun.

3. If we don´t have similar object that is handed down or still useful today, but some words can be found in the ancient text written corresponding to the characteristics of the ancient objects, we use the words to name the object. For example, a device in a specific form with three bags and feet was named "GUI"according to the explanation in the ancient dictionary "Shuo Wen Jie Zi".

4. In burial in Spring and Autumn Period there are often slips listing the name and quantity of the grave goods. According to the slips the unearthed and unknown ancient pottery from the burial in Spring and Autumn Period can be identified and named.

5. Finally, if the mentioned methods above do not work, the objects can be named according to their characteristics and shapes, such as the bottle from Yangshao culture time with sharp end and small mouth and Shang's big mouth device. The former device has been found as a device to draw water. But we don´t know, if the developed device later could still be used to draw water. As the same, someone believes that Shang's big mouth device was a gauge, but this needs further confirmation. It seems a little "pictographic", which means to make the objects graphical at first, and then into a symbol - the text.

That the objects can be named according to their characteristics and the name becomes then "signifier" (linguistics), is based on the fact: if two persons want to communicate with each other, the most important condition is that the speaker and the recipient's have the interlinked language skill, which is called the language sharing. That means people must have close reserve of symbols and marks. The naming of the objects according to their shapes and uses and the determination of "signifier" in this way may enable people to have the most reserve of close symbols and marks and the recipient's understanding of sentences, namely the information decoding procedure is simplest in this situation. Therefore, this language symbol is the most intuitive and popular. This language can be understood as a self-contained metaphor system based on the emotion, thought and belief expression of the Chinese prehistoric culture. Like the British archaeologist Rawson said, "The symbolic system was the ancient China's characteristic and was closely related to various people's activities. Here we use the concept of symbolism to describe that a thing or a mark is used to symbolize one kind of thing or a pile of different things.

3   Surface Decorations of Pottery, Patterns and Semiotics

Symbol is abstraction of perceptual and emotional materials and generalization of them into some kind of universal form; while language is constituted by a certain number of symbolic elements. The value and meaning of each symbol can be gotten through the comparison with other symbols. For example, the word "horse" (mǎ) holds certain space in the aspects of pronunciation and meaning. From the perspective of pronunciation, the own boundary of the word can only been defined, when the boundary of its space has been determined, such as distinguishing the word "horse" (mǎ) from the other very similar sounded words "mother"(mā) "numbness"(má) and "scold"(mà). In the complex real life people use not only the language to communicate with each other not only in use of language, they also use a lot of non-word methods. All kinds of art have formed their special languages. The ancient pottery naturally provides a wealth of information through the surface decorations and patterns.

The surface decoration of pottery has two categories: the first one is working on the pottery surface. The working on the pottery surface has two methods — polishing the surface and painting pottery clothing on it.

A.   Polishing the Surface

Before the pottery blank is fully dry, we can use bones, bamboos, stones and other hard and smooth tools to polish the pottery blank in the same direction, so that the clay particles will arrange along one direction and have dim luster. It was not an accidental action, it showed that the ancient human, based on possessing basic living material, have achieved a higher level of requirements for their own life —— aesthetic. This action symbolizes a leap of the quality of the human creation.

B.   Painting Pottery Clothing

If we reconcile thin pottery clays or porcelain clays with water into a mud, and paint them on unbaked pottery blanks or porcelain blanks, a thin layer of color paste will leave on the surface. The color could be white, red and gray etc. This color paste is called pottery clothing in the ceramic technology.

The method of painting pottery clothing is often used by making color-painted potteries. The pottery clothing is per se a prime place for painting pictures. Pottery clothing with different materials has different coloration. This is sufficient to inspire our ancestors to paint pictures on the surface of potteries. The modern scientific analysis have the limitation, they can´t fully explain why people at the time have chosen different colors. As a medium for loading and transmission of information, symbol is a simplified means for learning things. It is expressed as meaningful codes and code systems. As non-verbal symbols, symbols in the creation design have many similarities to verbal symbols. It makes the linguistics a practical guide for designing. Generally speaking, you can regard the elements and basic methods of creation design as symbols. Through the processing and integration of these elements you can express your feelings and meanings. When human focused their creative behavior no longer just on the functional purpose, the decorations were added because of the preferences of different people. In a certain sense, it reflects the personality of people in particular times.

The other way of surface decoration of pottery is to use patterns. The painted pottery mentioned above may also belong to this kind. In addition, there are patterns by shooting or rolling, additional heap profiling, picked patterns, piercing patterns, colored drawing patterns after firing.

Pottery patterns have a variety of different titles. However, these titles are not standardized. Some patterns have been given different names, because different people have seen them in different ways. The general used names are jomon, blue-pattern, square-pattern, string-pattern, spiral-pattern, mat-pattern, knitting-pattern; fish-pattern in Yangshao color-painted pottery and deformed fish-pattern etc. Some other patterns shared the name with the potteries. Potteries in Shang Dynasty also had imitated brass ware patterns like surface-patterns, Kui-patterns, cloud-thunder-patterns, Hui-patterns etc. The ordering, abstraction and technicalization of these patterns symbolized the improvement of the human´s productivity; they also reflected the aesthetic tastes of ancient people. In the "signifier" and "signified" of these patterns, from the perspective that people can express meanings with decorations, decorations can be regarded as symbols. Symbols have the function of loading information. So the "signifier" can be understood as patterns, which is carrier of information, and its formation. "signified" is the information self. The different decoration patterns transmit the different culture and become the symbolic characteristic of the objects from a certain period.

Although the use of pottery has never become an important indicator of the productivity developments like bronze and iron as a result of its supporting function for the production tools. The pottery is only used as the alternative supporting tools and materials in the Neolithic Age. But the invention and use of pottery have facilitated and consolidated the settlement of human life and expanded the meaning of the agricultural production to progress of human society. Though the pottery is not the language and text, it is the physical legacy of a bygone era and the fossil of human life from the passing era in another sense. It brings the descendants very valuable information of ancient history and makes us to know our great ancestors and glory civilization better.

References

1. Wu, R., Xin, A.: History of Chinese Pottery and Porcelain. Tuanjie Press (April 2006)
2. Chinese Ceramic Society (ed.): History of Chinese Pottery and Porcelain. Wenwu Press (September 1982)
3. Zhang, R.: Ceramic Technology. Chemical IndustryPress (July 2007)
4. Echo (Italian), Wang, T. (trans): Semiotics and Philosphy of Language. Baihua Literature and Press (July 2007)
5. Xu, H.: Design Semiotics. Tsinghua University Press (July 2008)

Research Overview on Apparel Fit

Yufu Shan[1], Gu Huang[2], and Xiaoming Qian[2]

[1] Institute of Art and Fashion, Tianjin Polytechnic University, Tianjin, China
[2] College of Textiles, Tianjin Polytechnic University, Tianjin, China
shanyufu@hotmail.com, qianxiaoming@tjpu.edu.cn

Abstract. Apparel fit is defined as the relationship between the size and contour of the garment and those of the human body. Apparel fit is one of the most important factors that influence apparel comfort and apparel sale, and it is also a basic requirement of wearing. This paper states the present research status on apparel fit.

Keywords: apparel fit, fit evaluation, ease, fabric property, body-type, body scanner.

1 Introduction

Apparel fit has long been regarded as the single most important element in clothing appearance which attracts the customers. Good fit is crucial to clothing comfort, appearance and consumer satisfaction. Fit is very subjective and each individual differs on what they describe as good fit and how they like clothing to fit their bodies.

Numerous factors contribute to consumers' clothing and fit preferences including comfort, aesthetics, and personal choice in assessing fit. Current fashion trends, cultural influences, age, sex, body shape, and lifestyle also influence personal fit preferences, and changes in these elements may result in changes in personal fit preferences over the life span. So the definition of fit varies from time to time. Some general definitions are introduced here.

2 Definition of Fit

Cain said that " fit is directly to the anatomy of the human body and most of the fit problems are created by the bulges of the human body" ;Chamber and Wiley said that "clothing that fits well, conforms to the human body and has adequate ease of movement, has no wrinkles and has been cut and manipulated in such a way that it appears to be the part of the wearer" Erwin and Kinchen said that" fit is defined as a combination of five factors: ease, line, grain, balance and set"; Hackler said that" clothing should fit the body smoothly with enough room to move easily and be free from wrinkles"

These divergent definitions of fit reflect the lack of agreement within the industry on the features which are responsible for a good fit. Therefore, a more detailed understanding of the factors contributing to clothing fit is necessary.

J. Luo (Ed.): Soft Computing in Information Communication Technology, AISC 161, pp. 39–44.
springerlink.com

3   Factors That Efect Apparel Fit

Physical comfort, psychological comfort and appearance all play in the consumer's perceived satisfaction of fit.

A.   Ease

Many researchers agreed that optimum apparel eases are important to the garment appearance and fit (Reich 1991).The optimum amount of garment ease is an important parameter for mobility, appearance and fit as too much ease of allowance may block the movement and create wrinkles on garment surface, too little nearly "just fit" may create pressure sensation and also affect the body movement.

The optimum level of ease allowance is important for achieving the high satisfaction on garment fit for body movement unless the product is stretchable enough for wearing or is a tight-fitted style of garment. Unfortunately, guidelines on the optimum range of ease allowance are very limited.

B.   Body Cathexis

Body cathexis is defined as positive and negative feelings toward one's body. Various scales evaluating body cathexis have been used to examine attitudes toward the body. Alexander et al. (2005) found that respondents who were dissatisfied with their weight preferred a loose fit in dresses, and respondents who were satisfied with Particular body parts (such as thighs, bust, or hips) preferred a closer fit in the area of concern.

C.   Fashion Trends

When fashion emphasizes the physical body (body primary), less ease is incorporated; When fashion places an emphasis on the clothing (clothing primary), more ease is incorporated to conceal the body beneath (Fiore and Kimle, 1997).

D.   Fabric Property

Fabric is also an important factor that affects the fit of the garment, especially mechanical properties of fabrics, such as bending rigidity, the shear resistance, elongation of fabric-self-weight, and so on.

4   Relevant Research on Apparel Fit

A.   Apparel Ease

Generally ease is divided in two types: (a) wearing ease—the amount of extra fabric allowed over and above body measurements to ensure comfort, mobility, and drape of the garment and (b) design ease—additional amounts of fabric added to achieve certain design effects by changing the line and shape of a garment. Because design ease is subjective, style dependent and not essential to the basic understanding of fit, so ease studying of apparel fit is referred to wearing ease. Makabe et al (1988) has attempted to determine the part and quantity of necessary ease of clothing in the upper limb motion. The change of length surface and relief of the human body during the upper limb motion has been investigated by means of dermatograph, moire topography and modeling by gypsum tape. Janice et al (1996) has evaluated the amount of garment ease in protective garments for allowing the worker to move uninjured and being recognized as comfortable by the wearer. Shu et al (1997) claimed that a lot of geometrical style lines could be drawn in accordance with the same measurement value from human body. That means the two-dimensional pattern may not fit the human body without considering the body shape. Musilova et al (2003) assessed the allowances on the looseness of ladies blouse, ladies polo coat and ladies coat. Alexander et al. (2005) studied fit problems of females with four body shapes: pear, hourglass, rectangular, and inverted triangular.

B.   Fit Evaluation

1) Subjective fit assessment

In order to understand the optimum garment ease for wearers and enhance the customer satisfaction in terms of fit and appearance, many researchers conducted a lot of subjective assessments on garment fitting. This not only ensures the quality of the garments, but also provides a yardstick to the manufacturers for product development. Moreover, it helps to generate precise garment patterns with the consideration of customers' preference.

Subjective fitting assessments were carried out by Shim et al. (1993), who conducted a study with respect to the satisfaction of apparel fit for the women 55 years and older. Huck et al (1997) evaluated the subjective wearing sensation of the protective overall. Wearers were required to follow the procedure of exercises such as kneel on left knee, duck squat, stand erect etc. Kohn and Ashdown (1998) found a positive correlation between fit analysis using a video image and traditional fit analysis using a live model. Both models resulted in a reliable analysis of the garment/body relationship. Fit researchers have used scanned images and expert judges to evaluate the fit of pants (Ashdown et al., 2004) and cooling vests (Nam et al., 2005). Nam et al. (2005) found using 3D scan images for fit analysis convenient and accurate because there were no constraints due to time, model availability, or fatigue. Although the subjective assessment tends to be inconsistent and unreliable, but it is still important for manufacturers to understand the fit requirements of the wearers to a particular garment in terms of personal perception and garment appearance when hanging on the body.

2) Objective evaluation of apparel fit

Different objective methods have been proposed or developed for assessing the garment fit more precisely and accurately. These include fitting index, summarized dot pattern, waveform of clothing, moiré topography and video image.

C.   Correlations between Body-Type and Apparel Fit

Most fit problems occur due to figure variations in body contour, posture, and proportion (Kwong, 2004). Traditionally, body-type can be classified into four types, that is pear, hourglass, rectangular, and inverted triangular. Alexander et al. (2005) studied fit problems of females with four body shapes, They found that the participants who identified their body shapes as rectangular, pear and hourglass were more likely to express fit problems at the bust area than those who perceived themselves as the inverted triangular shape (Alexander et al., 2005). Kind (1995) used the height measurements of 358 college females to classify their body types into petite, average, tall, and large sizes. Kind (1995) surveyed their fit perceptions and found that the college females, with exception of the average group, were dissatisfied with pant length (Kind, 1995).

D.   Application of Body Scanning Systems

The first element for the implementation of mass customization is the accurate body measurements by 3D body scanners. Istook and Hwang (2001) pointed out that a comprehensive and accurate set of measurements could help to customize garments for fit. The development of three dimensional body scanning technologies is key to the apparel industry in terms of mass customization.

According to Istook et al (2001), Blais (2004) and Fan et al (2004), the principle of three-dimensional body scanners is to capture the outside surface of human body without physical contact with the body using optical techniques or light sensitive devices.

A shadow scanning method developed by Loughborough University was used in one of earliest 3D body scanning systems, which was named The Loughborough Anthropometric Shadow Scanner (LASS) in the UK in 1987.

SYMCAD is Telmat's 3D body scanner and developed in France in 1995 (Russell 2000). The measurement data from Telmat body scanner can be integrated into apparel CAD systems such as Gerber Technology's Accumark system or Lectra System's Modaris software.

Now Lectra 3D body scanner is widely used all over the world. This system contains VitusSmart (used to scan the body) and ScanWorX (mainly used for data processing).

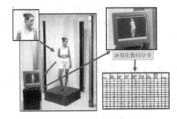

Fig. 1. Lectra 3D body scanner

E.   Fabric Properties to Apparel Fit

Material property is one of the important parameters related to garment fit. Fabric properties have been considered for pattern generation for enhancing garment fitting. Aisaka et al (1998) used the physical properties of fabric to develop an estimating method of the three-dimensional shape of made-up garment in order to develop a suitable planar pattern information processing method. Okabe et al. (2001) used the mechanical characteristics of fabric for three-dimensional dress simulations.

F.   CAD Technology and Computer Simulation on 3D Virtual Garments

Apparel CAD is a very important element in mass customization. CAD is a well-known tool in the clothing industry. CAD systems are widely used in clothing industries, such as textile design, grading systems, marker making systems, pattern design software, pattern generation software etc. The following review is focused on CAD for garment design.

Yamakawa (1989) developed the computer patternmaking system, which makes garment patterns automatically by the input of the fashion drawing, as there were no systems for patternmaking. Heisey et al (1990) developed algorithm for draping a basic skirt in three-dimensional form and convert the individually fitted skirt to two-dimensional pattern. Kang et al (2000) developed an apparel CAD system to perform automatic garment pattern drafting and the prediction of the final shape of designed garment dressed on the human body. Istook (2002) proposed a step by step guideline for using the CAD technology so as to enhance the custom fit of the designed garments.

5   Conclusions

In summary, apparel fit is a complex issue, which relates to many relevant fields, such as apparel construction, anthropometry, apparel comfort, apparel psychology, computer graphics, and so on. In this field, there are so much more for us to do further research, such as objective evaluation of apparel, the correlations between mechanical characteristics of fabric and the apparel fit.

References

1. Ashdown, S., Loker, S., Adelson, C.: Use of body scan data to design sizing systems based on target markets. National Textile Center (April 1, 2003)
2. Ashdown, S., Loker, S., Schoenfelder, K., Lyman-Clarke, L.: Using 3D scans for fit analysis. Journal of Textile and Apparel, 15–21 (May 2004)
3. Petrova, A., Ashdown, S.P.: Body Size and shape dependence of ease values for pants' fit. Clothing & Textiles Research Journal 26(3), 227–252 (2008)
4. Yu, W.: Subjective assessment of clothing fitness. Hongkong Polytechnic University, Hong Kong (2005)
5. Alexander, M., Connell, J., Presley, A.B.: Clothing fit preference of young female adult consumers. International Journal of Clothing Science and Technology, 52–64 (2005)

6. Huckabay, D.A.: Perceived body cathexis and garment fit and style proportion problems of petite women. Virginia Polytechnic Institute and State University, Blacksburg (1992)
7. Devaragan, P., Istook, C.L.: Validation of "female figure identification technique(FFIT) for apparel software". Journal of Textile and Apparel, Technology and Management, 1–23 (2004)
8. Loker, S., Ashdown, S.P., Schoenfelder, K.: Size-specific analysis of body scan data to improve apparel fit. Journal of Textile and Apparel, Technology and Management, 1–15 (2005)
9. Ashdown, S.P., Delong, M.: Perception testing of apparel ease variation. Applied Ergonomics, 47–53 (1995)
10. Ashdown, S.P.: Introduction to sizing and fit research. In: Clemson Apparel Research, 2000: The Fit Symposium (2000)
11. Ashdown, S.P., Schoenfelder, L.K., Lyman-Clarke: Visual fit analysis from 3D scans. In: Abstracts of the Fiber Society 2004 Annual Meeting and Technical Conference, p. 111 (2004)
12. Kong, M.Y.: Garment design for individual fit. In: Fan, J., Yu, W., Hunter, L. (eds.) Clothing Appearance and Fit: Science and Technology, Textile Institute, pp. 196–233. Textile Institute; CRC, Cambridge (2004)
13. Choi, J.H., Hong, S.A., Kim, S.A.: 3D sewing and virtual fit of stage costume. Paper presented at Seoul International Clothing and Textiles Conference, Seoul (2005)
14. Kim, S.M., Kang, T.J.: Garment pattern generation from body scan data. Computer-Aided Design 35(7), 1–8 (2003)
15. Simmons, K.P., Istook, C.L., Devarajan, P.: Female figure identification technique (FFIT) for apparel, part II: development of shape sorting software. Journal of Textile and Apparel Technology and Management 4(1) (2004)
16. Goldsberry, E., Shim, S., Reich, N.: Women 55 years and older: Part II. Overall satisfaction and dissatisfaction with the fit of ready-to-wear. Clothing and Textiles Research Journal 14(2), 121–132 (1996)
17. Simmons, K.P., Istook, C.L.: Body measurement techniques: comparing 3D body-scanning and anthropometric methods for apparel applications. Journal of Fashion, Marketing and Management 55(3), 26–32 (2003)
18. Carrere, C., Istook, C., Little, H., Hong, T., Plumlee, T.: 100-S15 Automated Garment Development from Body Scan Data, National Textile Centre Annual Report, National Textile Centre, Auburn, AL (2001)

Research on How to Incorporate Helping Mechanism of Students Who Have Learning Difficulties into Teaching Management

Wei Shao, Fuzhong Wang, Jie Huang, Jing Sun, Ru Li, and Qi Yang

College of Science, Tianjin Polytechnic University, Tianjin, China
Shaowei_tj@hotmail.com,
{Wangfuzhong,huangjie,sunjing,liru,yangqi}@tjpu.edu.cn

Abstract. This study combines with our university's actual situation, investigates helping and management system of students with learning difficulties from the view of teaching management. On the basis of analysis of causes for learning difficulties, we study the existing teaching management system reform, promote the hierarchical management, develop management systems for special students with learning difficulties pointing at different teaching processes, and strive to develop the potential of each student with the principle of promoting construction through management to truly improve the quality of teaching.

Keywords: management, students with learning difficulties, Helping Mechanism.

1 Significance of Research and Analysis of Present Situation

"National medium and long-term planning and outline for education reform and development (2010--2020)" clearly prescribes that "we should focus on individualized teaching, concern about the differences between students with different characteristics and personality to develop the potential of each student; We should simultaneously advance the reform of stratification of teaching, classes in turns, credit system, tutorial system and build up helping mechanism of students with learning difficulties."

Priority to the development of education proposed by the party and the country is a major policy which has to be adhered for a long-term. With the full implementation of China's strategy of science and education and building up the strength of the country with talented people, higher education gets a rapid development of an unprecedented scale. College enrollment has expanded each year, China's higher education formal entry into a stage of popularization, more students have access to higher education. But at the same time many problems have been magnified which we need to pay more attention. For example, regional differences in the quality of students, the major emotions caused by blind choice of major, the jobs confusion and anxiety caused by the increasing competition and pressure and so on. With the growing in the total number of students, the number of college students with learning difficulties

J. Luo (Ed.): Soft Computing in Information Communication Technology, AISC 161, pp. 45–48.

increases, some students can't even complete their studies and get failed in graduation. Currently all colleges are facing the problems of increasing number of students with learning difficulties, which has a great impact to the whole quality of teaching. Therefore, the situation of college students with learning difficulties has become a problem that should not be ignored in China's higher education [1].

Teaching management is an important part of education, especially for underachievers with high failure rates. The fundamental requirement of education is to educate people. Students are the main component of education, we should be concerned about each student, encourage the initiative and lively development of them, following the rules of physical and mental development and education. The objective and significance of this research is how to incorporate helping mechanism of students who have learning difficulties into teaching management to provide the most suitable education for each student and truly improve the quality of education. We should also make efforts to train high-quality workers, professionals and creative talents.

2   Analysis of Students with Learning Disabilities

At present the colleges are faced with a problem of growing number of students with learning difficulties which has an increasing influence on the overall quality of teaching.

We made an detailed investigation for existing students with learning difficulties[2], then we found that their deep-seated reasons of learning difficulties can mainly due to the following aspects:

- Unevenly in different parts of different education level can have different recruitment conditions, so students are of very different basis. Many students are born "congenitally deficient".
- Parents do not understand the development and content of their major, the kids just choose majors depending on assumption and preferences. Many students are still "confused about their majors after entered into university."
- The increasing employment pressure on students on campus can't be ignored, many students are very confused on the whereabouts after graduation, thus resulting in " step up with disabilities."
- In order to achieve good results in the college entrance examination, the centralized management of high school and too much intervention from parents have caused a lot of students less capable of self-management. Now facing more and more temptation, it is not easy for students to control themselves, on the contrary they indulge themselves in and resulting in " go down out of control. "
- Growing up with parents' over favorites, many students have weak character, bad psychological bearing ability of failure and pressure which easily lead them to extreme emotions, low self-esteem and desperation, thus result in "give up learning".

3 The Existing Research on Teaching Management System and Reform

Teaching Management goes through the whole process of teaching, it is a powerful assurance for maintaining the normal order of teaching, teaching research and reform, full implementation of quality education as well as the accomplishment of high effectively teaching task.

With the constant expansion of various colleges and universities, the college's educational level and strength is also rising. However, from the view of teaching management, they still have problems of lacking of modern management concepts, systematic, serious formalize pattern, serious neglect of the students with different characteristics and personality differences; they lack of stratification and service management thinking, failing in mobilizing the enthusiasm and initiative of students as well as setting a good style of study [3]. Therefore, our reform should highlight the following aspects:

- Combining with the actual situation of our university, we should implement teaching and management reform according to the characteristics and personality differences of students to develop potential of each student with the principles of promoting construction through management.
- University, college leaders at all levels should pay more attention to the psychological problems, learning and living problems of students who have difficulties in learning.
- Taking good use of the school forces at all levels to help students with learning difficulties, within the methods of reasonable and effective teaching management system to integrate teachers, teaching management, students management and parents together and implement the joint management.
- Promote the hierarchical management, humanized management, establish a special management practices for students with learning difficulties and then fully mobilize the subjective initiative of students.
- Learn the advanced experience from other universities.

4 The Present Research Findings

The total number of students in Science College of Tianjin Polytechnic University has been growing from the original 400 of 3majors to more than 1000 of four majors. In this process, both students work and teaching task have encountered so many problems, but they all got solved finally. We did a lot of attempts to the extent permitted in the school policy, and have formed a unique working methods with the characteristic of science college, that includes:

- Early warning system for failing students in high credit system
- Methods for student registration transaction in college of science (Revised Edition)
- Notification system for midterm exam results
- Management methods for special students in college of science

These systems and methods of work have achieved very good results. For example, in 2009, 83% students with learning difficulties in our college had well achieved learning task, of which 50% of them scored even more credits than the normal students and then skipped a grade.

5   Conclusion

Although our previous reform had very good results, the current reform still has the problems of low efficiency, large workload, inharmony of various departments and many other defects. With the increasing number of students with learning difficulties, we must work hard to continue to advance the future of teaching management reform, the formation of efficient, smooth, humanized and advanced management mode.

References

1. Gaosong: Causes and Countermeasures of Students With Learning Difficulties in Colleges and Universities. Hei Longjiang Education (Higher Educational Research and Evaluation) (12) (2005)
2. Jie, H.: The Establishment of Special Education Management Mode for Guiding Work of Underachiever. Higher Education in Chemical Engineering (4) (2004)
3. Jing, Z.: Build a Harmonious Relations Between Teachers and Students Eliminate "Learning Difficulties" of College Students. Journal of Hunan Institute of Engineering 18(4) (2008)

On the Influence of Online Education on Teacher-Student Relationship

E. Lai and Yansong Xue

College of Management, Tianjin Polytechinc University, TJPU,
Tianjin, China
printppp@sina.com, yansongxue@163.com

Abstract. With its distinctive features of openness, sharing and interaction, online education facilitates and strengthens a new teacher-student relationship which is characterized by teacher-led and student-centered characteristics. While providing an effective teaching that cannot be realized in traditional face-to-face classrooms, online education has some disadvantages in itself such as a lack of emotional communication, ineffectiveness in supervision. Therefore, to develop online education, it is necessary to draw some lessons from the traditional teaching model in maintaining a close and intimate teacher-student relationship. To that end, on the one hand, interaction between students and the teacher should be emphasized and an emotional connection between students and the teacher should be built; on the other hand, more emotional elements can be included into the online education resources so as to arouse students' emotional participation.

Keywords: online education, teacher-student relationship, leading role, subject status.

1 Introduction

Online education, which integrates such means of artificial intelligence, decision science and system science, is a computer- and network-enabled distance education in which learning content is delivered and stored with technologies of digitalization and multi-media [1-4]. With the emergence of a brand new teaching mode brought about by the transformations in teaching concept, content and pattern, a new teacher-student relationship is gradually taking its shape. It is of extreme significance to explore the characteristics and functions of online education, its influence on teacher-student relationship, and ways of establishing a teacher-student relationship under the context of online education. This article reviews teacher's leading role and students' subject status in online education mode and presents some new perspectives on and solutions to the influence of online education on teacher-student relationship.

2 Teacher's Leading Role in Online Education Model

In terms of teacher's status in a teaching relationship, in traditional face-to-face classes, the teacher is regarded as an authority whose responsibility is "impart truth, teach students and clear up difficult questions" and therefore takes a dominate role. Teacher controls nearly all the teaching resources and assumes the mission of maintaining cultural continuity. It has been thought as a perfectly justified fact that teacher is responsible for modeling students in overall aspects by imposing on them knowledge and values. Under the traditional mode, teacher learns knowledge from predecessors and then passes the knowledge on to students. By doing so, teaching becomes a simple process of knowledge –processing and students become passive recipients.

In terms of teaching model, the traditional face-to-face class is a teacher-centered one. First, teacher nearly controls all the key teaching activities such as selecting textbook, making curriculum plan, adopting teaching style and method, utilizing assessment instruments and means. When a teacher is enjoying his absolute right to design his teaching plan, he is likely to know what to teach and how to teach so as to help students learn knowledge about the world in an easy and quick way. It is also likely that when designing his teaching plan, the teacher will take students' interest and needs into account. However, it is not possible to meet each student's needs, for they are, after all, not students' choices by themselves. Second, under traditional model, teacher has to spend most of his time and energy on arranging and imparting knowledge. Meanwhile, students' enthusiasm of critical learning is killed and therefore their learning achievement is lowered, for this kind of cramming education leaves students no room and possibility to explore and appreciate knowledge world.

Online education has greatly impacted traditional face-to-face classes and therefore has brought new disequilibrium to the education system. In this context, teacher tends to take a leading role instead of dominating the whole classroom. The teacher-centered teaching model tends to become a student-centered one.

Access to sharing information is the most distinctive characteristic of network, which implies that as a kind of information, knowledge can also be shared. Education resources are expanded to an infinite knowledge bank in a network system rather than a limited knowledge body in a teacher's brain. Students have access to knowledge to meet their specific needs without turning to a teacher, which fundamentally changes the situation in which teacher owns knowledge and "know earlier than students", for students can learn more and earlier with the help of online resources. In online education, students have access to any digital library in the world and even can exchange ideas with those most respected in the specialized realm directly; therefore, they will not uncritically trust the teacher in his location. Thus, teacher's dominant role vested by owning knowledge is weakening.

In online education, teacher's main responsibility is supervision instead of imparting. The interaction and communication between teacher and students is not realized in a face-to-face classroom but in carefully designed online classes with effective learning-supportive services based on information and computer technologies. In this kind of interaction and communication, teacher becomes a designer, an organizer and a promoter of the online education activities instead of an instructor who only imparts and crams knowledge. Thus teacher is only one part of the whole online learning system, which requires that teacher should adopt brand new teaching model, teaching method and teaching plan. The traditional teacher-dominant teacher-student relationship is transformed into a more and more interactive and collaborative one.

Therefore, under the online education model, teacher should take the participant of his teaching into account and design the class with the help of information technologies. He also should make a teaching plan and syllabus that are more applicable in online education. Moreover, more work should be done such as introducing a textbook that can facilitate self-learning, making multi-media courseware, organizing daily learning activities and proving supportive learning services to students. Currently students have access to a large amount of online resources. However, not all of those resources are effective and high-quality. Under this context, while assisting in students' learning, teacher has to assume the responsibility of managing online resources. Online education imposes higher requirements on teachers. Teacher is not only a role who imparts knowledge and enlightens wisdom but a supervisor of vast online resources and a mentor of leaning content, learning way and positive value.

3   Students' Subject Status in Online Education

In terms of teacher-student relationship, in traditional face-to-face classes, students highly depend on teachers, for their main learning sources are teachers. Although a part of their knowledge comes from textbook, teacher is still the main lecturer of the textbook. Without teacher's participation, the whole teaching activity cannot be implemented. Without teacher's lecturing, mentoring and organizing, the teaching activity will be in a mess. Therefore, students have to depend on teacher and become passive recipients.

In terms of learning autonomy, the traditional face-to –face class leaves little room to student. Students have no autonomous rights in choosing teacher, teaching content, teaching method etc. In short, with so many limits, it is impossible for students to choose new knowledge and skills and choose learning timetable. However, online education gives back the autonomous rights to students and makes it possible to break up the limits of time, space and other limits in traditional face-to-face classes.

Online education is based on students' active participation, autonomous selection, for its teaching content, knowledge system, learning method and management is student-centered. Online education presents its advantages by providing student with autonomous learning environment and resources, which provokes learning interest and emphasizes autonomous learning mode.

Students gain a real subject status in online education. The content of traditional face-to-face class is impacted by teacher's interest and specialized knowledge. During the whole teaching process, students cannot gain a subject status and their needs are difficult to meet. Students' creativity and innovation are neglected. In online education, students are viewed as an independent individual, which makes it possible for students to develop their own learning habits on the basis of their talents and cognitive modes. Therefore, students become active participants instead of passive recipients. Online education focuses on learning environment that facilitates learning, on student-centered model and on collaboration and interaction in learning, instead of on controlling students' learning process. By doing so, students are saved from the simple and boring face-to-face class. Under the online learning model, students can bring their autonomy into full play and show their personalities by gaining information from diverse sources and then integrating it into their existed knowledge system and cognitive structure.

Network provides students with a free learning and communication environment. Teacher is no longer the sole source to turn to when students need information and reference. With more information available online, students' learning autonomy and enthusiasm are given into a full play. However, online education imposes new requirements on students. First, students should bear autonomous learning in their minds. To accomplish the learning goals, students should manage their behaviors in class-designing, learning assessment and information-retrieving. They should design their learning activities according to their own learning needs and goals. Second, students should be familiar with network technology and utilize all the learning resources to carry out their learning activities. The whole learning process is implemented by high interest, active thinking and great inspiration, which develops students' self-management and self-control abilities.

Of course, emphasizing students' subject status do not imply an underestimate of teacher's role in students' development. In online education, teacher is still a supervisor, manager and mentor. Students' subject status should be based on teacher's leading role.

4   Problems in Teacher-Student Relationship and Their Solutions in Online Education

As a newly-emerging teaching model and method, online education has been a general concern in education realm since its emergence. It changes the traditional teacher-student relationship into a new democratic, equal and interactive one[3, 6-7]. However, due to a lack of effective and efficient models and popularity of online education, it cannot be denied that there is a conflict between ideal teacher-student relationship and the actual one in reality. The problems in teacher-student relationship are as follow:

A.   A Lack of Emotional Communication

In traditional face-to-face classes, the face-to-face communication is not only a process of knowledge-imparting, but also a process of exchanging attitudes, emotions and ideas. Teacher can influence students by his unique personality, thirst for knowledge and appropriate outlooks, in which students can learn a lot that cannot be learned from textbooks. In particular, some excellent teachers can arouse student's learning interest and leave an expressional memory on them with humorous and interesting classroom teaching style.

In online education, the communication between teacher and students is realized by BBS, E-mail and other network communication tools, which lacks face-to-face and situational communication. In other words, the relationship in online education is a virtual one. Individual in the virtual network environment is deprived of emotion and becomes an abstract data or sign. When participating in online education, students are confronted with lifeless and emotionless objects. There is only a process of knowledge–imparting without any emotional communication. By doing so, students will only focus on learning itself and therefore, neglect developing their learning interest and learning attitude.   Thus, it poses a great challenge to the design of online classes in bringing emotional elements to the objective equipment.

B.   Absence of Effective Supervision

Currently, a lot of studies have shown that Chinese students lack autonomy, independence and self-control ability, especially the self-control ability. Online learning model is a highly autonomous learning model, which requires students own high self- control ability. Students should conduct a niche targeting autonomous learning based on their own learning characteristics and weakness. In other words, in online education environment, students should have high self-control ability and learning autonomy. In the current online education activity, teacher mainly focuses on reorganizing knowledge, designing classes, providing learning resources and teaching learning methods, while, neglect necessary supervision and management on students. Providing information to students without constraints and monitoring their effects is likely to make the online education a laissez-faire learning.

In order to overcome the aforesaid defects in the communication, it is necessary to draw lessons from traditional face-to-face classes in interactive teaching and integrate the advantages of the two.

To solve these problems, the following solutions can be adopted:

C.   Focusing on Emotional Education in Online Education

First of all, focus should be put on strengthening the interactive communication between teacher and students and on creating an emotional connection among the participants in the process of teaching. On the one hand, teacher should get himself known by students through such means as blog, micro blog, electronic resume etc. In this way, teacher can give students an idea that they are facing a real person instead of lifeless signs and words. Meanwhile, it is also feasible to reach this goal by improving the intelligent question-answering system and other intelligent systems. One the other hand, teacher should take an active part in the whole process of online learning, management and communication. Before the class, teacher should inform students of such information as learning goal, learning task and learning way. In the class, teacher should help students in critical thinking on the contents, information, attitude and values. Meanwhile, collaborative learning should be encouraged so that students can feel their counterparts' support in the learning process.

Second, online education resources should include more emotional elements and provoke more emotional participation. Online education resources are mainly made up of words, pictures, animation and images, which lack vocal communication and gestures. Students' emotional participation mainly depends on the emotional elements contained in these resources. Therefore, on the one hand, attention should be paid to the layout of webpages, the use of color and the arrangements of such elements as words, pictures, animation and sounds. On the other hand, designing a situational online class can also provoke more emotional participation. Teacher can create vivid situations, adopt simulative conversation strategy and develop conversational learning materials to add more emotional elements to the learning process.

D.   Strengthening Teaching Control in Online Education

In online education, teacher can offer students proper and reasonable guide through interactive control. For example, he can realize his supervision and macro-management on students by monitoring learning activities, creating electronic files, issuing notice and requirements and providing guidance. He can also make adjustment to the learning process according to students' learning circumstance such as the quantity of questions raised by students, quantity of students who participate in discussion and quantity of materials reviewed by students. When receiving interactive feedback from students, teacher should give a response as soon as possible to maintain a highly interactive learning process. Teacher should adjust his teaching contents according to students' learning dairy, test results or feedback, so as to help students accomplish their learning goals. Meanwhile, teacher should collect those problems and questions frequently meet by students and find solutions to them in time.

To help those students who lack self-management ability, teacher should offer them help in finding right learning methods and in developing their self-supervise t ability. For example, to avoid being lost in online learning and to make learning targets more clear, students can create a navigation map under teacher's guidance. Learners should be encouraged to manage and allocate their time properly. They can make a detailed learning goals list and then find the corresponding strategy to accomplish those goals. And they also need to make a timetable of tasks to make their learning more effective and efficient. Students should learn to assess learning activity by themselves, for, after an online test, they are required stopping to check their own tests and estimate their scores. Self-questioning method is also an important skill that should be mastered by students. By using self-questioning method, students use some questions that are raised before the learning to check their learning effects and then start the next phase of online learning. In the online learning process, students can improve their self-supervise abilities by repeatedly using self-questioning.

References

1. Harasim, L.: Online Education: An environment for collaboration and intellectual amplification. In: Harasim, L. (ed.) Online Education: Perspectives on a New Environment, pp. 39–64. Praeger, New York (1990)
2. Harasim, L.: Shift Happens: online education as a new paradigm in learning. The Internet and Higher Education 3(1-2), 41–61 (2000)
3. Lu, X.: A Study of Transformation in Teacher-student Relationship of Online Education: Its Reasons and Solutions. Liaoning Normal University
4. Xie, X.: Introduction to Distance Education. Central Radio &University Press, Beijing (2000)

5. Nan, G.: Information Technology Education and Creative Talents Cultivation. In: e-EDUCATION RESEARCH (2001)
6. Warschauer, M.: Computer-Mediated Collaborative Learning: Theory and Practice. The Modern Language Journal 81(4), 470–481 (1997)
7. Sorensen, K.: Design of TeleLearning: A Collaborative Activity in Search of Time and Context. In: Chan, T., Collins, A., Lin, J. (eds.) Proceedings of ICCE 1998: International Conference on Computers in Education. Global Education On the Net, vol. 2, pp. 438–442. Higher Education Press, Springer, Beijing, Heidelberg (1998)
8. Hongliang, M.: Rethinking of Some Notions in the Field of Web-based Course Building. Modern Education Technology, 21–24 (2003)
9. Xie, Y.: Development of Online Course in Higher Education. In: e-EDUCATION RESEARCH (2000)

Research on Foreign Direct Investment of Chinese Private Enterprises

Yan Luo and Ruiqi Qin

School of Economics, Tianjin Polytechnic University, Tianjin, China
272513278@qq.com

Abstract. This paper analyzes the status and characteristics of foreign direct investment of Chinese private enterprises. Also, the paper profoundly analyzes the motive and reasons for foreign direct investment of Chinese private enterprises. By the means to study case, the author analyzes the strategies of Haier's foreign direct investment.

Keywords: Private enterprises, Foreign direct investment, Motive and reason.

1 Current Situation Analysis of Foreign Direct Investment of Chinese Private Enterprises

At present, Foreign Direct Investment of Chinese Enterprises has grown into a certain scale, with a significant increase in the amount of fields related and increasingly diversified investment patterns. According to statistics of Chinese Commerce Department, more than 10,000 outbound China-invested enterprises had been approved by the end of 2006. The total amount of Chinese Foreign Direct Investment had been $16.1 billion in 2006, which was 31.6 percent higher than it was in 2005. Business turnover realized by contracted projects abroad amounted to $30 billion, 37.9 percent higher than it was in the previous year. The total amount of newly contracted projects abroad was $66 billion, increasing by 12.3 percent, and realized turnover of labor cooperation abroad was $5.4 billion, increasing by 12.3 percent. The details of current situation are illustrated as follows from 4 aspects:

A. FDI of Chinese Companies Are Small in Their Average Scale

Not so many companies have a registered capital of more than $5,000,000 among those private companies which are abroad. Take Zhejiang Province, which has a heavy proportion of private enterprises, as an example, the average amount of its foreign part is merely $587,800, while the average amount of Chinese sole proprietorship is only $96,700. It is far below the average of $6 million in developed countries and even the average of $2.6 million in other developing countries, for the funding ability and average scales of our private companies are still limited.

J. Luo (Ed.): Soft Computing in Information Communication Technology, AISC 161, pp. 57–62.
springerlink.com   © Springer-Verlag Berlin Heidelberg 2012

B.   Market of FDI of Chinese Companies Are Strictured and Too Concentrated

Even though FDI of Chinese companies are spread all over the world, they are actually regionally concentrated and mainly centre around the Southeast Asia, Russia, the East Europe and a few developed western countries. Moreover, Manufacturing and processing foreign investments are focused in the ambitus countries in Southeast Asia and Trading, Researching and Developing investments are mainly distributed in developed countries and regions, such as the USA, Japan and Germany.

C.   Majority FDI of Chinese Private Companies Are about Trading and Servicing Industries

Currently, although fields involved in FDI of Chinese enterprises have gradually enlarged from original trading and catering business to fishing, energy, clothing and retailing businesses, they still concentrate on trading and servicing industries. According to the statistics, trading and servicing weigh more than 60 percent of total amount of Chinese FDI, while other industries, like Manufacturing and Resource Developing, weigh merely above 30 percent. Majority private enterprises remain investing in small projects of frontier trades and they are still suffering spontaneous phase of transnational investments and economic cooperation.

D.   Most Modes of FDI of Chinese Private Enterprises Are in Joint Venture or by Acquisition

There are mainly 2 types of Chinese FDI, one of which is Greenbelt Investment. It means to invest capital to set up enterprises of sole proprietorship or joint venture. Another type of Chinese OFDI is transnational merger investment, which stands for investment outward via acquisition and merger. Because Chinese private enterprises have a lack of core competence, limited capital and resources and weak ability to handle risks and there are huge differences of oversea markets and also Joint Venture are more efficient and flexible on financing, techniques, management and marketing than Direct Investment, most enterprises prefer multinational operations by Joint Venture or Merger. For instance, Hisense invested $37,450,000 to open up a Joint Venture Factory in South Africa. Moreover, Indonesia invested $1,000,000 to set up a Joint Venture Television Factory, while Konka Invested $9,000,000 to build a Joint Venture Enterprise and so does TCL. In addition, trend of transnational acquisition and merger is intensifying, making it the most significant pattern of FDI at present. In 2006, the amount realized by FDI through merger occupies 36.7 percent of the total amount of investment.

2   Analysis on Motive of FDI of Chinese Private Enterprises

A.   Export Helps Private Corporations Expand Their Sales

This type of investments is called Trading and Servicing investment, through which Chinese private corporations are able to build product-selling networks abroad and to have much more consumers so as to expand sales volume overseas. To take Zhejiang

Province, where there are plenty of private enterprises taking outward investments, as an example, these enterprises invest capital to establish "Chinese Mall" in bustling business district in St. Paul in Brazil. Plenty of small-and-medium-sized Chinese private enterprises settle there, improving local commodity export of their companies a lot.

B.   Protecting Original Exporting Markets and Developing New Markets by Steering Clear of Trade Barriers Abroad

Chinese private enterprises are able to avoid meeting trade barriers abroad, to protect original exporting markets and to open up new markets via manufacturing investments. At present, developed countries set up tariff and non-tariff barriers like anti-dumping and anti-allowance to restrict importing from abroad, protecting vested interest and market shares of local enterprises. And so do developing countries, for the purpose of protecting and developing national economy. However, FDI acts as an effective means helping steering clear of these trade barriers. Private enterprises are allowed to build industries in the sales location directly and occupy local markets without touching trade barriers. Some of private companies would take indirect tactics by opening up subsidiaries in those countries which have not been set up with quota restrictions by developed countries and exporting products to developed countries in order to occupy markets in developed countries, considering expensive cost to directly set up factories there. In 1990s, many of spinning enterprises or private enterprises of shoes and hats took their FDI in this way. Of course, these investments, focused on avoiding meeting trade barriers, are reduced in the circumstances that our country performs entering promises comprehensively. Instead, many private companies nowadays invest their capital for the great development potentials of host countries and these investments concentrate in developing countries and regions of rapid-developing economy and heavy demands. For example, both investment of Skyworth in Mexico and that of Chongqing Lifan in Vietnam focused on potential of local markets and occupying those markets.

C.   Taking Advantages of Resources Abroad and Implementing Industry Gradient Transfer

Chinese private companies can search for new markets for national saturated products, can transfer excessive production capacity and can proceed on industry gradient transfer by making overseas investments. Taking our national Appliance as an example, the average capacity utilization rate is merely between 50 and 60 percent, which illustrates that national production capacity of Appliance are seriously excessive and their market demands are relatively saturated. Therefore, Appliance enterprises can take a consider of setting up factories abroad, utilizing relative advantages and invest excessive capacity and labor into developing countries in AFILA by transferring gradient of labor concentrated industry to abroad, in order to lay out spaces for enterprises to develop inside and progress high-tech industries using centralized resources. For instance, TCL invested in Vietnam, the Philippines and Indonesia in exactly this way.

D. Emulating Advanced Management Experience Abroad and Implementing Technological Innovation

Learning Investment concentrates on emulating advanced operation principles and progressive techniques of host countries. On the one hand, via investing in developed countries, enterprises are able to study their operation principles. On the other hand, via purchasing shares of foreign companies equipped with advanced technical equipments or via participating in cooperative research and development of new products, enterprises are able to absorb developed techniques abroad so as to enhance present engineering level or even achieve new high-techs. For instance, Holley Group in Zhejiang entered into the promising high-tech industry with less cost by purchasing CDMA Chip R&D Apartment of Philips set in the Silicon Valley. What's more, it's also a form of Learning Investment by opening up R&D organizations in regions of intensive scientific research abroad directly. For example, Zhejiang East Communication Ltd set up a researching apartment of Eastcom in the Silicon Valley by investing in $3,000,000, treasuring local talents and technical information, maintaining the same pace with advanced techniques of cellphones in the world. Accordingly, it enhances its own technique innovation ability and research and development ability and improves its international competence.

E. Enriching Funds and Providing More Ways to Combine Funds

Capital Utilization Investment are intended to find ways to absorb more funds and it is also one of the reasons of FDI of Chinese private corporations. It is insufficiency of funds and lack of capital that restrict overseas investments of our private corporations. We try to not only absorb capitals but also get out, investing limited funds into developed and other developing countries and absorbing more funds. Capital Markets of western developed countries, especially the USA, are standardized and their financing channels are unblocked. If our private companies invest in these countries, we can easily obtain loans in order to expand our operation scales, as soon as the investment benefits are favorable. Moreover, it is also an essential channel for private corporations to finance abroad by appearing on the market in capital markets of developed countries. As our private corporations know much better about overseas capital market, more and more of them are focused on those markets, especially the USA capital markets. So far, more than a hundred enterprises have appeared on the USA market, such as Holley, WanXiang and etc. By appearing on the market, Chinese private companies have financed considerable amount of funds and have promoted the expansion of scale of private enterprises.

3   Analysis on FDI of Haier

The key of successful FDI of Haier lies in its sufficient preparation for Internationality. The development of Haier can be divided into 4 phases. The first phase, from 1984 to 1990, is called Introversion Development Phase. The second from 1990 to 1996, is called Exporting Phase. The third, from 1996 to 1998, is Overseas Investment Phase. The last phase after 1999 is called Localization Phase, in which it invests in money, designs, manufactures and makes sales in South Carolina in the USA. It means that in the first phase, Haier positively promotes the development of inland markets and tries hard to make sure the markets become huge and strong. In the second phase, Haier is ready to push its competitive products abroad and starts exporting products. The third phase stands for overseas investments and Haier invests capital to build factories in Indonesia. At last, Haier steers clear of trade barriers in western countries like the USA and implements the localization tactics, investing, designing, manufacturing and selling in the USA, in order to realize internationality better. Haier goes abroad through a gradual way, which is to proceed on internationality on the basis of occupying inland markets step by step. From the analysis of phases of Haier going abroad, we can obviously know that it gradually promote the localization of FDI.

The internationality way of Haier comprehensively reflects the leaders' elaborate planning on enterprises:

The first is to found a worldwide brand. Haier sets many strict standards on its own, such as Non-Defect of products, Non-Distance of distribution, Non-Inventory of Storage and Non-Bother of consumers, etc. The quality management of Haier achieves the instant effective control, saves costs, improves benefits and strengthens customers' confidence in Haier. After becoming famous brand in China, Haier put forward the idea of No Famous Brands Inland, believing that in opening markets, only those international is national. If any brand wants to be national famous, it has to be famous brand in international competences first. Haier sets about expanding its brand impact all over the world and its products start entering into the globe. It first focuses on the quality and fights for Quality Approve of International Authority. It obtains many kinds of approves of the USA, Japan, Australia, Russia, Canada and countries or regions of the European Union and the America, becoming the Chinese enterprise with most foreign approves and making a guarantee of high quality for consumers inland and outland. When building the worldwide famous brand, Haier keeps the same pace with global standards and ranks in the top place with its brand value rapidly improving. The internationality of Haier is build upon improvements of its own products and management, which forms a solid foundation for Haier to enter into international market later.

The second is to build international marketing networks. Haier establishes the tactics of Difficult Ones First and Simple Ones Later. It first brings its products into fastidious markets in developed countries. After obtaining favorable reputation, it continues expanding in developed countries in West Europe, Japan and Australia and rapidly gets through markets in the Middle East Europe, Latin America and South Africa. During making sales abroad, it mainly counts on foreign monopoly franchisers to set up sales sations and to build international distribution centers, making guarantees of product supplies of franchisers abroad and cooperating with foreign franchisers in circumstances of lack of internationality experiences. Although Haier has already have products of high quality and advanced management experiences, the management realizes their own drawbacks of sales of international products sufficiently and try hard to make sure their products can enter into international market and obtain international reputation better by abundant experiences of professional franchisers abroad and impeccable channels to sell its own products.

The third is to construct international technique R&D networks. In the age of information, traditional R&D way of manufacturing indoor is not only lacking in efficiency but also more aimless and likely to break away from the market. Therefore, Haier insists to internationalize technique R&D objectives, to marketize technique R&D projects and to commercialize its R&D achievements. It first imports the advanced fridge technology in Germany. Then it fosters its own ability of technique research and development, cooperates with many inland scientific research institutions, linked by capital, and sets up its own R&D system under help of those institutions. It builds networks of communication or cooperation and information centers in Tokyo, Los Angeles, Montreal, Sydney, Amsterdam and Hong Kong with a lot of major companies and technology centers. It also opens up designing segments in Tokyo, Montreal and Leon. Based on information of international markets, it follows the international technology trend and develops local products. The R&D mode of Haier not only makes full use of advanced science and technology to optimize its own products but also develops different products according to characteristics of different markets, fits the market and satisfies consumer demands better using international technique R&D networks.

The last is to build overseas production system. Haier sticks to its gradual strategy to Go Abroad, taking product exportation as primary stage to Go Abroad. It steps into technology exportation and capital exportation investing capital to set up factories abroad, when it grows into a certain scale. In 1993, Haier opens up a fridge factory in South Carolina in the USA, forming local operation systems of designing center in Los Angeles, of marketing center in New York and of manufacturing center in South Carolina, and implements comprehensive localization of investing, designing and marketing abroad, reaping favorable benefits. In December 1996, Haier sets up Haier Sally Paul Ltd in Indonesia, which means that it starts international production. From then on, Haier seizes the great opportunity to invest abroad during finance crisis in Southeast Asia and speeds up investing there.

References

1. Chai, Q.: Current Situation and Problem Analysis on Chinese Foreign Direct Investment. International Trade (2008)
2. Fan, X., Bao, X.: Research on Countermeasures for Foreign Direct Investment of Chinese Enterprises of Small and Medium scale. Economic and Trade in Heilongjiang (2008)
3. Zhang, Z.: Basic Characteristics and Inspiration of Foreign Direct Investment in Developing Countries. Commercial Times (2008)
4. Chen, W.: Motivation Analysis and Mode Selection of Chinese Foreign Direct Investment. Times of Finance (2008)
5. He, J.: Motivation Research on Foreign Direct Investment of Chinese Enterprises under Globalization (2007)

Discussion on the Implementation of the Credit System in Institutions for Higher Education

Zhiyong Zhang and Yansong Xue

Tianjin Polytechnic University, Tianjin, China
jlltj2010@163.com

Abstract. The credit system is one kind of education modes. The paper compared credit system and academic year system, Credit system is a complex and coordinated process to carry out the credit system, and it is also a giant project. at last the paper gives some suggestions on carrying out the credit system.

Keywords: credit system, academic year system, suggestions.

1 Introduction

The credit system is one kind of education modes, it is a comprehensive teaching management system, which takes course selection as the core, teacher instruction as the auxiliary, and measures student learning in terms of quality and quantity through the GPA and credits. The credit system, the class organization system and the tutorial system call three major education modes. At the end of 19th century, the credit system originated in Harvard University .In 1918, Peking University first carried out "the course selection system" in our country; in 1978 some qualified universities began to trial the credit system, and now the credit system reform has been pushed in colleges and universities.

2 The Comparison of Credit System and Academic Year System

A. The Academic Year System

The main advantages

- The subject curriculum and class hour are uniform, which can ensure the consistency about majority students' cultivating specification.
- It is easy to manage the teaching organizations.
- It is helpful for the centralized education of political ideology, moral character and collectivism spirit.

The main disadvantages

- It is not conducive to develop the students' personality and potential, and to teach in accordance with their aptitudes.

J. Luo (Ed.): Soft Computing in Information Communication Technology, AISC 161, pp. 63–67.
springerlink.com © Springer-Verlag Berlin Heidelberg 2012

- It is not conducive to arouse students' learning initiative and enthusiasm.
- Time utilization is poor in the school roll management, and this is detrimental to make the excellent students outstanding.
- Subjects are fixed, which is against to raise creative talents and inter-disciplinary talents.

B.   The Credit System

The main advantages

- It is easy to teach in accordance with students' aptitudes, could consider their individual interests, ambitions, and favors the development of students' individual characters.
- The flexibility of students' course selection also contributes to the teachers' enthusiasm for teaching and scientific research.
- Resilient school system benefits more people to accept higher education and the higher education massification.
- It is helpful to allocate resources for running schools reasonably and fully exploit the potential.

The main disadvantages

- Teaching management is relatively complex, and the teaching order is not easy to control.
- Some students with poor inquisitiveness and low demands would "avoid difficulty and seek the easy" in choosing class, which can reduce the studies quality.
- It is not conducive to develop group activities as well as moral character concentrated education.
- It is easy to cause knowledge separated, and makes the learning process lack of systematic.

For several decades, under the fetter of the planned economy system and influence of traditional concepts, our school emphasized teaching in step, memo ring knowledge and repetitive practice, and the development of individuality, creative thinking and practical ability were weakened .Now, facing the challenges of the knowledge-based economy, it is able to foster creative and aggressive talents only through relieving fetter early, teaching in accordance with their aptitude and promoting individuality development of students. So the implementation of the credit system is best able to adapt to the demands of the rapid development of contemporary technology. However, we should consider our unique educational traditions and specific national conditions. it is difficult to carry out "the complete credit system" in our country at present, which is implemented in the western developed countries, we cannot copy it mechanically. In addition, it also needs supporting measures and environment to implement the credit system, we should create good conditions gradually to ensure that the credit system can run smoothly.

3   It is a Systematic Project to Carry Out the Credit System

It is a complex and coordinated process to carry out the credit system, and it is also a giant project.

- We have to redesign teaching programmers and curriculum system combining with the adjustment of specialty structure; to offer sufficient elective courses to meet the needs of students; to break the traditional teaching operational mode and to arrange courses all day. We should redesign the schedule, the use of classrooms and the class time to improve the utilization of teaching resources. Arrange courses dispersedly so that students can make full use of different times to choose classes and order their study and life scientifically.
- We should try in the standardization of the course and credit, that is to standardize the curriculum name, to unite the code and to recognize the credit scientifically. Those will make the credit comparable and directional. Secondly, we should to implement elastic school system and flexible enrollment management, which allow students to speed up the learning process resulting in early graduation, or to reduce the process resulting in delaying graduation. The school system should be a certain degree of relaxation; it is one year for BA and one year for HVE, not too short or too long. The enrollment management involves changes about temporary absence from school, dropout, suspension, reentry and transfer. Schools with the credit system allow students to suspend from school once or more times.
- We should establish a network system. Then students can choose classes through the internet and have clear understanding of each course such as the teaching content, methods, examination, credits, teachers, as well as the class schedule and place, etc.

So we can see the implementation of the credit system is effective response to the idea of "human-orientation" and the education thought about "the subject position of the student". With the credit system, learners choose courses independently on the basis of their own understanding of society and themselves. That is to say, for students, they realize "flexible learning". This is a new style of learning with high flexibility and independence emerging under the background of economic globalization, social informatization and educational lifelongness. The credit system will bring learners' independence and benefit into full play. Tn relatively loose conditions, they can choose courses independently at greater and greater degree, and determine their learning contents, methods, place, time and teachers etc. And finally they take responsibility for their selection and learning behavior.

However, it is not difficult to see that implementing the credit system is a great challenge for teachers. Because students can choose teachers independently who will give them great satisfaction, there is a great disparity in classes. Some classes those few or no students choose may be merged or cancelled, so some unqualified teachers must be washed out. Therefore, as a university teacher, you must have the ability for lifelong learning and educational innovation; make yourself become an expert teacher. A university teacher should not only have the teaching knowledge and skills, but also have the ability to explore and deal with wider problems such as the teaching purposes, behavior consequences, ethical background, methods and the curriculum principles. In the teaching process, the teachers have to change traditional education, make contents novel and precise to benefit the students. The teacher will also change the teaching method and the role, sets up the concept that a student is the protagonist; create a relaxed and harmonious atmosphere, listen to each student, and conduct flexible teaching methods according to individual differences. Finally, as a popular teacher, you must have elegant manners and immanent temperament and glamour. Because the influence of teachers is not only the surface of knowledge and skills, but also the soul and personality. So teachers should make great efforts to improve their ability to adapt the reform of the credit system. Nevertheless, it is not always smooth in the process of implementing the credit system. The main difficulties are that students are easily evasive, they only think highly of the credit, but give little attention to the actual effect and it will cause that students' knowledge structure is not balanced, and foundation is not solid. The inadequate reform of school administration cause disorder in the process of implementing the credit system, the traditional administrative class form is undermined, then affect daily ideological education and cultural activities, etc.

4   Some Suggestions on Carrying Out the Credit System

In short, the credit system is best able to adapt to the needs of the development of modern science and technology, we should explore hard to create conditions for it. The author thinks that we should carry out experimental reforms from several aspects with innovative:

We should encourage students' individual development; establish an equal and harmonious relationship between teachers and students. The traditional education thought which centered on inheriting knowledge restricted the development of young students' active thinking. The student is accustomed to "obeys", not accustomed to "doubt", let alone "the innovation". Even the relationship between teachers and students is also unequal, they cannot exchange opinions and discuss science knowledge equally, students' autonomy, independence and creativity are repressed. So we must accelerate the process of teaching democratization, teacher's first task is to create a kind of peaceful, relaxed and lively atmosphere. For example, as in teaching, the teacher should overcome chalk-and-talk teaching, and advocate heuristic way; the supervisors contact student frequently, interview them one-on-one; direct and absorb them to join in teachers' group to do scientific research. In a word, the teacher must lay down "dignity of the teaching profession", then the students can set up their subject position gradually in psychology and personality.

We should establish the tutorial system, which is the guarantee of the credit system. Under the credit system, the classes are loose organizations, students' idea of specialty is weak, they have many academic and ideological problems, require guidance and assistance. After the students enter a university, they do not understand the training mode and the knowledge structure; they are fuzzy about self-development targets. It is difficult to optimize the study plan for achieve goals. In addition, the credit system is holistic. If students don't understand its inner link, they will put one-sided emphasis on the credit and choose the courses which are easy to get credits. This will break down the knowledge structure. And the students are ignorant of the society demand of talents, this will cause great blindness. So there's a need build the tutorial system. Not only can the tutors guide the students' academic, but also they have subtle influence on students in ideological and moral aspects. Thus will improve the environment of communication between teachers and students.

References

1. Altbach, P.G.: Measuring academic progress: the course—credit system in American higher education. Higher Education Policy 14 (2001)
2. Shedd, J.M.: The history of the Student Credit Hour. New Directions for Higher Education (2003)
3. Ehrlich, T.: The Credit Hour and Faculty Instructional Workload. New Directions for Higher Education (2003)
4. Wellman, J.V.: Accreditation and the Credit Hour. New Directions for Higher Education (2003)

On Ecological Architecture Technology and Design Aesthetics

Wei Wang

Institute of Art and Fashion, Tianjin Polytechnic University, Tianjin, China
wwseaww@126.com

Abstract. There are greater and greater contradictory and conflicts of development of human and living environment we have to face with. The proportion of architecture's energy consumption is quite big. It is obvious in architecture energy, ventilation design, construction of natural light and artificial lighting. As the technology becomes mature, the demand of design aesthetics has been proposed. The necessity of combination between technology and art is analyzed from style beauty and structure beauty.

Keywords: ecological architecture, technology, aesthetics.

1   The Production and Development of Ecological Architecture

A.   Production of Ecological Architecture

Ecological architecture is produced and developed with design concept of ecologism. Ecological environment is root of human survival and development. In view of ecologism, the essence of architecture is adaptation or improvement of human for survival and development, such as Fujian Hakka Earth architecture, shown as figure 1. Tall and closed external form helps resist external invasions and internal communications. Social architecture reflects social human, the cause of producing

Fig. 1. Fujian Hakka Earth architecture

J. Luo (Ed.): Soft Computing in Information Communication Technology, AISC 161, pp. 69–79.
springerlink.com   © Springer-Verlag Berlin Heidelberg 2012

different functional architecture is the division of labor of human. The functions of cluster effect are to improve mini environment, make effective use of resources and prevent harm together. These are human's early practice and exploration on ecological architecture, but not forming conception of system and design.

With the presence of industrial revolution, more and more contradictory and conflicts are coming up. The early way of harmonious coexistence between man and nature becomes dream of human. The progress of technology and the development of industrial civilization, driven by exploiting nature, have paid too much. Figure 2, taken by latest remote sensing satellite maps, firstly proves the forest crisis. Large areas of earth's original forests are less than 10% of the land area and remain forests are missing rapidly. The disappearance of forests directly contributes to demise of biodiversity. At present, the speed of extinction of animals and plants is100-fold faster than that before human exists, which is estimated to increase to 10000-fold. While indulging in brilliant achievements in economic growth and technical progress, human's cultural development is impeded by population expansion, environmental deterioration, exhaustion of resources, the imbalance of relationship between human and nature, making human's living environment suffering from serious tries. People begin to realize that only if deal with the harmonious relationship among economic development of human society, technical progress and nature and strive to coordinate human social economy and environment well, their history can be prolonged.

In terms of energy consumption, the current statistics show that the proportion of architecture's energy consumption is great in total energy consumption, about 25%~40% (Figure 3), moreover is increasing on and on. If included the production and transportation of architecture materials and the energy consumption in process of removal, the proportion will be increased to approximately 50%.

So the development of social architecture is urgent. In the 1960s, Italy architecture planner PaoloSoleri firstly combined ecology and architecture, merging ecology and architecture into Arcology, creating a merging interdisciplinary subject--arcology.

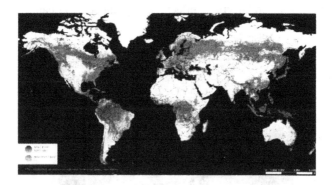

Fig. 2. Original forests satellite map announced by Green Peace Groups

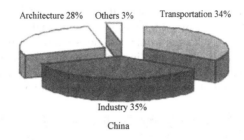

Architecture 28%   Others 3%   Transportation 34%

Industry 35%

China

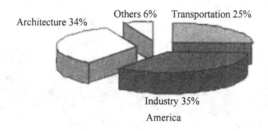

Others 6%   Transportation 25%

Architecture 34%

Industry 35%

America

Fig. 3. Statistics of energy consumption

B.   The Development of Ecological Architecture

With the development of modern science and technology, ecological architecture forms relatively independent system, which mainly combines leading technical advantages with avant-garde ideological concept, and produces parts of entities of ecological architecture in the whole world, such as hall 26 for the Deutsche Messe AG (Figure 4): the length of hall is 200m and width is 116m, which is planned averagely as 3. Its stretching and dynamic exterior appearance shows exclusive artistic technology. Its architecture structure and energy consumption relates to the effect on environment and sustainable development, which is the perfect unification of technology and art. Reduce 50% of pays on architecture air conditioning by ventilation design combined artificial and nature. Hall uses glass windows faced north to lighting. Meanwhile, reflex mirror area partly installed under roof can reflect artificial light and natural light into room. Following ecological features are realized in big size geometric modeling of this architecture: first, it has ideal form with long-span spatial adaptation; second, it has representative cross-section shape, which makes functional spatial height enough to large area of hall and supplies essential height of ventilation of hot air, considering Venturi effect on the roof; third, natural light can entering the large area of architectures and avoiding direct light, bright light without dazzling is the key to create the whole hall's spatial quality; fourth, the application of natural renewable materials. The roof's area is up to 20000 m^2. Use wood as a roofing panel not only for cheap cost, but also more consideration into advantages of reducing energy consumption. Steel structure is a light weight structure, while can be recycled and made use of numerously.

Fresh air can flow into hall along with transparent tube in service area, either through entrance in the height of 4.7m or through the ground. Then the contaminated heated and upward air is discharged from continuous folding on the roof. The folding opens alone in different angles according to different wind's direction to assure effective natural ventilation. It can be enhanced by horizontal strip components in the fixed exit and create a kind of Venturi effect, shown as figure 5.

Fig. 4. Hall 26 for the Deutsche Messe AG

(a) Venturi effect of ventilation

1: Natural wind into the vent from the front. 2: Cold air is diffused in hall through glass vent tube. 3: Air becomes heated and upward through absorbing room's energy. 4: Hot air is ejected from vent on the roof. 5: Warm air into room by mechanical vent devices and hot air is up and ejected

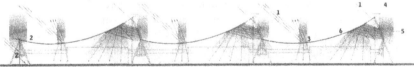

(b) Daylight lighting diagram

1: Direct daylight. 2 and 4: Diffuse daylight. 3: Skylights cast some light. 5: Visor reflects sunlight into the room. 6: Reflective roof

(c) lighting diagram

1: Lighting equipment in the ventilation ducts. 2: The roof reflective makes light even and soft. 3: Hanging light with the roof

Fig. 5. Lighting and ventilation diagram

We can directly know about the development of ecological architecture and its technology in the whole world through this analysis of representative of ecological architecture.

2 The Application of Ecological Architecture Technology in Architectural Design

A. Architecture Energy

Ecological architecture technology in architectural design is manifested in imitating and forming saving in architecture energy by use of modern science and technology. One of important objects of ecological architectural design is efficient and clean energy. Efficiency means improving efficiency of resources and energy in the whole life period as soon as possible, reducing cost material and energy cost. Use clean energy and renewable materials positively. Clean refers to reduce eject waste, waste water, waste gas and turn them into no harm and resources. Realize waste recycling and minimize the pollution and damage on the natural environment.

It is estimated that current renewable resources (solar energy, wind energy, hydrogen power, terrestrial heat, etc) only 2%-4% of total energy demand in the whole world. In 1994, WPA established that the rate of clean energy would be up to 15% in 2010. Solar energy is main method in architectural energy of ecological architecture. (1) Keep in consideration of height and position in various seasons when certain the direction of architecture and main design. (2) Keep in consideration of direction and gradient of ground. (3) Keep in consideration of land surrounding architectures' occlusion of the sun. (4) Keep in consideration of potential shading effects on surrounding plants. (5) Keep in consideration of glass types, distribution of glass windows and relationship between image and elevation. (6) Keep in consideration of relationships between internal spatial heat demand and solar heat. (7) Analyze the sunshine effect in computer system. (8) If possible, research on effect of natural lighting by models. (9) Keep in consideration of relationships between sun trance and main façade in a year, the direction and gradient, external wall's form and window's area, positions, etc. (10) Make full use of south façade to collect solar radiation, and store them in forms of electricity and heat, etc. (11) The fixed or removable devices can reduce direct daylight, prevent room being too hot, so as to reduce load of conditioning system.

B. Architectural Ventilation Design

Ecological architecture technology in ventilation is mainly manifested in forming air circulation by making use of circulation of natural wind. The greatest benefit of natural ventilation firstly is to improve internal air quality environment. Aside from the places where the contamination is too serious that external air can't meet the healthy standard, we should provide more fresh air by natural ventilation as possible as we can, in order to reduce incidence of architecture syndrome effectively. Another benefit is that it can reduce the dependence on conditioning system, so as to save conditioning energy consumption. If external temperature is lower that internal temperature, lower temperature by natural ventilation is best. However, external temperature can's meet the natural ventilation's need all the time. Therefore, except some experimental architecture in special places, it is almost impossible to full natural ventilation. The most common in modern architecture is to mix various methods, that make full use of natural ventilation while configuring mechanical ventilation and conditioning system. The four ways of hybrid ventilation are as follows: (1) Mechanical ventilation is used as aids only when needed. (2) Use different ways of ventilation according to practical requirement of various areas of architecture and practical conditions of different areas. (3) Natural ventilation and mechanical ventilation are used at the same time. (4) Natural ventilation and mechanical ventilation system is as a mutual alternative means. For example, lower room's temperature by natural temperature at night and meet acquirement by mechanical ventilation in the daytime.

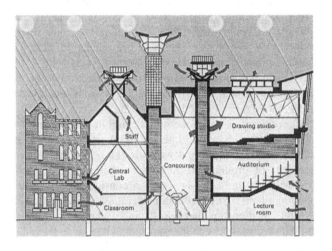

Fig. 6. De Montfort University

A typical example is De Montfort University (Figure 6). Due to the functional demand and limit of ground conditions, architecture plane is deeper into ground. So it is difficult to form draught by use of windows in the external walls; moreover, noise of transportation and exhaust pollution from surroundings affects architectures seriously. That makes it necessary to use closed windows. However, architects build four light wells in integrated plane, besides courtyard, to provide natural lighting and deflate hot air as ventilation wells, at the same time, to inhale fresh air into pipelines in the floors. After fresh air absorbing heat, it rises up and is ejected from chimneys or lighting wells. Wind cowls are installed at the air outlet, which promises the air in the basement to eject successfully and not be pressed into pipelines because of change of external pressure. These devices are controlled by BEMS control system, which can absorb fresh air into room at night and take away daytime's gaining heat of thermal inertia materials such as concrete floor. Besides, BEMS system has function of self-learning, which can learn how to regulate ventilation rate to meet people's demand by recording results of manual intervention in a day. The closed architectures don't need mechanical ventilation, with the energy consumption as 64 kwh$\diagup$m^{2}y, and reduce amount of carbon dioxide emission to 20kg$\diagup$m^2, whose energy consumption approximately to 85% of ordinary conditioning architectures.

C.   Natural Lighting and Artificial Lighting

Ecological architecture technology advocates improving power rate of artificial lighting as much as possible by natural lighting and technical ways. One of effective way to save energy is to make full use of possibilities of natural lighting. Besides, another important reason for making use of natural lighting is it more suits human's biological nature, which is especially important to healthy in psychology and physiology. Therefore, one of important index to explore internal environment qualities is natural lighting.

The factors of natural lighting level are such as direction of windows, gradient of windows, environmental shadings (plants' configuration, other architectures, etc), sunlight reflex of surroundings, areas of windows, depth of plane and height of section and shading devices indoor and outdoor, etc.

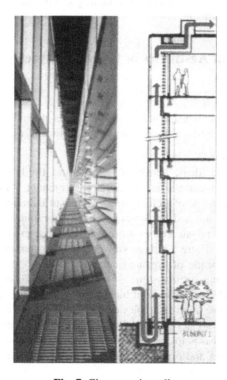

Fig. 7. Glass curtain walls

Shading devices are for shading and reflecting sunlight, reflecting sunlight from high places to low places, or from external to room depth. Multi-reflex by ceilings is an effective way to improve spatial illumination of rooms with a large depth.

The angles of sun visors need to be calculated meticulously, in consideration of incidence angles of domestic sunlight in winter and summer, preventing from too dim light due to low height of sun in winter. Sun visors and blinds are used to prevent too high temperature because of direct sunlight in summer, meanwhile, which can also meet the principle of requirement of natural lighting illumination. Of course automatic control and adjustable sun visors are ideal choice, but the cost is relatively high: sun visors in places where air pollution is serious should be cleaned frequently to keep the abilities of reflex. While the mutual relationship between solar energy and heat energy is the important way to solve ecological architecture technology, such as glass curtain wall (Figure 7). A typical design strategy of valuing technical energy and materials utilization is the strategy of integrating passive and active energy utilization, paying attention to external structure of architectures, utilizing modern computer technologies, change fixed external structure into self-adjustment structure whose change according to external climate conditions and internal functions. (parts of internal architectures can adjust themselves automatically with these adjustment). These external structures become "skins" of architectures, which can breathe, control energy and materials exchange between architecture control system and external ecological system and enhance ability of adapting to continuous change of external ecological system environment.

3   Embodiment of Aesthetics of Ecological Architecture in Design

A.   Beauty of Style

Many ecological architectures learn from appearances natural shapes and technical parameter, producing scientific architecture style in following ways:

1)   Systems analysis

While making bionic ideas, above all, we should think about the differences between natural environment and architectural environment. Although organisms in nature are origin of aspiration of architectures, we mustn't simply copy and we should lead to a deeper exploration and implementation on aspiration in methods of systems analysis. The systems analysis comes from one of three great theories of modern science---system theory. There are three viewpoints in system theory: 1 viewpoint of system, referring to principle of organic integrity. 2 viewpoint of dynamic, declares that life is self-organizing open system. 3 viewpoints of organizational hierarchies, thinks that there are different levels and hierarchies among things with different self-organizing capacities. The three principles of system theory are element, structure and hierarchy. Systems analysis not only helps us understand and hold the characteristics of organism itself but also helps us to seize the common essential features from the changing of architectures and organisms, and the common of structure, function and types.

2)   Analogism

Analogism is a cognition method based on forms, mechanics and similar functions, by which we can not only find the similarity among related organisms in same family, but also find form structures' similarities in totally different systems. A common architecture can be seen as being, with internal circulating system and nervous system. The similarities of human architecture activities and organisms and be found by analogism.

3) **Model test**

Model test combines theory and practice by quantitative experimental methods, on basis of certain qualitative understanding of bionic design. Architecture an effective bionic model helps us know more about structures of organisms and develop a new creative mode of thinking on basis of common regularities in comprehensive architectures and biosphere.

These three ways usually can form objective ecological architecture form, such as Jean-Marie Tjibaou cultural center, designed by Italy architecture Renzo Piano (Figure 8), honored as typical performance by use of wind pressure to natural ventilation. New Caledonic is a tropical island locating on the south Pacific in east of Australia. Its climate is hot, wet and windy all year around. Therefore, maximum natural ventilation utilization to decrease temperature and wet becomes the core technology to adapt domestic climate and value ecological environment. Renzo Piano found aspiration from traditional seaside huts in New Caledonic. These huts are generally built in multi-layers leaf, which provides cool ventilation and diffusing wind power. Outer wooden screens designed by Renzo Piano have the same function. Curved mulberry glued wood ribs can not only impede the direct sunlight but also divide wind power, forming a architecture shapes of strong aesthetics.

Fig. 8. Jean-Marie Tjibaou cultural center

B.   Beauty of Structure

All the structures of ecological architectures are distinctive, mainly classified as fiber structure, shell structure and spatial skeleton structure, etc. Through scientific researches on spider nets, shells, skeletons, crystal structure, form exclusive structural form of ecological architectures by reasonable and utilizable sections. We can feel strong aesthetics of form generated from orderly array forms in some representative architectures. The designs have both reasonable structures and architectural aesthetics

Fig. 9. Berlin Parliament Architecture

will be future development direction of ecological architectures. Such as the rearchitecture plan of Berlin Parliament Architecture (Figure 9) positively added technical methods of ecological architecture, utilizing new structural forms and concepts in rearchitecture cupolas. It becomes the representative of combination of technology and art because of its focus on structural sequential (Figure 10) and aesthetics.

Fig. 10. Structural sequential

4   Conclusions

As concept of sustainable development deepening into people's hearts, people will pay more attention to development of ecological architectures. Decrease in energy consumption will reduce obviously environment's burden. Human's production and life will gradually go back to orbit where human's development is harmoniously developed with the natural environment. Meanwhile, human's spiritual demand requires ecological architectures to combine with aesthetics to meet the functional need, aesthetic demand. Moreover, the organic combination between technology and artistic science is generated.

References

1. Li, H.: High-tech Ecological Architecture. Tianjin University Press (2002)
2. Zhou, H.: Ecological Architecture: the future-focused architecture. Southeast University Press (2002)
3. Liu, X.: Arcology. China Architecture & Architecture Press (2009)
4. Sue, R.: Ecohouse 2 a design guide. China Forestry Publishing House (2008)
5. Li, D.: On the ecological design. China Architecture & Architecture Press (2009)
6. Xiayun: Ecological and sustainable developing architecture. Journal of Architecture (2002)
7. Gong, C.: Considerations in developing ecological architectures. Architecture (2000)
8. Hanpei, Liangxue, Zhang, Y.: Analysis on spatial shapes of modern ecological architectures. World Architectures (2003)
9. Lizhe, Zengjian, Xiaorong: New Aesthetic Conception of Contemporary Arcology. New Architecture (2004)

Study on Digital City's Technique and Urban Planning

Wei Wang

Institute of Art and Fashion, Tianjin Polytechnic University, Tianjin, China
wwseaww@126.com

Abstract. Digital city's technique provides multi-level, high quality, high efficient information services and decisive support for governments, enterprises and individuals, accomplishes fast paced development of productivity and general improvement of people's living standard. GIS is built platform of total system. Provide objective overall information for urban planning design and three dimensional virtualized direction of development also provides technical support for process of urban planning design and achievements exhibition.

Keywords: digital city, GIS, three dimensional virtualization, urban planning design.

1   Constitution of Digital City

A.   Constitute Overview

Digital cities can be classified as broad sense and narrow sense. Generalized digital city is informational city, referring to integrating city's information resources and realizing social economic informational urban, by constructing basic device platform such as broadband multimedia information network and GIS, etc.

Narrow digital city refers to urban network information environment constructed by GIS, GPS, RS, etc, a informational basic facilities and applied system serving urban planning, construction and management, serving governments, enterprises and public, serving population, resource, environment and economic social sustainable development.

Specially speaking, digital city refers to make full use of digital information processing and network communication, in process of urban planning, construction, management, production, circulation and lives, regulate and utilize various digital information and informational resources, widely apply to various aspects of urban economy and social life, supply multi-level, high-qualified, efficient information service and decisive support for governments, enterprises and individuals and accomplish paced development of social productivity and general improvement of living standards.

B.   Constitution of Digital City

1)   Digital urban infrastructure

The construction of urban infrastructures include Metro broadband backbone network construction, broadband access network (BRAN), cities' existing integration

J. Luo (Ed.): Soft Computing in Information Communication Technology, AISC 161, pp. 81–87.

of computer, telecommunications and cable television networks and optical fiber communication and mobile communication construction. City infrastructural informational construction also includes city space based information access tools and infrastructure construction, including GPS data real-time receiving, processing, applied service facilities and other digital photogrammetry facilities construction.

2) **City construction and digital management**

Various areas of digital construction in city construction (such as aviation remote sensing, urban road, underground utility, environmental sanitation, urban landscaping and city supervision, etc) integrates the data from GPS or aerial photography and forms various city management information systems, urban disaster relief emergency command system, city planning and police information systems based on GIS.

3) **Digital government affairs**

From viewpoint of globe, it is a trend to propel governments to be office automation, networking, electronic and comprehensive information sharing. In other words, government digital information is the basis of social digital information. Electronic governments include three aspects: one is governments and staffs can achieve information from network, including internal workflow information and operating information from external departments; two is governments put information on network for social understanding and utilizing, namely, open administration. Three is government interactive with social public in network, namely electronic administration.

4) **Electronic business**

Electronic business is an economic revolutionary affecting whole society and times. It not only changes managing methods, but also changes our shopping ways even modes of thinking. Electronic business is a landmark, including electronic business comprehensive service system, technical standards for electronic transactions, CA, digital signature, electronic contract, electronic receipt, etc. Besides, electronic business platform, various electronic business application systems (such as B2B, B2C, G2B) and logistics systems construction based on GIS are included, too.

5) **Digital enterprise information management**

Digital enterprise information management refers to improve enterprises' economic profits and market competitive ability by changing and improving traditional industry in information technology. It mainly includes internet engineering company, enterprise informative management. Typical applications are like CAD, CAM, PDM, ERP, CRM, CIMS, etc.

6) **Digital Social business**

Social business information refers to information in education, culture, sanitation, tourism and publishing, etc, to promote service industry by remote education, remote healthcare and health-care service, digital library, etc.

7) **Digital community information**

Digital community information is integration of community information service and application function based on space information infrastructure, which refers to realize digital communications, lives, tourism, medical, educations, etc, realize informational subsistence mode, based on harmonious living environment and open information infrastructure, in networking and digital environment. Moreover, realize convenient and safe life, reduce consumption of energy and materials, to accomplish sustainable development of society.

2   Technical Development of Narrow Digital City

A.   Essential Technical Methods of Narrow Digital City

Narrow digital city mainly refers to 3S system integration technology. 3S is collectively called for GPS, RS and GIS. GPS conducted by American is generally used in the globe, which is a geodetic network comprised of 24 satellites with distance of 20200 kilometers to ground. It can be all-day, high-precision positioning, qualitative and practical on any point, linear and polygons. Record basic qualities and detected time, through ground signal receivers, and transform to digital information for storing and outputting. Remote sensing detects object properties and spatial forms by the characteristics that objects can send, reflect and absorb electromagnetic wave. Current remote sensing has function of all-day high-precision monitoring, more rightly and comprehensively holding up landscape regime and provide information resource for GIS. As space database management system, GIS can keep, manage and access landscape object space and properties data from GPS, RS and other ways. Analyze and mark landscape states and regulation of landscape functioning process by superposition, adjacent. Predict change and influence of landscape and digitally simulate virtual landscape.

Fig. 1. GIS geographic information system

B.   Technical Development of GIS Digital City

GIS (figure 1) is main content of digital city technology, data of series of city geography situation formed by GPS and RS, also main basis of city planning. Three dimensional GIS gradually becomes main direction of technical development, whose range including database, GIS, Visualization in Scientific Computing, virtual reality, etc. Three dimensional GIS can be defined as: a computer system conducting received three dimensional sample data input, memory, edit, select, spatial analysis, simulate and assist in decisive support, whose space coordinates join fixed displayed operation.

   With the fast development of computer technology in recent 20 years, it has become true that three dimensional visualized GIS generates, displays and operates data structures of three dimensional geometry characteristics and properties characteristics. It can be classified as two categories: surface representation and volumetric representation. Spatial data model is an abstract, classification and simplify description of real world. Three dimensional spatial data model is a community of geometry data organization, operation method and constraint rules, etc. Three problems have to be considered while defining and development a new three dimensional data model: certain target need to be described; memory of three dimensional data and description of logic relationship; how to display model. Three dimensions visualization have solved these problems. So called "visualization" is transform object properties data into graphic images which can be better understood. In general, visualization refers to the process of displaying objects' internal properties and mutual relationship, etc in information select, browser and demonstration from object properties to be sensed, acknowledged, imagined, deducted, integrated, abstracted and its changing forms and process, through imagery, simulated and actualized methods, by assist of computer software, utilizing various media materials and design methods.

C.   Technical Methods of Three Dimensional Visualized GIS

The visualization stresses simulation on real spatial relationship, which makes a straight feel in various states in three dimensional spaces. No matter a map which can be zoomed freely and selected in screen or a three dimensional surface model, we can know about real world spatial relationships more perceptually, specifically. We can also say "What you see is what you get". The visualization of spatial information also includes surface properties graphics visualization. For example, with an electronic administrative division map, we can extract population statistics data in 2010 in various provinces, municipalities and autonomous regions from spatial database, calculate population density and display corresponding graphics according to graded index in different colors and fill patterns. Thus, the spatial surface properties can realize visualization of spatially referenced information.

The fast development of three dimensional data collection provides large amounts of data for describing geometry objects accurately. However, due to limit in computer software and hardware level, how to clip and intercept, displayed rapidly, real-time roam these data is still waited to be solved. With application of VR, three dimensional visualized GIS are more and more next to our demand. VR has characteristics as follows: (1) immersion: no longer like traditional computer interface technology, interactive mode between computer and users is natural, just like the interactivity between human and nature in real world. Placed in a real multi-dimensional information space, people feel to be surrounded by virtual world, as if fully integrated, generating a feeling of entering into it, the passive observers become active joiners and join various activities in virtual world. (2) Interactivity: different from traditional three dimensional flash, VR no longer passively receives information from computers, but can operate virtual objects by interactive input devices (such as data gloves, power feedback, etc), so as to change virtual world. Users take interactivities with various objects in virtual world by natural actions, such as walking, rotating head, moving hands, etc, making users timely operation on virtual world and get feedbacks from virtual world. (3) Imagination: users acknowledge perceptual knowledge and rational knowledge form qualitative and quantitative comprehensive integrated environment by VR, deepening concept and generating new ideas, as a result of enlightening new ideas. So to a great degree, three dimensional virtual GIS system depends on virtual-reality technology.

3   Application of Digital City Technology in Urban Planning

A.   Demand of City Information in Urban Planning

The demand of city information in urban planning is quite overall and deeply, which is an important trademark, including city digital map, digital photography grammetry, remote sense video, building digital information, etc. These data can be accomplished with assistance of digital city information system, especially three dimensional visualized GIS, more conveniently providing related information. City digital maps include various digital maps, such as relief map, cadastre, land use map, house property map, etc. These digital maps not only include essential geometric information of 3D model, such as geographic object position, scenery contours or elevation points of digital ground model, roof forms, margin and area, lean, direction of margin, height and level, etc, but also include abundant properties semantic meaning information. Photographic grammetry generates geometric data of geographic relationship effectively and record semantic meaning information, provides fine three dimensional model accuracy for buildings with an obvious outlines. Elevation and ground also can be measured by digital aerial graphics. Through remote graphics, under support of GIS, we can get large area, comprehensive, dynamic, timely information, combining and comprehensively analyzing which with ground qualities, earth physics, earth chemical, earth biology, military application, etc. With such information, current objective basis of analysis can be generated.

B.   Three Dimensional Expression of Urban Planning

Large amounts of CAD have been used by urban planning, in the process of considering designing, three-dimensional urban scenery model (Figure 2) is an important condition for regulating plans. Three dimensional visualized GIS not only provides related three-dimensional urban scenery model, but also provides surface and climate situation in any position. Moreover, we can know about corresponding infrastructure, plants distribution, population, etc. So three dimensional urban planning expression is urban planning's important process and display mode of final result. Big-scale three-dimensional urban landscape model is basis of building three dimensional landscape planning, with a widely and diversified objects and data. It includes both manual geometric objects, such as construction, road, bridge, subway, underground pipelines, sculptures, etc and natural objects such as relief, water system, mountain, grass, trees, etc. Besides, some significant spatial phenomenon is also included in city, such as post tagged, cadastre, temperature field, rainfall and population distribution, etc. Relief is basis and spatial positioning carrier for other city surface models. Therefore three dimensional urban landscape models can be classified into two categories.

Fig. 2. Three-dimensional urban scenery model

C.   Development of Three-Dimensional Design

Technology in urban planning in large-scale city digital simulation, the number of thin pieces of polygons in three-dimensional scenery is generally large, often reaching several hundred thousand. Calculation is large when generating three-dimensional scenery. Practicalness of three dimensional images' generation and display is essential condition for real feeling. In three dimensional landscape, practicalness in magnificent in updating speed more than 24 f/s in order not to make eyes to feel flash. The image generating speed depends on software and hardware system structure in image process, especially image processing capacities of hardware speed devices and various speed devices used in image generating.

LOD technology develops along with operating large-scale three dimensional digital models. LOD mode refers to, to the same scenery or objects in scenery, get a group of models in different description methods for drawing. As the distance increases, the objects become vaguer, so as to not able to distinguish many detail structures in objects. Besides, when object's cast area in screen is small, no complicated model can describe its details. Otherwise, it will result in huge waste of CPU and processing time. In condition of not affecting image visual effects, reduce geometric complication by gradually simplifying superficial details to improve efficiency of drawing algorithm. The technology usually builds several different geometric models approaching accuracy to every original polygon model. Compared to original models, every model saves certain level's details. While drawing, according to different standards, show objects in proper model. LOD has a wide application domain. At present, it is applied to practical image communication, interactive virtualization, virtual reality, relief representation, flight simulation, collision detection, image drawing with time limit, etc, having been a key technology, Many model software and VR developing system start to support inserting LOD model.

4   Conclusion

Digital city is a huge complicated system which has to be mutually supported by multi-class technology. GIS is a core comparatively, a platform for total system.

Three dimensional virtualized GIS system is major method of researching and results expression. Urban planning support it in this process, to a degree, however, urban planning require more information supplied by digital city technology, to form scientific design reference and design plan. Finally, demonstrate the design process and design result by three dimensional technology.

References

1. Cao, C.: Study on virtual city three dimensional modeling theories and method. Master Thesis Central South University (2004)
2. Cheng, J.: How to digitalize city-study on digital city construction. China City Press (2002)
3. Chen, Q.: Study on city landscape virtual reality application Master Thesis Jilin University (2005)
4. Chen, S.: City urbanization and UGIS. Science Press (2000)
5. Dengyu: Ideology and historic development of virtual reality. Journal of Donghua University (1), 7–8 (2002)
6. Gong, J.: Object-oriented spatio-temporal data model in GIS. Journal of Surveying and Mapping 26(4) (1997)
7. Liu, X., Linhui, Zhanghong: Virtual city construction principles and methods. Science Press, Beijing (2003)

Path Analysis on How Attention Impacts Investment Decision Making

Zongsheng Wang

School of Economics, Tianjin Polytechnic University, Tianjin, China
wzs0822@163.com

Abstract. Attention bottleneck exerts double restraints on information set in the course of the information receiving. As a kind of subjective resource, attention determines the condition of information set which leads to decision-making expectation deviation and fuzzy condition in action and status space. This further influences his economic behaviors. Attention has become a key link in the decision-making chain from information to attention to decision-making and then to new information. The author analyzes why and how attention influences economic behaviors by using the set theory under the framework of decision-making. The paper discusses the relationship between attention and economic behaviors from the angles " attention dark box " in the course of information of normal and practice in order to discover the receiving.

Keywords: Information set, decision-making chain, action space, double restrains.

1 Attention-Based Decision-Making Chain

The essence of economic behavior is decision-making problems and decisions are driven by information[1]. Information and attention flow towards the opposite way, and information acquisition will be focus on the basis of investment of attention resources. Information - Attention - Economic behavior have formed a ring-shaped relationship which is shown in Figure 1.

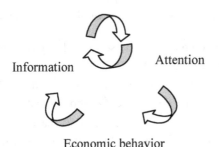

Information   Attention

Economic behavior

Fig. 1. The ring-shaped relationship

J. Luo (Ed.): Soft Computing in Information Communication Technology, AISC 161, pp. 89–96.
springerlink.com   © Springer-Verlag Berlin Heidelberg 2012

Attention must be allocated in order to get information, how to reflect the value of information depends on the allocation of attention; Attention allocation will certainly affect the quality of decision-making, whereas rational economic behavior will be focus on the basis of rational attention allocation. Consequently information - attention - decisions - new information have formed our decision-making chain, attention has become an integral part of this chain, or a key part. This decision-making chain will determine the impact that attention exerts on economic behavior.

With the related knowledge of set theory, we discuss how attention influence economic behavior in the framework of decision theory. Because the essence of the economic behavior is decision-making problems, in order to facilitate the analysis of the problems, we focus on the analysis of how attention influence decisions to explain the concrete impact that attention exerts on economic behavior.

2   The Elements of Decision Making

For policy makers, the decision-making purpose is to solve the problems, the policy makers get the benefit after solving problems, which is often referred to as payment in the process of decision-making.

From the perspective of decision-making, the payment of decision-making depends on two factors, first, the one is background state of the decision makers, and we call it the natural state in the process of decision making, it is beyond the control of policy makers, is exogenous, decision-makers can do nothing but to accept it. According to previous and present information knowledge, decision makers estimate the state space and decide which natural state will appear as well as the possibilities. That is to say we can get a subjective estimate of state space probability distribution. The other one is that decision-makers can control behavior with interchangeability. Generally speaking that policymakers can choose certain programs and have a choice of action within a certain autonomy.

Therefore, we can abstract decision-making into four elements: the state space, the probability distribution of state space, action space and payment (can called sample space) [2].

State space is formed by all the sets of natural state which is denoted as Ω.

Action space is composed by all the sets of alternative actions which is denoted as Γ.

The sample space is composed by the sets of all payment and can be denoted as K, there obvious exists $\Omega \times \Gamma \to K$.

The probability distribution of the state space is denoted as Θ.

When an element β in the state space Ω appears and decision-makers choose a particular action α, $\alpha \in \Gamma$ from the action space Γ, it will generate a corresponding result R, $R \in K$, this relationship can be expressed as:
$$R = f(\alpha, \beta) \qquad \alpha \in \Gamma \qquad \beta \in \Omega.$$

3 Process of Attention Allocation

Given t the set of unknown external state as Ω, policy makers understand Ω (state space) by noticing some information. When policy makers observe some of the information, in fact, they transform a certain state from the outside world into information input by entering function $\xi : \Omega \to \Theta$, and ξ reacting processing ability of information. Due to the attention consumption of the process of information input, decision-makers judge the state space on the basis of information set obtained by attention configuration. Thus this input function is actually attention configuration function, which can be written as: $A : \Omega \to \Theta$, Θ is the subjective estimate for the distribution of the external state proposed by decision-makers. When the outside state is β, he observed that the input is $A(\beta)$ and then he output the attention 'black box' as $A^{-1}(\theta)$.

Supposed that some kind of transcendent distribution $\pi(\theta)$ may exist in decision-makers' knowledge for the external state, the decision-makers modify his understanding of external state according to the output $A^{-1}(\theta)$, and then define this belief modification as $\vartheta : \Theta \to \Delta$, Δ is a set for all probabilities distribution defined by Ω.

On the basis of the modification of recognition for external world, policy makers will take some kind of action α. Assume that decision makers' unknown replaceable action collection as Γ, decision makers should first identify action space while identification process is the process of information acquisition, the identification process can be described as $\varepsilon : \Gamma \to M$, M is a knowable action space, ε reflecting the decision makers' processing capacity for observing information, that is, the ability of the rational configuration of attention. It can be described as : $A : \Gamma \to M$.

Decision-makers have some value judgment system, such as revenue maximization, thus forming mapping from the action and external state to earnings: $u : \Omega \times \Gamma \to K$. This value function can be regarded as a law f common determined by external state and action, $f : \Omega \times \Gamma \to K$, $f(\alpha, \beta) = R$.

f reflects the processing ability of decision-makers in the whole process of information processing. Because information processing is carried out based on attention allocation, so information processing capacity is the rational allocation capacity of attention[3], that is, common ability for efficient attention allocation of the process of both state space identification and action space identification and choose. Above ξ represents processing and recognizing capacity of information for state space, ε represents processing and recognizing capacity of information for action space, so that this law can similarly describe as: $f(A) = f(\xi(\beta), \varepsilon(\alpha))$.

4   The Boundary of Optimal Solution

The purpose of our decision-making is to optimize the results by the choice of alternative action. Simply saying is that to select an option to implement from all kinds of alternative programs to get the best results.

Starting from the decision-making purposes, we are seeking an action under which $f(\beta,\alpha)$ can be maximized and meet certain evaluation criterion. In other words is that when $\beta \in \Omega$ is determined, seek α^* $\alpha \in \Gamma$ from $f(\beta,\alpha^*) = Maxf(\beta,\alpha)$, α^* is the optimal action. When we process β as a variable, $f(\beta,\alpha^*)$ constitutes the optimal solution border.

Optimal solution border becomes a set of all the best actions, it is $f(\beta,\alpha^*)$ formed from $\beta \in \Omega$, that is curve $f(\beta,\alpha)$ shown in Figure 2.

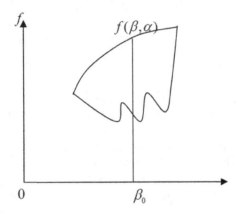

Fig. 2. Optimal solution border becomes a set of all the best actions

5   Decision-Making Based on Attention

Payment concerned by policy makers are determined by action state and external state, this is $f(\alpha,\beta) = R$. The policy-makers' expectation of action and the external state are built on information sets while information sets are the result of attention allocation, in other words information sets is the one based on attention allocation. "What kind of information sets you get depends on what kind of information you notice", therefore we can describe information sets as : $I = g(A)$ or $I(A)$.

If the decision-making environment is simple, the information flow for describing change will become small and slow. Thus attention will not become the "bottleneck" of obtaining information, policy-makers can obtain all information for judging both state space and action space. The policy-makers can correctly identify the optimal solution boundary that is the sets for all optimal action. At this point the decision-making is relatively simple and clear.

However, some action of the practical market often has a huge amount of information which flows quickly and requires short-cut of decision-makers, of which the securities market is the most typical market with the above features. In the stock market, attention "bottleneck" will show in the process of receiving information for decision-makers: "surgent information torrent will absolutely generate exceptional information flow in bottlenecks, either we are inundated and overwhelmed by information, or our brain rapidly processing without distinguishing, then information processing bias will appear. "

Attention "bottleneck" cause double constraint of attention: First select constraint, we choose some information to note at the same time means that some information is discarded[4]. The next is allocation constraint, attention is a scarce subjective resource, too little attention allocating can only excavate too little information value, while too much configuration can cause uneconomical. Therefore, policy-makers are often faced with following situations: (1) Decision-makers simply do not find reliable information to invest and determine; (2) The information is selected to be noticed, but policy-makers miss some important information. For example, we might ignore that the central bank has raised interest rates, some public opinions about the government plans to prop up the market, the listed companies adjust operating projects, and other information, whereas all these information plays a key role in the process of the policy makers forming correct expectation and developing scientific investment decisions[5]. (3) Pay attention to a variety of information, particularly key information, but in the process of attention, attention resource distribution is unreasonable, such as we focus less on key information while focus more on information that impacts less on policymaking.

Dual constraint caused by attention "bottleneck" make decision-makers have an understanding deviation in policy-making space and action space. In this case, policy makers may not fully understand the state space and action space, that is to say the state space which is known to decision-makers is only a subset of the state space. If the state space changes, the probability distribution of state space must change. As policy makers' understanding of the state space itself is ambiguous, that is to say it is not clear how many states may occur. The expected adjustments for probability distribution of state space are based on information set $I(A)$. Therefore, the probability distribution of state space known for decision-makers is necessarily vague. Similarly, the action space known for decision-makers is a subset of fully actional space, state space and action space known for decision-makers are respectively describe as $\Omega^{'}$ and $\Gamma^{'}$, then there exists $\Omega^{'} \subset \Omega$ $\cdot$ $\Gamma^{'} \subset \Gamma$.

We know that policy makers' understanding of $\Omega^{'}$ and $\Gamma^{'}$ is based on the original accumulation of knowledge and new information collection integrated by new information, thus $\Omega^{'}$ can be described as $\Omega^{'}(I + \Delta I)$, that is to say the known state space is a function of new information set. Since the new information ΔI is also the function $I(A)$ of attention A, therefore, $\Omega^{'} = \Omega^{'}(I(A))$ can simply described as $\Omega'(A)$. Similarly action space is also the composite function of attention which can be described as $\Gamma'(A)$.

The policy-makings of investors are built on the basis of $\Omega'(A)$ and $\Gamma'(A)$, even if the policy-making results is optimal in $\Omega'(A)$ and $\Gamma'(A)$, they are not necessarily optimal in Ω and Γ.

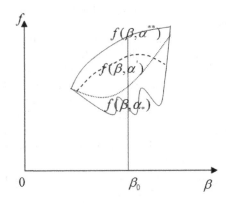

Fig. 3. Under the situation of complete information and decision-makers' knowing about β_0, they select α action, as α is the optimal action

As shown in Figure 3, under the situation of complete information and decision-makers' knowing about β_0, they select α action, as α is the optimal action.

Because of dual constraint of attention configuration, it may appears: (1) in β_0, but the decision-makers don't know about α action, that is $\alpha \in \Gamma$ but $\partial \notin \Gamma'$. Only if decision-makers choose α' action from Γ' to make $\alpha' \in \Gamma'$ optimal in β_0. Due to that Ω' is a proper subset of Ω, there may exist $\alpha^{**} \in \Omega$ but $\alpha^{**} \notin \Omega'$, and $f(\beta_0, \alpha^{**}) > f(\beta_0, \alpha')$, it is obvious that α' is just "satisfactory action" rather than the optimal action. Therefore curve $f(\beta, \alpha')$ constitutes a "satisfactory" boundary, and the relationship between satisfaction boundary of actual policy-making and theoretical efficient boundary is shown in Figure3. If the decision-making is hasty, and the decision-makers select α_* according to a small amount of information without any satisfaction, thus the decision-makers select a very bad action. (2) β_0 is not included in Ω', $\beta_0 \in \Omega$ but $\beta_0 \notin \Omega'$. Decision makers do not know the state β_0 will happen, when β_0 appears, decision-makers' choice of action α in the condition that state β_0 is not going to occur might be the worst or near worst action, at that time the decision-makers will face a greater risk[6] [7].

Even in the case of $\Omega' = \Omega$ and $\Gamma' = \Gamma$, decision-makers know all state that they will face, and they also know they can choose all of the action by themselves, in other words decision-makers know all of the state space as well as the whole action space. But the decision makers also estimate the probability distribution of state space based on attention allocation. With the information update of policy makers, policy-makers' probability distribution for state space will also be adjusted continuously. Policy makers may choose different actions because of different attention configuration in the same state space and action space.

First of all decision-makers seek expected return for each action of action space, $EMV(\alpha) = \int_{\Omega} f(\theta, \alpha)\pi(\theta)d\theta$, then decision-makers select the action with the maximum expected profit. As long as $\pi(\theta)$ changes, $EMV(\alpha)$ will change along with the turnover of action satisfying the maximum expected profit. Therefore, as long as the probabilities distribution of final state is not fully consistent with expectations, policy-makers will be at risk.

If the policy-making is made in a similar environment, policy makers will achieve optimal state after a series of attempts. However, if the decision-making environment is complex and can't be repeated, or this decision needs a fast decision of decision-makers, then the policy-making will be full of chance, action will be with the random choices. In this environment a large number of decision makers will cause the diversity of economic behavior.

6   Conclusion

The essence of economic behavior is decision-making problems, decisions are driven by information while the access to information is based on the investment of attention resource. So our economic behavior is based on attention. In the background of relative surplus of information, the attention constraints make action space and state space of behavioral agent unclear. The behavior agent will adopt satisfactory policy instead of optimal decision-making in the framework of sense. The uncertainty of satisfactory program makes behavioral agent's attention-based economic activities full of risk.

References

1. Li, L.: Uncertainty in Decision-Making and Investment in Financial Assets. Guangdong People's Publishing House (August 2002)
2. Li, L.: Modern Statistical Analysis of Financial and Investment. China Statistics Press (December 2004)
3. Zhang, L.: Attention Economics. Zhejiang University Press (April 2002)
4. Zhang, Y.: New Progress of Research on Selective Attention Mechanism–Negative Priming Effect and The Inhibition of Distracting Information. Psychological Science 21 (1998)

5. Liqi, Z., Gang, H.: Psychological Decision-Making Under Uncertain Situation-Adaptation and Cognition. J. Advances in Psychological Science 5 (November 2003)
6. Zhao, D., Liu, M.: New Progress of Research on Attention Visual Search. Advances in Psychological Science 1 (October 2002)
7. Zhang, L., Fu, X., Sun, Y.: Cognitive–Ecological Sampling Approach of Judgement on Biases analysis. Advances in Psychological Science 3, 601–606 (2003)

Positive Analysis of Allocation of Attention

Zongsheng Wang

School of Economics, Tianjin Polytechnic University, Tianjin, China
wzs0822@163.com

Abstract. An investor's limited capacity of attention is in contradiction with the vast amount of information in modern financial markets, which makes the investing people focus their attention selectively. This paper makes such a positive study so as to ascertain indirectly that influence does exist. and that comments on stocks, the way of their release and even the representation of the comments are all influential factors to affect man's selective focus of attention.

Keywords: selective attention focus, comments on stocks, security investment.

1   Introduction

Stock market is one that pushed forward by capital and information. And it's acceptable that allocation of attention does have effect on the decision-making of investors, theoretically and perceptually. However, we face lots of difficulties and obstacles in trying to make a positive analysis of it, and it seems absolutely impossible to do it directly. The paper tries to analyze the changes of the prices and trading volume of the recommended stocks with the investors' adjustment of allocation of capital after their noticing the securities analyst' stocks comments, thus to test indirectly the basic conclusions of the theory of allocation of attention and its influence on the decision-making of investors.

2   Ideas for Testing and Descriptions of Samples

A.   Ideas for Testing

Our ideas for testing are very frank: if investors notice the information released by security analysts and make a reaction to the stock comments, there must be obvious differences between the opening prices, trading volume and the number of shares traded of the day before and after the stock comments, otherwise, there's no change. Basing on the ideas, we can not only test our theory through the following hypothesis and test whether the contents of the stock comments have effect or not, but also test whether the release time, form and location of the stock comments have effect on the investors' attention as its contents.

H_0: No Obvious change in opening prices (trading volume, the number of shares traded) after the stock Obvious

J. Luo (Ed.): Soft Computing in Information Communication Technology, AISC 161, pp. 97–105.
springerlink.com   © Springer-Verlag Berlin Heidelberg 2012

H_1: Obvious change in opening prices (trading volume, the number of shares traded) after the stock comments

B.   Descriptions of Sample

We chose three authoritative newspapers which have a significant market share and wide coverage: China Securities Journal, Shanghai Securities News and Securities Times. The detailed columns are as follows: "investment grade presentation" of the Markets section, China Securities Journal, Mon to Fri and "the potential species" of sections B07 and B08, Sat.; "recommended stock" of B3 section, Shanghai Securities News, Mon to Fri and "golden weekend stocks" on Sat.; "thoughts of choosing stocks", "the competition on Huashan Mountains" and "the cradle of black horses" of the Markets section, Securities Times, Mon to Fri, "close attention on individual stocks" of sections Chosen Individual Stocks and Crisscross Market, Sat.

We chose all the 3139 stock comments on the three securities newspapers of the months July and August of 2008. After wiping out incomplete datum, there're 1109 left in July and 1083 in August.

3   Process of Positive Analysis

A.   Variables Choosing

The 'Candle line' in the first 5 minutes is formed through call auction, and it contains all the effective information the day before and it's a description of the short-term tendency of individual stocks. Therefore, it can reflect the adjustment of attention of the investors. The paper chose the information contained in the candle line in the first 5 minutes of the sample term after the opening quotation at 9:30. With the help of the system software of Eastern Securities, we chose the opening prices, trading volume and the number of shares traded of the recommended stocks reflected in the candle line in the first 5 minutes of the sample term after the opening quotation as analysis variables[1] [2].

To get rid of the influence of stock grail, we did the following thing: the change rates of the opening prices, trading volume and the number of shares traded of the sample stocks reflected in the candle line in the first 5 minutes minus the pricing limits of the trading volume and the number of shares traded in the candle line in the first 5 minutes of index (the Shanghai Composite Index of SSE and SZCZ of Shenzhen Stock Exchange) of the according date. That is

$$\Delta P\% = (P_{t+1}^i - P_t^i)/P_t^i - (P_{t+1}^s - P_t^s)/P_t^s$$
$$\Delta Q\% = (Q_{t+1}^i - Q_t^i)/Q_t^i - (Q_{t+1}^s - Q_t^s)/Q_t^s$$
$$\Delta M\% = (M_{t+1}^i - M_t^i)/M_t^i - (M_{t+1}^s - M_t^s)/M_t^s$$

The superscript i stands for individual stock, s stands for the big board, and $\Delta P\%$, $\Delta Q\%$, $\Delta M\%$ stand for pricing limits of the stock prices, trading volume and the number of shares traded without the influence of grail. We assume that other interference factors are normally distributed, and the positive-negative action is zero. Therefore we can say that $\Delta P\%$, $\Delta Q\%$, $\Delta M\%$ are cause by information, and we call them analysis variables[3].

B. Testing the Value of Stock Comments

Regard the stock comments from Mon to Fri and those of Sat. as two different samples and we get the data of the samples independently. And we assume that the change rate of the number of shares traded ($\Delta P\%$, $\Delta Q\%$, $\Delta M\%$) is nearly normally distributed with the change rate of prices. We test the following hypothesis with t testing of two independent samples

$$H_0: \quad \mu_1 - \mu_2 = 0 \qquad H_1: \quad \mu_1 - \mu_2 \neq 0$$

Using SPSS12.0 to test it and the output is shown in Table 1. The change rate of the trading volume and t the number of shares traded can refuse null hypothesis under the significant level $\alpha = 0.01$. Therefore, we have reason to believe that there're obvious differences between the stock comments from Mon to Fri with those of Sat.

Table 1. Testing differences between stock comments from Mon to Fri and those of Sat

Variables tested	variance test	Test for Equality of Variances		t-test for Equality of Means			
		F	Sig.	t	Sig.	99% confidence	
$\Delta P\%$	Equal	3.535	0.060	0.939	0.348	-0.2229	0.4783
	not			1.305	0.192	-0.1247	0.3801
$\Delta Q\%$	Equal	65.663	0.000	-6.425	0.000	-7.0503	-3.0124
	not			-6.349	0.000	-7.0753	-2.9874
$\Delta M\%$	Equal	66.904	0.000	-6.428	0.000	-7.2442	-3.0967
	not			-6.363	0.000	-7.2662	-3.0747

In order to test that whether are most of stocks very active on Monday? We choose a controlled group every Saturday. To reduce the subjectivity in choosing samples, We sequentially chose 25 stocks each from Shenzhen Stock Exchange and Shanghai Stock Exchange. In this way, we chose 450 stocks on 9 Saturdays, which makes 1/3 of all the 1370 stocks of the whole market, and having representativeness to some extent.

After getting rid of the effect of the big board using the way shown above, we find that the numbers of rising stocks and falling stocks of the control group are almost the same, rising samples covering 49 percent of the whole sample group. And it seems that the market perform of those stocks that were not recommended is random. To do further research on the "Effect of stock comments of Sat.", we regard stocks recommended on Sat. and the control group as samples from two normal populations, and do hypothesis testing.

$$H_0: \quad \mu_1 - \mu_2 \leq 0 \qquad H_1: \quad \mu_1 - \mu_2 > 0$$

Do the t-test (Finite Sample) of the differences of means of two independent populations with different variances and the results are shown in Table 2. Table 2 shows that both the change rate of the trading volume and the change rate of the number of shares traded betray the original hypothesis H_0 under the condition $\alpha = 0.01$.

That is to say, there're obvious differences between stocks recommended on weekends and those not recommended on weekends, which can't be explained with "effect of stocks recommended on weekends". That is to say, the "effect of stocks recommended on weekends" is limited, even if there is. Stocks recommended on Sat. are valuable, which accords with the results of empirical studies of Xiang Lin (2001), Baoxian Zhu (2002) and Liang Ding and Hui Sun (2001). Therefore, we can conclude that stock comments of Sat. do affect the allocation of attention of investors.

Table 2. Testing differences between stocks recommended on Sat. and the control group

Variables tested	variance test	Test for Equality of Variances		t-test for Equality of Means		
		F	Sig.	t	Sig.	99% confidence
$\Delta P\%$	Equal	0.626	0.4290	7.048	0.0000	0.0129	0.0278
	not			7.181	0.0000	0.0130	0.0276
$\Delta Q\%$	Equal	16.945	0.0000	2.693	0.0070	0.2560	12.0082
	not			3.485	0.0010	1.5903	10.6739
$\Delta M\%$	Equal	18.897	0.0000	2.814	0.0050	0.5497	12.6780
	not			3.662	0.0000	1.9517	11.2759

C.   Effects of stock comments's form on attention allocation

Dose issue form and position of stocks comments, like the contents affect attention allocation? Next, we answer this question.

1) Effects of simultaneous recommendation and high-frequent recommendation on stock comments

Generally speaking, investors are not likely to buy all of three newspapers mentioned above in a day. We call it simultaneous recommendation if a same stock is recommended on different newspapers by one particular organization on the same day, and high-frequently recommended stocks which are recommended on a same newspaper by different organizations and stock reviewers. The effect of stock comments is achieved by many investors together answering to it. High-frequently recommended stocks raise investors' degree of recognition of the stock review while simultaneously recommended stocks are given more attention by investors.

Table 3. Testing differences between simultaneous recommendation and sole recommendation

Variables tested	Variance test	Test for Equality of Variances		t-test for Equality of Means		
		F	Sig.	t	Sig.	95% confidence
$\Delta P\%$	Equal	0.228	0.6330	1.735	0.0830	-0.0008	0.0137
	not			1.897	0.0580	-0.0002	0.0131
$\Delta Q\%$	Equal	19.286	0.0000	3.024	0.0030	3.6959	17.3757
	not			2.379	0.0180	1.8190	19.2526
$\Delta M\%$	Equal	20.636	0.0000	3.14	0.0020	4.2408	18.3887
	not			2.46	0.0140	2.2614	20.3681

Therefore, they both attract more attention. We have ground to suppose that the $\Delta P\%$, $\Delta Q\%$, $\Delta M\%$ of high-frequently recommended stocks and simultaneously recommended stocks are larger.

In this way, compare the simultaneously recommended stock comments of Sat. and the stock comments of Sat. recommended on a single newspaper, and the results are shown in Table 3. Table 3 shows that different from sole recommendation, simultaneous recommendation does attract more attention, which accords with our supposition.

Compare the high-frequently recommended stock comments of Sat. with other stock comments of Sat., and the results show that the high-frequently recommended stocks are stronger in the stock market than the others, attracting more attention, which also accords with our supposition. The detailed data are shown in Table 4.

Table 4. Testing differences between high-frequent recommendation and ordinary recommendation

Hypothesis testing	Variables	T value	P value	Significance level	Significant or not	Sample size
High-frequent recommend dation	Change rate of prices	4.223	0.000	0.1	significant	103:629
	Change rate of the trading volume	1.716	0.089	0.1	significant	103:629
	Change rate of prices	1.788	0.077	0.1	significant	103:629

2) **Location and stock comments**

On the basis of people's reading habits, we get the virtual value 1-6 in Figure 1 of the possibilities of differently located stocks' being noticed. Thinking of the possibility of different locations in the three securities newspapers of one particular stock, we take the mean value. Do hypothesis testing of different locations and the results are shown in Table 5. For $\Delta Q\%$, $\Delta M\%$, position's mean bigger than or equal to 4.5 are more significant than those with position mean smaller than 4.5. Besides, position's mean smaller than 3 is not significant compared to the control group, i.e. if the stock comments are poor position, they don't make more difference than ones which dose not be recommended.

6	6	6	6	5	5
6	6	Chart		4	4
5	5			3	3
5	5	3	3	3	3
4	4	2	2	2	2
4	4	1	1	1	2
3	3	1	1	1	2

Fig. 1. Virtual value

Through the testing we can conclude that the locations of stock comments affect to what extent they are noticed, thus attracting investors' different levels of attention.

It is thus clear that stock comments appearing in different locations are noticed to different degrees. Information in the top-left is noticed most easily, and the routine of people's attention is generally left→up→middle→down→right, which goes with "厂" (Chinese character, means "factory" in English) pattern. As shown in Diagram 1, the top-left headline stock comments with virtual value 5 and 6 attract the most attention obviously, the left stock comments with virtual value 4 are comparatively in the shade, and those with virtual value 1 and 2 are hardly noticed, which proving the limit of people's attention in another way. According to the "frame" theory of Tversky and Kahneman, the wording of information affects the decision. In the paper, we find that together with the wording of information, the location of information and time of reading the information affect how valuable a piece of information is, thus influencing the decision of investors.

3)   **Effects on attention of recommending agency and level**

We think that when investors' studying and judging stock comments, the fame of the stock reviewers, the strength of the recommending agency, and the ultimate assessment are all comparatively key points which attract much attention to an investor.

To test the hypothesis, we classify the stock comments into three groups "attention", "close attention" and "explosion" according to key words and tone of stock reviewers' ultimate assessment of a stock. For example, if a stock reviewer assesses a stock with expressions like   "breakthrough soon, close attention", "exploding quotation", "short-term speeding ascending",   "strong rebound, close attention" and etc, we classify the stock's recommending level as "explosion" and evaluate it 3; if the stock is said to be "rebound in afternoon market, close

Table 5. Contrasting effects of stock comments with different locations

Hypothesis testing	Variables	T value	P value	Significance level	Significant or not	Sample size
positional value≥4.5 versus＜4.5	Change rate of prices	1.326	0.185	0.05	insignificant	200:407
	Change rate of the trading volume	1.999	0.046	0.05	significant	200:407
	Change rate of prices	1.993	0.047	0.05	significant	200:407
positional value≥5 versus＜5	Change rate of prices	0.986	0.324	0.05	insignificant	180:427
	Change rate of the trading volume	2.089	0.038	0.05	significant	180:427
	Change rate of the trading volume	2.080	0.038	0.05	significant	180:427
positional value of the control group ≤3.5	Change rate of prices	4.394	0.000	0.05	significant	403:301
	Change rate of the trading volume	1.595	0.111	0.05	insignificant	403:301
	Change rate of the trading volume	1.940	0.053	0.05	insignificant	403:301
positional value of the control group≤4	Change rate of prices	5.444	0.000	0.05	significant	403:406
	Change rate of the trading volume	2.043	0.041	0.05	significant	403:406
	Change rate of the trading volume	2.437	0.015	0.05	significant	403:406

attention", "great rebound, active attention", "tendency getting strong, close attention" and etc, we classify the stock's recommending level as "close attention" and evaluate it 2; and other key words like "attention", "hopeful breakthrough", "hopeful ascending, worth noticing", "hopeful rebound" and etc are associated with recommending level "attention" and evaluate it 1.

Group the stocks recommended on weekends with different recommending levels, we do a cross hypothesis testing of the stocks. And the results are shown in Table 6.

From Table 6, we can conclude that though the difference between "close attention" and "explosion" isn't significant under the significance level 0.1, there is significant difference between "close attention" and "attention" as well as between "explosion" and "attention". The tone of stock reviewers does affect the allocation of attention of investors.

Referring to the strength and fame of recommending agencies and stock reviewers, we choose six authoritative organizations, and the top-5 stock reviewers of the ranking on the p5w.net(website of Securities Times). We do hypothesis testing of the stock comments released by them with those by other organizations and stock reviewers, and the results are shown in Table 7.

Table 6. Contrasting effects of stock comments with different recommending levels

Hypothesis testing	Variables	T value	P value	Significance level	Significant or not	Sample size
"attention" versus "close attention"	Change rate of prices	2.159	0.032	0.05	significant	499:188
	Change rate of the trading volume	2.271	0.024	0.05	significant	499:188
	Change rate of prices	2.316	0.022	0.05	significant	499:188
"attention" versus "explosion"	Change rate of prices	2.394	0.021	0.05	significant	499:45
	Change rate of the trading volume	2.025	0.049	0.05	significant	499:45
	Change rate of the trading volume	2.015	0.050	0.05	significant	499:45
"close attention" versus "explosion"	Change rate of prices	0.599	0.550	0.1	insignificant	188:45
	Change rate of the trading volume	0.047	0.963	0.1	insignificant	188:45
	Change rate of the trading volume	0.157	0.875	0.1	insignificant	188:45
"attention" and "close attention" versus "explosion"	Change rate of prices	2.114	0.035	0.1	insignificant	687:45
	Change rate of the trading volume	1.446	0.149	0.1	insignificant	687:45
	Change rate of the trading volume	1.509	0.138	0.1	insignificant	687:45

Table 7. Contrasting effects of stock comments of different organizations

Hypothesis testing	Variables	T value	P value	Significance level	Significant or not	Sample size
Contrast between different recommending agencies and individuals	Change rate of prices	2.851	0.005	0.01	significant	201:530
	Change rate of the trading volume	2.974	0.003	0.01	significant	201:530
	Change rate of prices	3.030	0.003	0.01	significant	201:530

The results refuse the null hypothesis(no significant difference) under the significance level of 0.01, and we can conclude that stock comments by authoritative organizations and senior stock reviewers are more worthwhile, attracting more attention of investors.

4   Conclusion

As an important part of second-hand information, Investors allocate attention on this information, and the information work together with the information already existing in the investors' mind, and through proving and testing, investors get more new knowledge, which leads to investors' adjusting his investment portfolio   in short time. With many investors' joint efforts, some stocks recommended by stock comments do much better on the second trading day than the day before the investors' noticing the stock comments[4], raising great fluctuation in trading prices, volume, and the number of shares traded.

Because high-frequently recommended stocks and simultaneously recommended stocks attract more attention of investors, they raise greater fluctuation on this stocks compared to other ways. Investors are used to paying more attention to stock comments located in top-front eye-catching places. These stock comments are identified with more, thus having more effects on the market than those located behind. The allocation of attention goes with "⌐". As the reading time goes by, investors get more and more sensitive to the loss of attention, they get more concerned with the key words like the conclusive assessment of stock reviewers and who recommended the stocks, and they rely on feelings rather than detailed comparing of different stock comments to adjust the investment portfolio. This is part reason why strongly recommended stocks and stocks recommended by authoritative organizations do better in the market[5].

In conclusion, Stock comments are valuable and the forms, locations, wording and etc of the stock comments influence the effects of the allocation of attention just like the contents of the stock comments themselves.

References

1. Liu, P., Smith, S.D., Syed, A.A.: Security Price Reaction to the Wall Street Journal's Securities Recommendations. Journal of Financial and Quantitative Analysis 25, 399–410 (1990)
2. Lloyd Davies, P., Canes, M.: Stock Prices and the Publication of Second-Hand Information. J. Journal of Business 51, 53–56 (1978)
3. Lavie, N., Hirst, A.: Load theory of Selective Attention and Cognitive Control. Journal of Experimental Psychology: General 133(3), 339–354 (2004)
4. Lin, X.: Analysis of China securities consultancy forecast. Ecnomic Reserch Journal (2001)
5. Ding, L., Sun, H.: Chinese stock market stocks recommended effects. Management World (May 2001)

The Information Processing Based on Attention Allocation

Zongsheng Wang

School of Economics, Tianjin Polytechnic University, Tianjin, China
wzs0822@163.com

Abstract. From relative quantity analysis of attention allocation, this paper has deduced the rule that tolerance of attention consumption reduces with time. Then, we can explain that the information searcher tends to selectively pay attention to information. Via anatomising study mechanism between information searched and priori knowledge, we have inquired into optimized paths during multi-periods dynamic allocation process. So this paper opened out entire course of information processing.

Keywords: attention allocation, information processing, utility.

1 Introduction

Attention is an ability that subject pays a certain degree of strength and continuance attention to object under excitation conditions. It is a subjective ability and belongs to Economics. The information collector pays attention to getting information, so the allocation process of attention goes along with the information collection and processing. If we want to understand the investors' investment decisions based on information, we must reveal the process of attention allocation.

2 The Static Model of Attention Allocation

Simon analogizes humans' information processing as a computer. Humans' minds, like information processing systems, receive, encode, store, extract, transform and transmit symbols. The information processing of humans can use some knowledge of Information Engineering: If we divide the whole process of persons dealing with information into these parts: (1) information receiving, (2) information processing, (3) information sending, the strength of attention is the same as the width of information channel. Therefore, according to the traditional theory of resources allocation, we can allocate the attention like this:

To a topic, the information researcher chooses n pieces of information $x_1, x_2 \cdots \cdots x_n$ and makes 'description' of each topic. According to information engineering, each piece of information takes up w_i (share) of the information channel[1],

J. Luo (Ed.): Soft Computing in Information Communication Technology, AISC 161, pp. 107–115.

and if the whole amount of attention is A, each signal will take up $w_i A$, and the Vector of attention is $(w_1, w_2 \cdots\cdots w_n)$.

The information collector achieves utility maximization by matching different attentions with different kinds of information. And the decrease of the searched topics' uncertainty will reflect the searching utility. While according to the information entropy theory, the decrease of uncertainty can be replaced by $-\Delta S$, so when the information collectors choose different kinds of information, the $-\Delta S$ to the topic is

$$U = -\Delta S(w_1, w_2 \cdots\cdots w_n; x_1, x_2 \cdots\cdots x_n) \tag{1}$$

Then, the solution of the attention resource allocation is

$$Max(U) = Max[-\Delta S(w_1, w_2 \cdots\cdots w_n; x_1, x_2 \cdots\cdots x_n)]$$

$$st \cdot \quad \sum_{i=1}^{n} w_i = 1 \tag{2}$$

$$A = A^*$$

Here, A^* is the attention maximum of the searcher, which is decided by emotion, physical strength and environment during the process. To a single-period, it is a fixed quantity. From above we know $\Sigma w_i A = A^*$. Because $\sum_{i=1}^{n} w_i = 1$, the constraint condition $A = A^*$ is an identity. Thus, the condition $A = A^*$ can be removed and only the solution of the model needs to consider during the equation solving

$$Max(U) = Max[-\Delta S(w_1, w_2 \cdots\cdots w_n; x_1, x_2 \cdots\cdots x_n)]$$

$$st \cdot \quad \sum_{i=1}^{n} w_i = 1 \tag{3}$$

Use Lagrange multiplier method to solve this optimization problem, and Lagrangian function is

$$\Phi = -\Delta S(w_1, w_2 \cdots\cdots w_n; x_1, x_2 \cdots\cdots x_n) - \lambda(\sum_{i=1}^{n} w_i - 1) \tag{4}$$

Next, Do derived function between Φ and w:

$$\begin{cases} \dfrac{\partial(-\Delta S)}{\partial w_1} = \lambda \\ \quad\vdots \\ \dfrac{\partial(-\Delta S)}{\partial w_n} = \lambda \\ \sum_{i=1}^{n} w_i = 1 \end{cases} \tag{5}$$

To solve this equation, we can get the perfect allocating vector $(w_1, w_2 \cdots\cdots w_n)$. We can see the best allocating result is that each marginal utility is equal. We call this model static model of attention processing.

It is obvious that if we use tradition resources disposition theory to process attention resources, some deviation may appear: (1) according to filter theory of attention, that we choose to pay attention to information is sequent, that is, the searcher chooses n pieces of information to describe the topic according to time rather than choose n pieces of information to go into one channel at the same time and every signal takes up w_i part of the channel. So time factor must be introduced into the model. (2) According to the model, the searcher reasoningly allocates A^* into every chosen piece of information. However, we discovered that before referring to information, we do not know which piece of information is closest to the topic, so we can not allocate our attention rationally. (3) To an important topic, people often do not finish it at a time, but arrange the allocation of attention during many times. (4) Actually, the searcher does have some prior knowledge about the topic. And the former searched result and the former prior knowledge will be combined into the next prior knowledge. That is, every process of information search is not independent.

So, we need to modify the static model of attention. First, according to the realistic rule of attention, we suppose:

- During a short time, attention decreases accelerately as time passes, namely $dA/dt < 0$ d^2A/dt^2.
- During a relative long time, attention will regenerate and the full amount is increased, namely $TA \approx A_1 + A_2 + \cdots\cdots A_t$ Regeneration makes attention allocated across time.

Next we will make use of these two supposes, combining the prior knowledge, to make a series of modification to the attention processing.

A.   The Modification of Time to the Single-Period Attention Allocation

ormation, which compel information searcher to weigh between the amount of information and the depth of information:

If we pay too much attention to number i piece of information, another piece of information behind i may be missed though we can get more information about the number i. So the searcher must weigh the most cost of attention to n is C, which makes C/A measure the relatively cost of attention. We call it tolerance of attention cost (psychology base line), which is represented by T:

$$T = C/A, \quad \text{T-tolerant, C -cost, A-attention} \qquad (6)$$

To see from formula superficially, the tolerance of attention cost decreases with attention increasing, that is, it seems $dT/dA < 0$. It means that, the less searcher's attention is, the more the tolerance is, which disobeys the reality. After analyzing, we find that when T is A's function, C is A's and t's function as well, namely $C = F[A(t)]$.

In order to get the change relationship between T and A, we make a derivation between the two sides of $T = C / A$, then we can get:

$$\frac{dT}{dA} = \frac{A \cdot \frac{dC}{dA} - C}{A^2} = \frac{1}{A}(\frac{dC}{dA} - \frac{C}{A}) = \frac{T}{A}(E_c - 1) \tag{7}$$

Here, $E_c = \frac{A}{C}\frac{dC}{dA}$ is the flexibility.

We can see that with A decreasing, how T, the searcher's tolerance of attention cost, changes depends on E_c: if $E_c > 1$ then $dT / dA > 0$, T changes with A in the same direction, that is, when A decreases, T increases; if $E_c < 1$ then $dT / dA < 0$, T changes with A in the opposite direction; if $E_c = 1$, T changes without A .

Therefore, we can see that there seems no a consistent conclusion. The change of T depends on the flexibility of C to A, which is accepted by the information searcher. And this flexibility depends on the searcher's preference to risk. Better he likes risk, more the tolerance of attention cost will be, and otherwise it will be less. Now, we can use E_c to classify the preferred features that the searcher's attention to dangers.

- If the most allocation quantity C that the searcher can accept has much flexibility to the storage quantity A, namely $E_c > 1$, the searcher is a risk-avoider.
- If $E_c < 1$, the searcher is a risk-seeker.
- If $E_c = 1$, the searcher is indifferent to risk.

Let's see the time track of T. Because both C and A are functions of time, we make a derivation between the two sides of $T(t) = \frac{C[A(t)]}{A(t)}$ and t:

$$\dot{T} = \frac{dT(t)}{dt} = \frac{A(t) \cdot \frac{dC[A(t)]}{dt} - \frac{dA(t)}{dt} \cdot C[A(t)]}{[A(t)]^2}$$
$$= \frac{C \cdot \frac{dA}{dt}(E_c - 1)}{A^2} \tag{8}$$

$dA / dt < 0$ has been decided in suppose 1, if $E_c > 1$ then $dT / dt < 0$ and T decreases with time passing; if $E_c < 1$ then $dT / dt > 0$ and T increases with time passing; if $E_c = 1$ then $dT / dt = 0$, so t will not change. So the time path of T depends on E_c and then it is decided by searcher's preference to risk as well.

To a risk-seeker, because of $E_c < 1$, $dT / dA < 0$ and $dT / dt > 0$, which means that when A acceleratively decreaseswith time, T increases simultaneity. Little by little the searcher will lose sensitivity to attention cost. Then, the searcher's sense of 'gambling' will be stronger and stronger and he will pay more attention to the last information. As a result, the 'recency cause" phenomenon will appear obviously. This kind of person is often too confident in the process of investing, believes his own capability of acquiring information and depends on 'recency cause" information that he chooses too much. The narrow choice of information makes some information affect the decision too much, so the searcher will over reflect to this information[2]. If this kind of information comes from others, the searcher will easily follow the herd to enlarge or shrink the capacity of stock exchange, which increases volatility of market. There, when the searcher is in charge of a bigger risk voluntarily, he deviates from the track of behavior ration at the first minute.

To a risk-neutral invester, he treats all pieces of information equally and allocates his attention to each piece of information equally, rather than makes necessary adjustments according to different topic. So the searcher deviates from the track of behavior ration at the first minute, too.

To a risk aversion invester, because of $dT/dA > 0$, and $dT/dt < 0$, as time increases, he will be more and more sensible to the attention cost. Relatively familiar information, and the information whose style coincides with the searcher's habit and lying in front page costs less attention, it will more easily go into the area of his attention allocation. The searcher starts to observe and pay attention to different information selectively. As time goes by, the degree that the searcher chooses the "available" information will be stronger.[3] So, from the above analysis, we can say that irrationality deviation is a natural reflection of the searcher to the accelerated decrease of attention. To some extent, Kahneman and Tversky's Frame theory is also a theory about the allocation of attention resources[4] [5].

B.   Prior Knowledge's Influence on Attention Allocation

People don't search information without any aims, but always have some constructed prior knowledge. next, we will discuss prior knowledge from the dimensions of divergence and abundance.

1)   Divergence σ_I

That is, the deviation degree of prior knowledge, which is the degree of clarity for searcher's understanding to the given topic

2)   Abundance R_I

That is, the richness degree of prior knowledge, which is the degree of enrichment for searcher's understanding to the topic.

If σ_I is small, the searcher's understanding to the topic is relatively clear. Unless there is strong information attracting him, he will not leave the former understanding way. The searcher easily trends to 'Path-Dependence' and to short cut understanding information selectively. At the same time, he deduce results directly depending on experience. At this moment, usually T is small.

But if σ_I is large, namely the searcher doesn't have a clear understanding of the topic, he will not have successful experience to follow and not have obvious tendency. At this time, the searcher tends to pay attention to the selecting information carefully in order to have a clear mind. And usually T is large.

Therefore, there exists a weak positive correlation between T and σ_I, namely $dT/d\sigma_I > 0$.

The R_I is small means prior knowledge is poor and the searcher needs to refer to a lot of information to understand the topic better. Every piece of information's description of the topic is new to the searcher, So T is usually larger.

But if R_I is large, searchers can leave out many details depending on rich information. So, T is usually small.

Therefore, there exists a weak negative correlation between T and R_I, namely $dT / dR_I < 0$.

3   The Dynamic Model of Attention Allocation across Period

Attention can regenerate during a relative long time, because we can recover our attention and concentrate on the next search. In many cases, especially some complex topics, the information search cannot be finished at a time. At this time, the rule that attention is of scarcity is still not broken. The curved line of long-time attention resource is showed in Figure 1: allocation period and rest period take turns; because of the differences of the searcher's health and environment, the attention during each period maybe different but not obvious; the time of allocation period and rest period is different as well.

The regeneration of attention resource allows the searcher to plan his attention allocation for a long rime. Because attention is scarce, the multi-period of attention allocation doesn't mean that attention moves around many periods, but the planned content of the topic is allocated during each period. To a decision needing long time and much attention, we do the information search many times and start learning mechanism, which can reduce irrationality considerably.

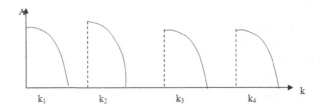

Fig. 1. The curved line of long-time attention resource

Now we will discuss the relationship among T, periodic numbers k and prior knowledge I.

In multi-period attention allocation, with k increasing, I will be richer and obvious more and more, namely $d\sigma_I / dk < 0$ $dR_I / dk > 0$. With learning mechanism, the searcher knows the topic more specific. The searcher tends to pay more attention to these concrete conditions and then T will be lager and larger.

Here T and k changes in the same direction, namely $dT / dk > 0$. It tells us that during a long period, because the generation of attention, we can allocate more attention to important information. Then, two effects will turn up: in single-period, $dT / dt < 0$ while in multi-period, $dT / dk > 0$.

The specific changes of T is influence by k and time process t_k. But what we are sure is, when the searcher needs to make judgments quickly during a short period (usually during the 1st period), $dT/dt < 0$ matters a lot and irrationality takes up most of the process; if he needs to search for a long time during multi periods, later $dT/dk > 0$ will take its place and rationality will turn up.

So, T is also the function of k, and the function form will be changed to $T = g(A,C,t,\sigma_I,R_I,k)$.

Now, we can combine time, prior knowledge and other factors to describe the across-period attention allocation. Suppose during k period, the information searcher, who is affected by prior knowledge I^{k-1}, chooses n pieces of information $x_{k1}, x_{k2} \cdots\cdots x_{kn}$ to describe the topic. In the t_k attention allocation period, prior knowledge and the searched information during number k period have an organic combination. At the end of k period, the searcher edits and processes the information into I^k as his new understanding of the topic and I^k becomes the prior knowledge of $k+1$ period. The searcher gets the utility during number k period like this:

$$U_k = -\Delta S\left[w_1^k(t_k), w_2^k(t_k)\cdots\cdots w_n^k(t_k); x_1^k, x_1^k \cdots\cdots x_n^k \mid I^{k-1}, A^k(t_k), T^k(t_k, k, I^{k-1})\right]$$

$$st. \qquad \sum_{i=1}^{n} w_i^k = 1 \qquad\qquad (9)$$

$$A = A^k(t_k)$$

$$T^k = g\left[A^k(t_K), C^k(t_K), t_K, \sigma_I^{k-1}, R_I^{k-1}, k\right] \qquad (10)$$

$$I^k(\sigma_I, R_I) = F\left[\sigma_I^{k-1}, R_I^{k-1}; w_1^k(t_k), w_2^k(t_k)\cdots w_n^k(t_k); x_1^k, x_1^k \cdots x_n^k\right] \qquad (11)$$

It can be simply described as:

$$I^k = f\left[w(k), x(k), t_k\right] \cdot I^{k-1} \qquad (12)$$

We can find that in the attention allocation during k period; both the width that each piece of information takes up in the information channel and the attention degree A during number k period are the functions of time. T is still very useful, which sometimes makes the attention allocation during number k period irrational; the understanding at the end of this period is the effect of the prior knowledge and the searched information.

4   Utility Analysis in Information Processing

In the period of $k+1$, Influenced by I^k, the searcher selects m information $y_{k+1,1}, y_{k+1,2} \cdots\cdots y_{k+1,m}$, the utility of attention   allocation   is:

$$U_{k+1} = -\Delta S\left[w_1^{k+1}(t_{k+1}), w_2^{k+1}(t_{k+1})\cdots w_m^{k+1}(t_{k+1}); y_1^{k+1}, y_2^{k+1} \cdots y_m^{k+1} \mid I^k, A^{k+1}(t_{k+1}), T^{k+1}\right]$$

In the later period of $k+1$, search people's recognition for subject as :

$$I^{k+1} = f[w(k+1), x(k+1), t_{k+1}] \cdot I^k \tag{13}$$

It can be named learning mechanism. We can see that the prior knowledge in the period of $k+1$ is more optimized than in the period of k. General speaking, I^{k+1} is more detailed than I^k, and is more closer to the truth of description of the subject.

If we complete the attention allocation in the period of $k+j$, then total utility is the sum of respective utility from 1 to $k+j$ period. For a searcher, the sooner he know the theme , the better for reducing the uncertainty Hence we need to discount the subsequent utility of each period. The total utility of such access is :

$$U = U_1 + \sum_{\tau=2}^{k+j} U_\tau e^{-\rho\tau} \qquad \rho > 0 \tag{14}$$

Information searchers allocate attention resource by searching the theme in a relatively long period to maximize his utility:

$$Max\left(U_1 + \sum_{\tau=2}^{k+j} U_\tau e^{-\rho\tau}\right) \tag{15}$$

$$\sum w(k) = 1$$

$$I^k = f(I, w) I^{k-1}$$

$$T = (I, t, k)$$

This not only is the model of attention allocation across multiple periods, but also is a process of searchers' gradually change from irrational to rational.

If the attention allocation completed in the period of $k+j$, the searcher has formed a complete understanding of topic in the later period of $k+j$, in other words his knowledge for topic converged in $I(end)$, at this time $I^{k+j} = I(end)$.

Make divergence and abundance as two dimensions, we divide the prior knowledge into four categories, they are shown in Figure 2. Let us look at the path and speed of these knowledges' convergence.

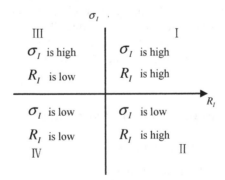

Fig. 2. Make divergence and abundance as two dimensions and divide the prior knowledge into four categories

Information searcher with different types of prior knowledge obtains different converging path and speed of cognition by searching. Generally speaking, prior knowledge of high abundance ratio and low divergence is easy to achieve learning mechanism with the fastest converging speed; Whereas the prior knowledge of low abundance ratio and high divergence is hard to achieve learning mechanism with the slowest converging speed; The converging speed of other two types of prior knowledge exists in the middle of the previous two types.

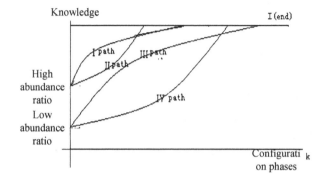

Fig. 3. The knowledge converging paths of four types of seekers

References

1. Wang, D.: Economics Description of Attention. Economic Research (October 2000)
2. Xue, Q., Huang, P.: Behavioral Economics - Theory and Applications. Fudan University Press (November 2003)
3. Zhang, J., Zhang, L.: Practical Ration: A Framework for Understanding Economic Behavior. Advances in Psychological Science (November 2003)
4. Tversky, A., Kahneman, D.: The framing of decisions and the psychology of choice. Science 211, 453–458 (1981)
5. Tversky, A., Kahneman, D.: Rational choice and the framing of decisions. Journal of Business 59, 251–278 (1986)

Basic Analysis on Molecular Behavior of RMB Settlement

Zongsheng Wang

School of Economics, Tianjin Polytechnic University, Tianjin, China
wzs0822@163.com

Abstract. Cross-border RMB settlement is important content of RMB internationalization and regionalization. In reality, it features as two practical types as "Yunnan model" and "Shanghai model". This paper analyzes deeply their situation, operational mechanism and development. Analyze trigger basis of enterprises choosing people by game frogs and its replicator dynamical equation of state, propose successful experiences of copying "Yunnan model", and improve the development routine of "Shanghai model" by initiative and clear gradual systematic accumulation.

Keywords: RMB cross-border settlement, accumulation system, game frogs.

1 Introduction

In the item of expanding trade, the experimental point of RMB cross-border settlement is the basis work of RMB gradual regionalization. On Apr 8th, 2009, executive meeting of the State Council decides, under the background of international financial crisis, to help enterprises away from exchange rate risk, improve terms of trade and keep foreign trade steadily growing, that to open RMB settlement experimental points with ASEAN countries, Hong Kong and Macao in Guangzhou, Shenzhen, Zhuhai, Dongwan. On July 2nd, according to cross-border trade RMB experimental point management measures, the cross-border trade RMB settlement experimental point started informally. On July 6th, 2009, Shanghai silk group started first RMB cross-border trade pricing and account settlement business. Until Feb, 2010, statistics of Shanghai government webs show, the cross-border trade RMB settlement business accumulates approximately to 900, and the total monetary value is up to 12.6 billion.

2 The Situation of RMB Settlement

A. The Exposed Problems of RMB Settlement "Shanghai Model" in Practice

RMB settlement business in Shanghai, Guangdong, etc is called Shanghai model, which starts its business up and down: through initial several attempts, the deals were frozen in next months, then countries intensified propaganda and proposed some supporting policies at the same time, RMB settlement demonstrated gradually warm situation since the end of 2009. But current business is in the state of warm water

J. Luo (Ed.): Soft Computing in Information Communication Technology, AISC 161, pp. 117–123.
springerlink.com   © Springer-Verlag Berlin Heidelberg 2012

again, and most enterprises are waiting to see. The investigation shows us some reasons as follows:

- The willing of using RMB to final consumer of exports is low. One important aspect of foreign trade is to freely exchange mutual monetary in international market. RMB isn't international monetary, so it can't be converted freely in international market. Such is a main reason why this business can't be used in a large scope.

- International trade habitually sustains strong competition of dollar, euro and yen, which makes low acceptance of RMB. Ever since a long time ago, the trade exports destinations of Shanghai model centered on America, Europe or Japan, etc, which are strong monetary mother land. They need a gradual process to adjust to RMB settlement. According to practice, enterprises choose which monetary is more convenient, safer and minimum in risk. So such will gradually reverse in practical advantages of exchange RMB and the shift among RMB, dollar, euro and yen.

- Export rebates are detailed but not clear. Unclear follow-up system anticipation leads to low willing of RMB settlement. The lag of data association mechanism of customs, tax office, complicate cancelling procedure, rebate operations detailed but not clear stops the export rebates, not high rebate ration, meanwhile, which affects cost of export enterprises. Besides, there are no definite sayings about follow-up relative policies of goods trade RMB settlement to tax office, which makes low enterprise system anticipation and not willing to invest in relative complete set service.

B.   The Practical Situation of RMB Settlement "Yunnan Model"

Sharply contrast to shanghai model, cross-border RMB settlement practice with ASEAN is warmer and warmer. After operating policy of border trade RMB settlement export rebates in 2004, the weight of RMB in border trade export settlement increased from 22% to 90%. The border trade RMB settlement is 2.7 billion in Yunnan province, increased by 265%, RMB settlement ratio increased from

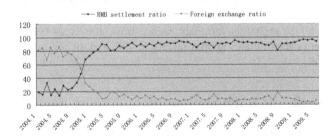

Data resource: according to month's statistics of Center for the People's Bank of China branch in Kunming

Fig. 1. Yunnan Province border trade exports settlement currency proportion from January, 2004 to June, 2009

average 30% to 85%; in 2006 and 2007, Yunnan RMB settlement of border trade keeps increasing in a high speed, which is 3.417 billion and 3.869 billion, whose proportion is about 91 percent. Influenced by global economic crisis in 2008, the border trade RMB settlement was 3.696 billion, with a slight decline, but the RMB settlement proportion was still up to 90.2%. RMB settlement between January and June in 2009 was up to 1.982 billion, with increasing by 5.5 percent, with proportion 95.2% of border trade in the whole province.

Meanwhile, Center for the People's Bank of China branch in Kunming shows, Yunnan border RMB input and output was up to 169.964 billion. Inflow was 33.018 billion and outflow was 36.946 billion, net outflow was 3.928 billion. The first half of year in 2009, Yunnan border RMB inflow and outflow were up to 22.980 billion. Inflow was up to 11.789 billion, outflow was 11.191 billion and net inflow was 0.598 billion.

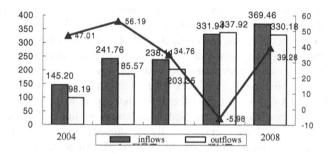

Data source: according to year's statistics of Center for the People's Bank of China branch in Kunming

Fig. 2. Yunnan RMB cross-border inflows and outflows

3   Comparison and Choice of RMB Settlement Spot Model

Now let us analyze the secret of success in Yunnan model. In general, with reference to area, the foreign trade can be classified into border trade and generally normal trade, the formal is called as border trade, the latter is called as normal trade. Both sides of trade in border trade are near, with many transaction rounds. RMB or foreign currency commutes familiar both sides many times, most of which exchange in cash. But normal trade includes large quantities of capital turnover, credits are needed, and complex payment system is needed to support. Due to these differences, the operations are different.

Researches show that the success of Yunnan model is to open a possible way to propel RMB settlement from border trade to general trade process. Specifically speaking, on the basis of spontaneous cross-border trade, eliminate differences between RMB and dollars through exports debate, so as to make enterprises to get true benefits from RMB settlement. Meanwhile, by use of reasonable cross-border settlement models, cultivate enterprises' RMB settlement habits and form steady settlement ways. Multilaterals supporting measures of regional governments build advantages of RMB settlement and out and in exchange platform, form a good start of RMB settlement and trigger positive feedback to form fine circle.

Therefore, Yunnan model can be described as to cultivate RMB settlement habit, link RMB settlement channels, solidify RMB settlement profits, optimize RMB settlement environment and support RMB settlement system. Yunnan model uses ideas of flexible intervention, propelling step by step, which can be consider as a kind of causative system, to realize coupling between system and market.

But Shanghai model is fast paced development, which uses RMB settlement in general trade without border trade inflows. It is a practicable settlement way without Chinese and foreign approvals. Shanghai model is inflexible policy implantation, which can be consider as compulsory intervention. The loss of settlement habits stimulates fierce competition from dollars and lag of foreign exchange platform construction makes RMB impeded reflow, going out but not return. RMB settlement of enterprises is a short-time trial behavior on the condition of debate policy. But in the long term, in the state of obscure follow-up supporting system policies, the expected yield is not obvious and the more consideration is in the state of watching.

Only from operational effect, undoubtedly, Yunnan model is better than shanghai model. However, proceeding from our national conditions and development, although there are deficiencies of nature, shanghai model is still the point of improving and propelling: basic hinterland provinces use Yunnan model, whose economic development is relative weak, ratio of import and export total, and the ability of enterprises which spontaneously use RMB settlement is limit. But the provinces which use shanghai model are economically developed, whose import and export shares are about 66 percent of the whole country, shown as table 1. This is real representation of imbalanced economic development.

Table 1. Import and export share (0.1 billion) and ratio of Yunnan model coverage and shanghai model coverage

模式	地区	2005 进出口	2005 占比	2006 进出口	2006 占比	2007 进出口	2007 占比	2008 进出口	2008 占比
云南模式	内蒙古	487625.3762		596082.0083		773588.5		891848.3	
	黑龙江	956601.7204		1285654.672		1729659.3		2313059.3	
	青海	41330.6953		65172.4082		61207.3		68882.3	
	新疆	794048.6754		910326.8399		1371582.8		2221736	
	云南	474344.0952		622483.6479		879356.7		959691.6	
	西藏	20546.839		32837.657		39346.4		76582.9	
	广西	518150.3754		666756.1614		925899.7		1323616.8	
	合计	3292647.777	0.023	4179313.394	0.024	5780640.7	0.027	7855417.2	0.031
上海模式	上海	18633673.79		22752419.59		28285387.8		32205531	
	江苏	22792276.24		28397838.36		34947178.6		39227193	
	浙江	10738965.57		13914160.75		17684736.8		21113373	
	广东	42796496.66		52719909.55		63418595.4		68496879.8	
	合计	94961412.25	0.668	117784328.2	0.669	144335898.6	0.664	161042976.8	0.628
	全国	142190617.2		176039646.9		217372601.7		256325522.8	

Data resource: China Statistical Yearbook

In the short term, these two models are not permitted to deform. But from developing aspect, if wish to truly propel RMB settlement deeply, we must develop shanghai model. We only copy partial successful experiences of Yunnan model, enhance policy supporting power of shanghai model, build fine environment through policy accumulation and gradually eliminate restraints of shanghai model.

4   Deepen Basic Analysis of Micro-behavior of RMB Settlement

After analyzing system and model of RMB settlement, we consider micro-implementation of RMB cross-border settlement. Enterprises are micro carriers and final motive force of RMB cross-border settlement, to specific enterprises, beyond the habit which consider dollars as settlement currency, RMB settlement is a kind of trial and innovation. Under what environment, will enterprises try a new currency? How to cultivate this trial to be a habit? We shall analyze these problems with reference to replicator dynamics of frog game.

To most of Chinese enterprises, the choice between RMB and dollars is not only influenced by self-settlement habits, but also by settlement habits of other enterprises. RMB settlement can dodge exchange rate fluctuations risk loss, meanwhile it is possible to enjoy state refund (exemption) tax concessions. Besides, spillover value (if most of enterprises choose RMB settlement, the probability for foreign enterprises to choose RMB settlement increases, so as to decrease difficult degree of negotiation) can be get.   Three points above comprises extra benefits V; due to complex formalities, bank negotiation while applying debate and financial self-adjustment, enterprises need to pay certain systematic frictional costs, there are also menu cost such as change prices, etc and saliva cost with foreign trade. They comprise cost C, so the extra benefits are V-C.

If choose RMB settlement, both two enterprises of random mating game get V-C extra benefits; if one enterprise chooses RMB, the other one chooses dollars, due to decreasing in overspill prices, the enterprise choosing RMB settlement gets a smaller U (U<V), with the same cost C, and takes on market cost M resulting from foreign clients dollar settlement inertia, this M is got by part; if all the mating enterprises keep original habits, they can't get extra benefits. Such constructs gain matrix shown as figure 3.

To specific enterprises, whether to choose RMB settlement depends on balancing results:

1) **Quantities and original ratio of anticipation model enterprise**
If most of native enterprises and foreign trade object enterprises use RMB settlement, in consideration of settlement environment and exchange convenient degree, they are apt to choose RMB settlement.

2) **Enterprises use extra benefits of RMB settlement, which is successful degree of model enterprises**
If anticipation enterprises get practical benefits in the process of RMB settlement, enterprises will keep using RMB settlement and instruct this experience to other enterprises and foreign trade enterprises.

No matter choosing RMB settlement or dollar settlement, supposing the ratio of enterprises choosing RMB settlement is x, then the ratio of enterprises choosing dollar settlement is 1-x, according to game matrix above, the replicator dynamical equation of state:

$$\frac{dx}{dt} = x(1-x)[x(V-C-M)+(1-x)(U-C-M)]$$

Limited by passages, this paper does not discuss complex relations deeply, only referring to analysis thought of game theory. When V>U, U-C-M>0, most enterprises will form $x^* = \dfrac{U-C-M}{V-U}$ steady-state point shown as figure 4, so as to form a fine situation of half of the world between RMB settlement and dollar settlement.

If V is big, C is small and V-C is certain and continuous, than after a course of game, enterprises will choose RMB settlement, namely steady-state is $x^* = 1$; but if V-C is not definite and expectation is not uncertain, either, most enterprises will convergent to steady-state $x^* = 1$, namely, although a few enterprises try RMB settlement, it will soon stop and watch.

Enterprise 1

		RMB settlement	Dollar settlement
Enterprises 2	RMB settlement	$V-C$, $V-C$	$U-C-M$, M
	Dollar settlement	M , $U-C-M$	0 , 0

Fig. 3. Extra benefits matrix of enterprises choosing RMB settlement currency

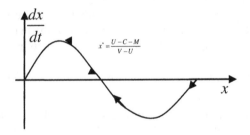

Fig. 4. Replicator dynamic phase diagram of two kinds of currency settlement game

Therefore, it has to form a certain number of enterprises to join and get practical benefits in settlement, so as to open situation of RMB settlement and get fine development. From the analysis above, it is easy to find that it is also the successful core of Yunnan model. All the systematic arrangement of Yunnan model are constructed and designed around this core.

In general, a currency's regionalization and internationalization can be realized in two ways: market spontaneous evolution and active and specific systematic arrangements. The representative of market spontaneous evolution is early pound internationalization, but success in dollar and euro internationalization can fully prove systematic arrangement or constructing order has a great force to currency internationalization. In the process of RMB going out of nation, the past experiences demonstrate most RMB circulation among border countries is the result of market spontaneous effect. We pay too much attention to depend on spontaneous spillover of RMB, which is passively regionalized in the long term, the propelling main of regionalization, is micro-economic main. We are short of actively systematic arrangements, and propel RMB to step into ways of regional development through the systematic construction center on government.

Yunnan model is a successful model utilizing active system to intervene market spontaneous state and developing process. So in the process of further promoting RMB regionalization, we shall improve Yunnan model, meanwhile, borrowing ideas from its experiences, induce and conduct behavior by actively arranging policy and systems, and propel RMB settlement work of shanghai model.

References

1. Zhou, Y.: Study on currency cooperation and RMB regionalization in Sino-ASEAN. Research of Finance (May 2008)
2. Shi, W., Ding, Y.: The basic principles and path of RMB regionalization at this stage. Journal of Economic Outlook (July 2009)
3. Han, M., Yuan, X.: Study on RMB regionalization based on trade. Global Economy (February 2007)
4. Qiu, Z., He, L.: The feasibility of regionalization of RMB. China Finiance (October 2008)
5. Suning: Steady progress in cross-border trade RMB settlement of the Yangtze River Delta. China Finance (August 2010)
6. Gong, Q.: Several thoughts in grass-root People's Bank to promote steadily and fast economic development. Zhejiang Finance (April 2009)
7. Feng, D., Xie, J.: Study on RMB regionalization strategies based on currency competition. Guangxi Finance Research (January 2008)
8. Cao, Y.: Discussion on systematic construction of RMB regionalization. Zhejiang Finance (April 2009)

Investment Portfolio Model Based on Attention

Zongsheng Wang

School of Economics, Tianjin Polytechnic University, Tianjin, China
wzs0822@163.com

Abstract. Markowitz portfolio theory is the core of microfinance, but modern stock market is increasingly reflected in a attention market. The modification and development of original theory model is to generate attention constraints from portfolio model, which is the theoretic basis of regulating stock market.

Keywords: attention constraints, investment portfolio, efficiency frontier.

1 Introduction

Markowitz portfolio theory is one of main contents of microfinance, since proposed in the 1950s, having been developed rapidly and become essential contents of financial investment and financial risk management. But to any viable theory without exception, which is improved by continuous modification of new phenomenon, new problems in economic development. Compared to Markowitz time stock market, although external form of stock market varied little, market operation is strongly supported by information technology development and the form of gathering information is innovated. To the whole, it has to be faced that limit attention constrains financial choices. So adding attention to modify on the basis of Markowitz is the premise of classical model's strong explanatory ability.

2 Model Hypothesis

Portfolio model based on attention is developed with attention constraints on the basis of Markowitz mean-variance model. A big change happens to modern hypothesis. Hypothesis 1- hypothesis 3 is different from Markowitz hypothesis:

Hypothesis 1: stock market isn't effective. Attention is limit in information. Every investor estimates expected return and standard deviation of each kind of stock by configuring attention selection and asset price. Information isn't total, information held by investors is determined by quantities of attention paid to. Different attention of investors is different in quantity but not in quality.

Hypothesis 2: the objects of stock investors: under certain attention input standard, given risk benefit hoped the biggest, or given benefit standard lowest, that is to say, investors hate risk.

Hypothesis 3: investors look on expected return and variance of yield through attention regulation as foundations. If select proposals with higher risk, they need extra investment benefits as compensation.

J. Luo (Ed.): Soft Computing in Information Communication Technology, AISC 161, pp. 125–132.
springerlink.com

Hypothesis 4- Hypothesis 9 is same as standard Markowitz mean-variance model.

3   Model Construction

A.   Model Construction Ideas

1)   Bring in attention in portfolio

Next we deal like this, regard each stock as a piece of information. Select n stocks in "stock forest" to constitute a portfolio, equals to select n stock information in "sea of information", a stock is a piece of information. This dealing has no effect on essence and can simplify the analysis. Proposing investors choose n stock notes in some time and make use of such to construct portfolio. The configuration attention in stock I is A_i.

2)   Multi- objective investment decisions

Subjective resource attention is a kind of scarce resource. Investors configuration attention selects stock and constructs effective portfolio. So attention is a kind of input. Investor configuration attention gets information set, on which form expectation of future stock prices and estimate expected yield and yield variance. The mean and variance of expected yield can be regarded as outputs, forms a kind of inputs with attention--- input –output relationship.

Investor input attention gathers macro-level and company-level information, analyzes factors influencing changes of stock prices, judges tendency of such stock prices and forms price expectation. Undoubtedly, much attention paid to some stock, will form more accurate expectation. But attention is limited, the competitive configuration relationship exists among stocks--- to pay more attention to some stock means configuring less other stocks or making them ignored .Investors wish to improve efficiency of the whole of attention resource, and that is to get information set as much as possible on the basis of certain attention input. So investors must balance this kind of input-output relationships.

At the same time, investors have to face with balance between risk and benefit, which will make the choice can't only set the maximum expected return as only criterion. If require higher investment return, they have to face more risks.

Under the background of relative surplus information, investors have to face two levels' weighing and three objects. Two weighing is: risk-benefit weighing and input-output comparison.

Three objects are: risk is as small as possible, benefit is as big as possible and input-output efficiency is as high as possible

They are described in math:

$$
\begin{cases}
\min \sigma_p^2(A) = (w_1, w_2 \cdots w_1)\Omega(A)(w_1, w_2 \cdots w_1)^T \\
\max R_p(A) = w_1 R_1(A_1) + w_2 R_2(A_2) \cdots w_n R_n(A_n) \\
\max F(A) \\
st. \quad w_1 + w_2 \cdots + w_n = 1
\end{cases}
\tag{1}
$$

$\sigma_p^2(A)$ --- Variance of portfolio. Because such is estimated on the basis of paying attention to gather information, variance is a function of information even a function of attention.

$R_p(A)$ --- Expected yield of portfolio, likely, a function of attention.

$\Omega(A)$ --- Covariance matrix based on attention portfolio.

$F(A)$ --- Attention input-output function.

So the stock choice is a multi-objective decision. It can merge objects; transform to a two-stage decision-making; transform other objects into constraints conditions, forming one object multi-constraint decision. This paper uses first method, certain replacement and amendment on expected yield and variance in attention frame.

B. Basic Metric Elements

1) Measurement of yield
Constrained by attention selection, investors choose n portfolio. Firstly, investors configure Ai attention in stock i, estimate expected yield $E(r_i)$ and standard deviation σ_i of stock i, then decide allocate w_i shares of finance to finance i.

In investment decision, the influence of attention on information gathering has been discussed detailed above, which can be summarized into three key points: first is limit constraint. The attention of investors is quite limited, to complicate decisions, they are always made on basis of incomplete information; second is selective constraints. The limit attention essentially demands selective notice on information. Investors usually select information easy to remember, conversant and fit their interests to notice, which always leads to directivity deviation on information gathering; last is configuration constraints, which is resulted from limit attention. In the process of information gathering, attention loss tolerance exists with time decreased, which will generate deviation in the process of attention configuration, for example, less attention paid to allocate key information but more attention paid to secondary information, the deviation surely leads to deviation on information gathering.

So, limit attention input is the key to form information deviation. When estimate portfolio expected yield based on biased information set, error cognizance always exists and systematic deviation will exist in expectation. To investors, stock i's expected yield equals original benefit and systematic deviation, that is $E(r_i)+\Delta s_i$.

To portfolio, the yield of portfolio equals the weighted average of yields of various asset in the portfolio

$$R = \sum_{i=1}^{n} w_i E(r_i) + \sum_{i=1}^{n} w_i \Delta s_i = \sum_{i=1}^{n} w_i E(r_i) + \Delta S \qquad (2)$$

The deviation ΔS is resulted from limit information, but information gathering relates to attention. In general, systematic deviation ΔS is inverse ratio to attention input standard. So yield of portfolio can be described as

$$R = \sum_{i=1}^{n} w_i E(r_i) + \eta/A \qquad (3)$$

It needs to be pointed that, due to attention resulting in systematic deviation, according to characteristics in attention configuration, this deviation may be positive deviation or negative deviation, that is investors highly estimate or lowly estimate yield. So parameter η is transfer coefficient and ± 1(positive deviation +1, negative deviation - 1) multiple.

   2) **Measurement of portfolio risk**

   In attention model, we also measure risk by expected benefit standard deviation. Standard deviation estimates possible deviation extent between practical return rate and expected return rate in some way. Because investors can't overall hold information related with asset prices, investors don't know various possible probability distribution of yield, which makes many unknown parameters. In practical, we statistically treat observed numbers of n risk assert payment, then select a group of subjective values to replace these parameters according to estimates or experience.

   In the process of estimating expected yield of stock I, due to the biased attention information set, the expectation produces systematic deviation. But according to definition of variance, $\sigma_i^2 = E[(r_i - E(r_i)]^2$, no matter whether highly estimates or lowly estimates expected yield, will get big variance, which can be described as $(1+\Delta)\sigma_i^2$

   The risk of portfolio can be described as $\sigma_p^2 = \sum_{i=1}^{n}\sum_{j=1}^{n} w_i w_j \sigma_{ij} + \Delta\sigma^2$. In the same way, $\Delta\sigma^2$ is produced by incomplete information, which is to result from attention limit input. Increasing information quantity to improve decision contributes to shorten deviation. So we can consider $\Delta\sigma^2$ is inverse ratio to attention input standard. Such portfolio risk can be described as

$$\sigma_p^2 = \sum_{i=1}^{n}\sum_{j=1}^{n} w_i w_j \sigma_{ij} + \varepsilon/A \qquad (4)$$

ε is transformation coefficient, which is multiple of ratio coefficient and unit conversion factor.

C.   Effective Frontier of Portfolio Based on Attention

Investors consider yield and standard deviation after regulating as reference while choosing plan. If want to choose plan which can get higher benefit, they have to take on extra investment risk as the cost.

   Such effective frontier can be get in following way:

$$\min \frac{1}{2}\sigma_p^2 = \frac{1}{2}\sum_{i=1}^{n}\sum_{j=1}^{n} w_i w_j \sigma_{ij} + \frac{\varepsilon}{A}$$

$$st. \quad \begin{aligned} & \sum_{i=1}^{n} w_i E(r_i) + \frac{\eta}{A} = E_p \\ & \sum_{i=1}^{n} w_i = 1 \end{aligned} \tag{5}$$

Describe it in matrix:

$$\min \frac{1}{2}\sigma_p^2 = \frac{1}{2}W^T \Omega W + \frac{\varepsilon}{A}$$

$$st. \quad \begin{aligned} & W^T E(R) + \frac{\eta}{A} = E_p \\ & W^T e = 1 \end{aligned} \tag{6}$$

In $W = \begin{pmatrix} w_1 \\ w_2 \\ \vdots \\ w_n \end{pmatrix}$ $\Omega = \begin{pmatrix} \sigma_1^2 & \sigma_{12} & \cdots & \sigma_{1n} \\ \sigma_{21} & \sigma_2^2 & \cdots & \sigma_{2n} \\ \vdots & \vdots & \vdots & \vdots \\ \sigma_{n1} & \sigma_{n2} & \cdots & \sigma_n^2 \end{pmatrix}$ $e = \begin{pmatrix} 1 \\ 1 \\ \vdots \\ 1 \end{pmatrix}$ $E(R) = \begin{pmatrix} E(r_1) \\ E(r_2) \\ \vdots \\ E(r_n) \end{pmatrix}$

Solve model in Lagrange multiple, suppose

$$L = \frac{1}{2}W^T \Omega W + \frac{\varepsilon}{A} + \lambda(E_p - W^T E(R) - \frac{\eta}{A}) + \mu(1 - W^T e) \tag{7}$$

First-order conditions for optimal solution are:

$$\begin{cases} \dfrac{\partial L}{\partial W^T} = \Omega W - \lambda E(R) - \mu e = 0 \\[2mm] \dfrac{\partial L}{\partial \lambda} = E_p - W^T E(R) - \dfrac{\eta}{A} = 0 \\[2mm] \dfrac{\partial L}{\partial \mu} = 1 - W^T e = 0 \end{cases} \tag{8}$$

This is a linear system of equations with n+2 unknown numbers. From first equation, $W = \lambda\Omega^{-1}E(R) + \mu\Omega^{-1}e$'s transpose can be get as:

$$W^T = \lambda E(R)^T \Omega^{-1} + \mu e^T \Omega^{-1} \tag{9}$$

Let (4.23) right multiple E(R), get $W^T E(R) = \lambda E(R)^T \Omega^{-1} E(R) + \mu e^T \Omega^{-1} E(R)$

Let (4.23) right multiple e, get $W^T e = \lambda E(R)^T \Omega^{-1} e + \mu e^T \Omega^{-1} e$ and constitute equations

$$\begin{cases} \lambda E(R)^T \Omega^{-1} E(R) + \mu e^T \Omega^{-1} E(R) = E_p - {}^{\eta}\!/\!_A \\ \lambda E(R)^T \Omega^{-1} e + \mu e^T \Omega^{-1} e = 1 \end{cases} \qquad (10)$$

This is a system of linear equations of two unknowns on λ and μ, to convenient calculation, we set

$$B = E(R)^T \Omega^{-1} E(R), \quad C = e^T \Omega^{-1} E(R) = E(R)^T \Omega^{-1} e, \quad D = e^T \Omega^{-1} e,$$

$$\Delta = BD - C^2$$

Then the equation changes into $\begin{cases} \lambda B + \mu C = E_p - {}^{\eta}\!/\!_A \\ \lambda C + \mu D = 1 \end{cases}$, we can get

$$\lambda = \frac{E_p D - {}^{\eta}\!/\!_A - C}{\Delta} \qquad \mu = \frac{B - E_p C + {}^{\eta}\!/\!_A}{\Delta}$$

Then bring $W = \lambda \Omega^{-1} E(R) + \mu \Omega^{-1} e$ into σ_p^2, get

$$\begin{aligned} \sigma_p^2 &= W^T \Omega (\lambda \Omega^{-1} E(R) + \mu \Omega^{-1} e) + {}^{\varepsilon}\!/\!_A \\ &= \lambda W^T E(R) + \mu W^T e + {}^{\varepsilon}\!/\!_A \\ &= \lambda (E_p - {}^{\eta}\!/\!_A) + \mu + {}^{\varepsilon}\!/\!_A \end{aligned} \qquad (11)$$

Take λ and μ into formula above:

$$\begin{aligned} \sigma_p^2 &= \frac{(E_p D - C)}{\Delta} E_p + \mu + {}^{\varepsilon}\!/\!_A \\ &= {}^D\!/\!_\Delta [E_p - ({}^C\!/\!_D + {}^{\eta}\!/\!_A)]^2 + {}^1\!/\!_D + {}^{\varepsilon}\!/\!_A \end{aligned} \qquad (12)$$

This is a hyperbolic line, write which as standard form:

$$\frac{\sigma_p^2}{{}^1\!/\!_D + {}^{\varepsilon}\!/\!_A} - \frac{[E_p - ({}^C\!/\!_D + {}^{\eta}\!/\!_A)]^2}{{}^\Delta\!/\!_D ({}^1\!/\!_D + {}^{\varepsilon}\!/\!_A)} = 1 \qquad (13)$$

Due to non-negativity of variance and covariance, covariance Ω is a positive definite matrix, and covariance Ω^{-1} is also a positive definite matrix. So $B = E(R)^T \Omega^{-1} E(R) > 0$, $D = e^T \Omega^{-1} e > 0$, $\Delta = BD - C^2 > 0$. We know in (σ_p, E_p) plane, minimum variance set $\sigma_p^2 = \frac{D}{\Delta}[E_p - (\frac{C}{D} + \frac{\eta}{A})]^2 + \frac{1}{D} + \frac{\varepsilon}{A}$ is right branch of hyperbolic line.

4   Choose Specific Portfolio

Suppose investor effect function is

$$U = U(\sigma_P, E_P) \tag{14}$$

So the equation of indifferent curve is

$$U(\sigma_P, E_P) = m \quad \text{(m is constant)} \tag{15}$$

Calculate slope of indifferent curve

$$\frac{dE_P}{d\sigma_P} = -\frac{\partial U}{\partial \sigma_P} \Big/ \frac{\partial U}{\partial E_P} = -MU_{\sigma_P} \Big/ MU_{E_P} \tag{16}$$

Because the optimal plan is to choose contact point between indifferent curve and effective frontier, there is such as follows:

$$\frac{\partial \sigma_P^2 \Big/ \partial \sigma_P}{\frac{1}{D} + \frac{\varepsilon}{A}} d\sigma_P - \frac{\partial (E_P - \frac{C}{D} - \frac{\eta}{A})^2 \Big/ \partial E_P}{\frac{\Delta}{D}(\frac{1}{D} + \frac{\varepsilon}{A})} dE_P = 0 \tag{17}$$

After arranging,

$$\frac{dE_P}{d\sigma_P} = \frac{2\sigma_P}{\frac{1}{D} + \frac{\varepsilon}{A}} \Bigg/ \frac{2(E_P - \frac{C}{D} - \frac{\eta}{A})}{\frac{\Delta}{D}(\frac{1}{D} + \frac{\varepsilon}{A})}$$

$$= \frac{\sigma_P}{(E_P - \frac{C}{D} - \frac{\eta}{A})} \frac{\Delta}{D} = \frac{\Delta \sigma_P}{DE_P - \frac{\eta D}{A} - C} \tag{18}$$

(4.39) is slope of effective frontier.

From nature of indifferent curve and effective frontier contacting, the slope of indifferent curve equals effective frontier curve at contact point.

$$-\frac{MU_{\sigma_P}}{MU_{E_P}} = \frac{\Delta\sigma_P}{DE_p - \eta D/_A - C} \tag{19}$$

So, the essential condition of contacting between indifferent curve and minimum variance curve is

$$\Delta\sigma_P \cdot MU_{E_P} + MU_{\sigma_P}(DE_p - \eta D/_A - C) = 0 \tag{20}$$

Besides, the contacting point surely crosses both indifferent curve and effective frontier. So the optimal state (σ_P^*, E_P^*) of portfolio is decided by equations as following:

$$\begin{cases} \dfrac{\sigma_p^2}{1/_D + \varepsilon/_A} - \dfrac{[E_p - (C/_D + \eta/_A)]^2}{A/_D(1/_D + \varepsilon/_A)} = 1 \\ \Delta\sigma_P \cdot MU_{E_P} + MU_{\sigma_P}(DE_p - \eta D/_A - C) = 0 \end{cases} \tag{21}$$

(σ_P^*, E_P^*) is a kind of portfolio according with investor risks. Because the indifferent curve of investors of risk aversion is positive slope and cima, it means the contact point is unique.

References

1. Li, Z., Wang, S.: EaR Risk Measurement and Dynamic Investment Decision. Quantitative and Technical Economics (1) (2003)
2. Chenshou, Deng, X., Wang, S., Liu, W.: Impact on the Efficient Frontier of Portfolio Varying Capital Structure. Chinese Journal of Management Science (1) (2001)
3. Tang, X., Pan, J.: The effective frontier assurance of portfolio on no condition of cost of short selling. Forecasting (5) (1995)
4. Liu, H., Fan, Z., Pan, D.: The optimal selection of portfolio with transaction costs. Journal of Management Science (2) (1999)
5. Liu, S., Qiu, W., Wang, S.: Universal Portfolio selection with transaction costs. Theory & Practice (1) (2003)

Development of Electronic Clothes and Thinking about Its Value

Sun Jing

Institute of Arts and Garments, Tianjin Polytechnic University, Tianjin, China
sunjing@tjpu.edu.cn

Abstract. In light of the development of the electronic clothes, its technology forms are analyzed from five aspects which includes the microminiaturization of device, device flexibility, connect technology, function fibers and power supply. The electronic clothes are classified as the entertainment clothes, communication clothes, monitor clothes, remote control clothes, locating clothes and clothes which can adjust temperature and pressure. Its current situation of development is introduced. And its cultural value and progressive meaning are expounded.

Keywords: clothes industry, electron industry, electronic clothes, clothes technology.

The development of Information Technology impels the informationized era to come into human's life. People will find that the electronic industry has been influencing their lives all around, from TV set, cell phone and digital camera, to credit card, VCD, DVD and MP3; from office automation facilities, to mechanical automation, enterprise IT application, digital architecture, multi-media instruction and e-commerce, especially, the popularization of computer and the application of the internet make people experience radical change in the fields of technology and culture. At present, with the further developing of electronic technology, the produces become smaller and lighter, with their capacity and security being improved. Including the computer, these kinds of produces are not only moveable, but also have being embedded into the clothes which serve as the vinculum between human body and environment. This endues the information technology with highly convenient handling, thereby, the electronic clothes appear in the fashion field as a maverick. Electronic clothes have become market-oriented in developed countries. It is really a wonderful domain with flowing information. This paper focuses on the direction of the new production. The author analyzes and concludes the technical composition and trend based on the mass of information and reports, and discusses the positive social value from the view angle of culture.

1 The Technical Composition of Electronic Clothes

The so called electronic cloth is a kind of clothes productions which have the functions of transferring, saving and processing the information. As a typical

J. Luo (Ed.): Soft Computing in Information Communication Technology, AISC 161, pp. 133–140.
springerlink.com   © Springer-Verlag Berlin Heidelberg 2012

high-tech thing, it's development recurs to the power of the latest modern science and technique. It's technical composition are as following:

A.   Tiny Components

People have made many facilities and computer systems based on the main stream techniques ---- electronic technique, sensor technique and communication technique. The original relationship between these techs and clothes were so simple, such as sewing a cell phone pocket on the clothes, covering a belt on the neck to tie the mp3 or putting a satchel on the back to contain laptop. These combinations obviously can not meet the social needs. The real electronic cloth should be with the components embedded or knitted in and fabricant should make them as a whole. Nowadays, the valve computer which once was 30 tons has been improved into a tiny piece of chip. Besides, as the developing of the information technology, chips are more highly integrated and smeller, while they have larger solidity and be lighter. The new technique, just like nanotechnology, speeds up the process which makes the components tinier.

B.   Flexible Components Joint Technology

The hard frames of the electronic components breach the comfort of clothes, therefore, exploit the flexible devices is necessary. These flexible components include flexible monitor, soft keyboard and flexible switch. Several pieces of these things have produced. Pressure-sensitive materials are the first choices to make flexible switch and soft keyboard. The pressure from fingers can reduce the resistance from hundreds of million Ω to 1Ω. The enginery of electric conduction is a appearance called 'quantum tunnel effect in induction field'.

People use luminescent fiber and special knitting technique to make the flexible monitor. Auburn University and Clemson University of USA are developing this kind of product. VISSON of Israel has made the monitor model by using the 0.2mm pellicular textile. It's frame is ranged electrode net and be inflicted different voltage which is controlled by chip. This set of things can bring patterns and characters through excitating the luminous material in the conductive fiber coat. At the Japan Fashion Show in 2001, some designer sewed the EL scene onto the cloth to be the monitor once won favorable comment. It is said that the flexible monitor made from nano composite materials has high intensity and wonderful optical quality. WRONZ in New Zealand and PERALECH in UK have the cooperation in exploring the soft switch. They use existing weave technique ---- coat and embroidery, into the producing of clothes and the outputs can be quantity produced and washed.

C.   Joint Technology

There is so different between textile and electronic technology. The connection between chip's micron and textile's millimeter is the difficulty in the progress of designing electronic clothes. Currently, two scenarios are supported: one is joint the chips into the complicated circuit fabric, just like connect the metal leads; the other one is embedding the electronic elements into the electric plastic pellicle which can twist like fabric and is only 0.5 micron. These two scenarios replace the flexible circuit board formed by printed circuit board. Electronic clothes can not be clipped as the same method as tatting clothes are done, but people should design the flexible circuit according to the style characteristics. Besides, the connecting methods of the devices also relates to blue-teeth wireless communication, mobile communication and auto image recognition.

D.   Functional Fiber Material

Several kinds of functional fibers are involved in electronic clothes, and the electric fiber is the key point. People can find many metal electric fiber and polymer electric fiber. For example, the producer spread electric coat on the polymer fiber or fabric, or, mix the electric substance into fiber or plastic pellicle, also, people can spin the stainless steel fiber or bronze fiber together with terylene fiber to make the yarn. All of these are the good ways to obtain electrical conductivity. New functional fibers also include optical fiber, euthermic fiber and color turning fiber, etc. Optical fiber can transfer the light and has been widely used in optical fiber communication; euthemic fiber can give off different quality of heat according to the changing current The latest production is made from carbolic nano-tube and nano carbolic fiber. And the usage of the color turning fiber is mixing the substance that can change color in the electromagnetic wave visible light into the fiber or accrete on the fiber, and different colors can be regulated under static or dynamic electric field.

E.   Power Supply

Supplying the power to electronic clothes is a great challenge to modern science and technique. The early productions were equipped the rechargeable batteries, and now this power supply is still being used and improved. As the processing of the usage of solar energy, solar batteries have been equipped into the electronic clothes step by step. The French scientists take measures to put the tiny solar batteries into fibers. It is said that solar battery yarn with scarfskin-core structure has already been quantity produced. Especially the outer wears can obtain power through sun light. INFINNEON in France has excogitated a new silica-based heat energy electric power generation core. It can generate electricity through the difference in temperature. Between common clothes and electronic clothes, there is a difference in temperature of 5°C on the surface and the technique can output voltage more than $5V/cm^2$. This can startup the electronic facilities and can be used in special medical probe head and to be the power of micro-electronics core. Moreover, there is a kind of tiny battery can give electricity depends on the wearer's amount of exercise, that would be the trend.

2   Status Quo of Electronic Clothes

Electronic clothes started in 1990s and based on the study of spacesuit and army uniform. But now electronic ones have been developed much further. From the electronic clothes exhibition of the whole world, people can find that the designers have the strong consciousness to syncretize the electronic productions into clothes and series of productions are presented which can be classified as following:

A. Entertainment Clothes

The development of electronic clothes is grounded on people's sweet imagination, for example, singing clothes and color changing clothes. These dreams have become at present. At the Internet Digital Science and Technique Fashion Show in Hong Kong, 2000, a black jean jacket was showed. The designer put the optical fiber into cloth and designed different patterns. After touching the tiny switch, the pattern began to shine as electric light with the glint like neon light. Another one was a T-shirt printed with pattern. The secret was the designer used the high-tech paint. When people press different patterns, the T-shirt uttered different sounds. This kind of technique can be involved in the design of bride's wedding dress and costumes of singers, because it has brilliant colors. French Communications Company has studied out an optical fiber screen whose power can be supplied by batteries. It can release AD words or can be used as a security facility; Massachusetts University explores fabric pictorial which can change the colors to frame patterns according to heat change. The example, or the representation, of sounding clothes is the one with MP3 in it. The designers include INFINEON in French, Phillip in Netherlands, MIT in USA and NIKE, etc. The MP3-clothes can cooperate with skiing, swimming, running and daily relaxation to play music, switch music and adjust the volume, and all of these just need the users to press the buttons on the sleeve. A German company has exploited singing jacket, called "MP3 BLUE". All the musical devices are jointed by blue teeth. The design is very skilled that when the collars stand up, the music only can be heard by use him/herself, but when the collars down, the people surrounding can also hear it.

B. Communication Clothes

As the modern communication tools, telephone and cell phone have become the parts of human's lives. Electronic clothes, at the beginning, combined with these functions, have already come into the market. The production from INFINEON in Germany equipped flexible key board and MP3. It can receive and send messages and has the batteries which can keep work for 8 hours. The pictures of this cloth are published in several magazines. Customers like it and the price is € 600 per piece. BILLON in France has developed the clothes can send e-mail. The computer on it is as small as a decoration. The scientists of MIT are trying to embed phone and computer into clothes. This kind of clothes with key board and computer can play music, make records and receive and send messages. Phillip in Holland cooperates with Levi's in USA to exploit the clothes that can send ICD messages. They design the structure to contain cell phone, MP3 and earphone. Inner the jacket, LAN has been prepared. If people wear it on, they can talk through the phone and listen to the music without using their hands to control something. When they want to dial a number, they just need to speak out the target person's name, then the microphone will send the information to cell phone, and at once, the number will be dialed up. The army uniform with flexible key board and fabric antennae enhance the soldiers' ability of communication. At the Electronic Exploration in Berlin, a bicycle suit was showed. The key board and flexible screen were accreted on the sleeve with sharp tuning elements in it and it can receive messages. The power was from the solar battery on the shoulder.

C.   Monitor Clothes

Electronic clothes are more frequently used to monitor human's body. This kind of clothes has sensors embeded at collar, oxter, colpus and stomach and can record blood pressure, pulse and temperature. It also can be adopted to test the health of athletes and patients. Many companies in Europe, American and Japan have developed this production. It can link to PC and to reveal heartbeat and breath, furthermore, it can analyze the patient's condition. Medical monitor clothes are not only fit to patients, but also be the same with elder citizens and children. They can monitor physiological changes, and mental changes, as well. Once the clothes diagnose that the users' emotions are turning bad, they will play music or give off some fragrance to relax the customers. These clothes also do good to test the physical ability of athletes, just like tennis suit does. They can monitor the changing of resistance during activities and send the related data to help people to study the shoot technique; the skiing suit can send the signal to seek help when the wearer in dangerous. Finland University and another 2 universities discovered a monitor cloth once was called 'intelligent electronic cloth'. It not only had monitor function, but also found the change of athlete's heart rate, and increase the temperature to 30°C under the urgent situation. A kind of soldier's monitor cloth has the optical fiber knitted into suit. It has the ability to estimate soldier's wound degree and wound position through the abnormity of the signals.

D.   Location Clothes

The so called 'location cloth' has GPS and wireless communication technique in it. The cloth can location the longitude, latitude, distance and scene of the target in order to assistant the rescuer to help. As long as make the target person to wear this, people can follow him. No matter where he is, the cloth can locate him with the error which is only few meters. This production has come out. Phillip has a series of children's wear equipped GPS and digital camera. It can chase the whereabouts of children and take pictures; the productions of Bristol University in USA are more often used in adventure and tourism; a skiing suit which is discovered in Europe has an inductor on the back to hint the user to dodge other skiers. Locate person is on one hand, on the other hand, INFINEOU company discovers a intelligent sign to locate cloth productions. This sign is composed with several data which can save a plenty of information. It administers to anti fake and control logistics info to optimize the storage. Besides, US Army are studying on a kind of cloth called 'SENSORNET'. The army can explore the distance of vehicles which belong to both enemy and our army. Moreover, it can chase the sound of blast and locate the position.

E.   Telecontrol Clothes

Electronic cloth uses blue teeth to achieve on-off control onto surrounding electronic facilities. The concealment makes the control be mystic. The cloth has a tiny wireless reflector, and its work can make other electronic facilities answer. The controlling scope is few meters, 50meters or much larger. It can dominate the light and dark of lamps or high and low of temperature, also, it can turn on the hided security equipment to give an alarm or control PC and other electrical equipments to save electricity. Of course, it can communicate with surrounding environment through remote sensing devices. This kind of productions is so convenient and has special purposes. For example, the bicycle suit has the function to lock the bike. People can lock the bicycle from a distance through press the button on his suit.

F.   Temperature and Press Adjusting Clothes

These electronic clothes equipped temperature and press sensors and can apperceive the change of temperature and press outside so that to adjust them to a suitable level. The development of temperature adjusting cloth has a long history. The resistance wire in it has low efficiency in heating up and the temperature controlling elements are too big, therefore, it's not ingenious enough. Nowadays, people have many kinds of euthermic fiber and fabric. The temperature adjusting system with electronic core in it has been processed. It provides with possibility of further advancement. However, there is a product dose not work depend on the euthermic fiber, but startups a tiny adjuster to do the gas charging to make the thermal barrier become thicker in cold weather and to deflate the gas to make the thermal barrier thinner in hot weather. Press adjustment is another need when people are under special situations. US Army is doing the research on a pilot suit which is fir for supersonic jets. This suit use electronic pneumatic anti-G press adjusting to anti centrifugal force. When the jet is speeded up, the pilot won't be ischemia seriously in head.

3   Values of Electronic Clothes

The reason why the electronic cloth catch people's eyes is that its cultural values and potential. It is not only the mixture of electronic things and clothes, but also the amalgamation of personal informationization and social informationization. The far-reaching senses are as following:

A.   Create the New Fashion of Scientific and Technical Clothes

Electronic cloth communicates imagination and real life, widen the train of thought of cloth taste, explore the new direction of digital cloth and bring fresh air to cloth market. Similar as the traditional cloth has the functions of beautifies and decorates to make customers happy, electronic cloth will arouse some people to uphold and pursue. 'MP3 BLUE', the writer mentioned above, developed by German company cater for the mental characteristics of youths who are love science and technique and advocate fashion. Thought it is a very new production to come out, it's active and has large amount of demands. Because the electronic cloth is the inevitable result of the development of human's science and technique, and because of the special charm and functions of itself, the writer can foresee the better trend. The electronic cloth put new life to the upgrading of textile and electronic fields.

B.   Elevate People's Life Level

The presentation of electronic cloth brings fresh feelings to human's life. Such as the appreciation of entertainment cloth, the mystical effect of controlling the switch on cloth, the adjustment of temperature and press and the convenient of communication cloth. All of these make us enjoy the results of electronic civilization. Keeping one eye on style design, color change, material and comfort, electronic cloth is being created and traditional functions are being advanced. It gives people satisfaction on substances and gives spice on mental. View ahead, electronic cloth will achieve more ---- auto dust elimination, recording on the spot and translation. It will be the personal assistant of people. It is imaginable that people can make the myth become true when they put on this cloth.

C.   Strengthen People's Ability to Conquer the Nature

The science and technique before information era, no matter handcraft, machine, electric power or magnetic energy, these all enlarge physical force, but electronic technique and computer enlarge human's abilities of apperceiving, thoughts and adjustment. This will make people have the capabilities to explore and change nature, form new thoughts and exploit new fields. For instance, the location cloth for adventure, the temperature adjusting cloth for lustrate cold and spacesuit for weightlessness in outer space are all strengthen people's abilities to conquer the nature and develop society.

D.   Strengthen the Relationship between Single Person and Network

Electronic culture, represented by computer, infiltrates into society fully, signed by digitalization and based on networking. It is contacting all the people in the world closely. Electronic cloth is born under this background. It supplies the convenient to people's lives, and increases people's awareness of creation. Every single person is really in the network world with the electronic cloth on. It changes the life styles, break the confine of terrain and space, and widen the connection among groups. All of these can promote the development of informationization powerfully.

4   Conclusion

Electronic cloth is a new question for study in China. It will arouse the hot competition in the future textile and cloth industry. Currently, electronic cloth trends to higher level, more functions and lower costs. It is believable that as the perfecting of information technology and the increasing of human's riches, electronic cloth will turn to come into more people's common lives, but not be the individual things for youths or specifical groups. Our country must pay attention to this closely. China needs to follow and even to overtake. In this way, China can be the large textile and cloth producer creditably and to become the real great power in this field.

References

1. Gui-xin, C.: The development of electronic information intelligent textile. Transaction of Tianjin Polytechnic University 23(4), 51–54 (2004)
2. Li-xia, C.: Study and development of electronic cloth. Knitting Industry (1), 53–56 (2005)
3. Li-xia, C.: Scientific and technical cloth in information era. New Textile (7), 10–11 (2003)
4. Xi-fan: Development and the foreground of textile intelligentize. Silk (11), 41–43 (2004)
5. Deflin Koncar, V., Weill, A., et al.: Bright optical fibre fabric: A new flexible display. Textile Asia (5), 25–28 (2002)
6. Gang, Yun-jie, J.: The exploration and thinking of computer culture. Transaction of Qujing Normal University 21(3), 79–82 (2002)
7. Bao-shu, W.: Enjoy the high-tech cloth. Textile in China (5), 15–16 (2003)
8. Adhan, W.: IT evolution at infineon. Textile Month (7), 24–25 (2002)
9. Ju-xia, X.: Intelligent clothes–— the engine for development of cloth industry in 21Centry. Textile in Shanxi (3), 25–26 (2003)
10. Xiu-min, Y.: The Channel for the fashionization of intelligent cloth. Textile in China (7), 15–16 (2004)
11. Xiu-min, Y.: Future clothes. Tex Leader (2), 50–51 (2001)

Discussing Three-Dimensional Design Cheongsam from Clothing Human Engineering

Yun Wang[1], Haiwei Yu[1], and Xiao Zhang[2]

[1] Tai Shan University, Fashion Department of Taishan University, Taian, China
[2] Tianjin Polytechnic University, Tianjin, China
{wangyun04,zhangxiao612}@163.com, yuhw2000@yahoo.com.cn

Abstract. From traditional and modern cheongsam, compared with the human body, the author analyzed three-dimensional design cheongsam in body measurement, loose measure design, structural design, and other human engineering functions. And further elaborate method of body measurement, important parts of loose design, method of structural design. The study of further improvement of cheongsam to adapt to the aesthetic demand of contemporary significance.

Keywords: clothing human engineering, cheongsam, body measurement, loose design, structural design.

1 Introduction

Clothing human engineering should adapt itself with clothing design and human body, making human body matches up clothing interface inside and outside, reflecting optimum condition and the most superior achievements. Clothing design is the people's trunk, limbs and other body parts, reflecting the beauty of the human body and clothing. While the best to show this is cheongsam.

Chasing the route of cheongsam, from the main line of traditional cheongsam style to the human body which can better reflect the structure of the modern decent dress, cheongsam came from the planar to a three-dimensional process. While chasing cheongsam three-dimensional design way from human engineering, analyzing human measurement, loose design, structural design. Undoubtedly they are helpful to modern clothing designers in modeling concept and creating structure from the ancien.

2 Analysis of the Performance between Cheongsam Form and Human Space

A. Analysis of the Performance between Traditional Cheongsam Form and Human Space

Traditional cheongsam is with straight lines, slits on both sides, tide oxter, almost the same width of bust and lap. Difference in lapel styles, with lute lapel, oblique side

J. Luo (Ed.): Soft Computing in Information Communication Technology, AISC 161, pp. 141–148.

opening, wishful lapel and so on; Sleeves are connected with the whole body; Garment pieces are left and right symmetry on fore and after pieces (except cardigan), sleeves are of fore and after symmetry (except the depth of the neckline). There is no difference between chest and back in structure, neither emphasis tall on the back nor curve of the chest, fore and after pieces have no difference expect cardigan and depth of the neckline. The early stage of collar structures are round and without collars, until Qing Dynasty and the Republic of China emerged collar structure. The structural design of the whole cheongsam is also without waist and hip, there is no waist line, hip circumference and waist line of the arc, only by contacting people can reflect the shape of the body. It can be seen, the performance of traditional cheongsam is not the body curves and surfaces, in addition to show a little part of three-dimensional collar status, the chest, shoulders, waist and hip are presented of the state plane, lack of three-dimensional feeling. Loose straight line and plane design make it difficult to recognize the change of human body from the shape of garment and sleeves, fore and back poor state and three-dimensional expression of hands (figure 1).

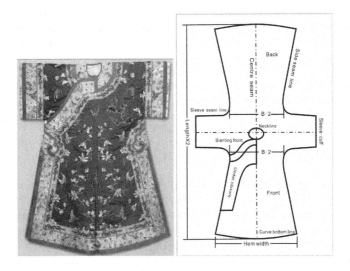

Fig. 1.

B.   Analysis of the Performance between Modern Cheongsam Form and Human Space

In the early 20th century, Chinese and Western culture fusion, cheongsam learned in the Western three-dimensional shape the way of clothing in cutting, new cheongsam came into being, which is characterized by a more relaxed waist, straight cut, the length of the foot or leg, wide cuff varied, and do the trim edge. From 1930s, under the influence of Western design, functional requirements of increasingly prominent, cheongsam broke the structural characteristics of a flat plane, began to stress the body lines, and learned the structure of lessons in three-dimensional structure from western clothing style. It changed chest, waist and hip state of usual, matched the side seam and human curve trend. Chest, waist eliminated surplus space in order to save allocated form, and gradually replaced the traditional reference to Western-style in sleeves, cuffs close to the arm, making which more simple, smooth, aptamer in style, cheongsam began to be three-dimensional (Figure 2).

Fig. 2.

It can be seen from the change of eongsam; the relationship between cheongsam and human is from loose cover to the performance of semi-tight and tight three-dimensional. It is maintaining the Chinese traditional culture while gradually learned Western clothing human engineering elements, united Chinese and the western structural. It can be said that the end of millennia of Chinese women's history, a sign of flat shape, is the form of clothing into three-dimensional model of the first step, it's shaping clothing design industry in China and also opened up a new way (figure 3).

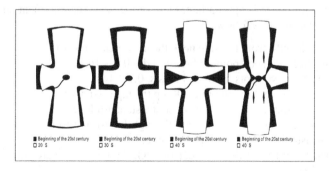

Fig. 3.

3   The Application of Body Measurement in Cheongsam Three-Dimensional Design

In clothing design process, from the requirement of comfort, function, improve the function of the body of engineering of clothing, more clear guarantees should be made in clothing design of human measurement, in order to match body and clothing. "Tailored" brilliantly summed up the relationship between body and clothing; it measures the length, width, circumference size, for the basis for tailoring. Therefore, size of the human body is the basis for costume design, as in 1930s, the famous designer Elsa Schiaparelli said: "fashion designer in any case should not forget the body shape." Therefore, as the representative of the modern cheongsam clothes, reasonable and accurate body size is the basis for the formation of three-dimensional, while the size of the acquisition can not do without the support of body measurement. Measurement of body shape and size of clothing is an important part of human engineering, is the main theoretical basis for scale.

Modern cheongsam as a representaive of junction clothing, its shape matches the body very well, so the choice of measurement sites is rich and delicate.

When measuring, in order to measure the exact scientific data so that measurements are comparable, sometimes it is necessary to identify the body surface as the measuring point and the base line. Cheongsam reference point can be measured separately from the body's fore and back. The fore main reference points are: side neck, shoulder, fore oxter, knee and extramalleolus; Back reference points are: back neck, back oxter, elbow and wrist. Baselines mainly are: bust line, waist line, and hip line. Among them the most basic points and lines are: side neck, shoulder and waist, because these are the basis sizes of producing length. Side neck point is at the root of the neck, viewed from the side of the human body; center horizontal line translates 2cm and meets the root of the neck. The point of shoulder is the most prominent point of scapula on the edge. However, due to the design of fitting cheongsam is reasonable, the endpoint located in the scapula up at 2 cm. The waist line is generally the thinnest, in order to make the cheongsam look more comfortable and meet the needs of eastern people. Waist measurement can be put on the 1.5cm ~ 2cm, it is to lengthen the lower body ratio and also make the side of the curve smoother. To ensure accurate measurement of the follow-up, bandlet with flexible can be put on it, no affecting the tightness of the girth size is appropriate. Parts of the basic measurement and methods of fitting cheongsam are as follows:

a)Cheongsam length: from the side neck through breast to skirt.
b) Breast height: side neck point to breast under 0.7 ~ 1cm.
c) Breast spacing: two spots between breasts, 0.7 ~ 1cm.
d) Bust: measure a lap through breast. Keep the tape standard.
e)Waistline: measure a lap from the waist at a standard leve.
f) Fore waist: start from the side neck through breast and to press the tape under breast, until measure the length of the waist line.
g) Back waist: start from the shoulder through the back, measure straightly to the length of waistline.
h) After the shoulder and waist: from the shoulder in the back measure straightly to the length of the waist.

i)Before the shoulder and waist: from the shoulder across to the breast, measure until the waistline.

j)Abdomen: start from the side neck through the breast to press the tape until the waistline.

k)Abdominal circumference: measure a lap from abdomen to a little higher

l) Hip: measure the most wrapped human butt part, move this part to the front of the body, from the side neck through breast and to press the tape under breast, until measure the length of this part.

m) Hip circumference: measure a lap cross the butt.

n)Fork High: from the side neck through the breast to press the tape under breast, until measure the fingertip

o)Shoulder: measure from the upper back to shoulder.

p)Back width: measure the width of the back from the shoulder to 2/3 oxter.

q)Chest width: measure the width of the roots of the two arms from the shoulder to 2/3 oxter.

r)leeves length: from the shoulder to the required amount of length.

s)Cuff: measure a lap from the level of sleeves

t)arm circumference: measure a lap from the roots of the arms.

u)collar length: measure a lap from the root of the neck.

v)collar height: measure according to the back of the neck, collar height is usually 4 ~ 6 cm, can be based on personal preference.

w)breast height: from the chest line through the breast to press the tape under breast, denoted as L. Measure from fore chest and sleeve cage wire across the breast until the length of the waist, denoted L1, the distance between L and L1 is called breast height.

x) Back dart center: from side neck through the back to the hollowest part of waist, upper 2cm length.

In addition, according to the different styles, root arm circumference, wrist circumference, knee circumference and ankle circumference can also be measured. Differences of individual hips can also increase the amount of people at its widest point of the legs and waist circumference at the smallest circumference.

To make a more accurate size measurement, the measurement should be measured by wearing tight-fitting underwear, reflected correct, natural standing position, not to have extra movements, gauger should measure in the decentralization of a finger of loose tape volume, so the measured values basically include the relaxation function of the amount of human activity. In the measurement of "circumference", we should pay attention to maintaining the level of tape, not too loose or too tight.

4   Analysis of Cheongsam Sample Design

A.   Analysis of Cheongsam Loose Measurement Design

Loose measurement design is the key element of cheongsam design, it decides whether cheongsam fits the shape and body or not. In the basis of measuring correctly, fully taking human activities, style, fabric and other elements into consideration, and matching with appropriate loose measurement. Loose measurement design of cheongsam is mainly reflected in three aspects: length, width, and circumference.

- **Length:** the length of the cheongsam is mainly cothing and long sleeve. Length size requirements based on measurement of long clothing styles with fabric shrinkage; Sleeve size requirements based on the measurement Sleeve style fabric shrinkage plus, cheongsam generally do not need padded shoulder, so sleeve size does not consider shoulder pads.
- **Width:** cheongsam that consider the main width

ts are the chest and the back. According to the kinematics point of view, the movement of the back are better than that of the chest, so back width are larger than the chest width. In case of the bodies of humpback or thoracic, one should also consider the chest width and back width of the change make adjustments on loose measurement. Shoulder is the site of many restrictions of Cheongsam design, the changes need to attach the body, generally do not increase the amount of relaxation.

- **Circumference:** the circumference is the main content of cheongsam loose design, including waist circumference, butt circumference, abdominal circumference, collar, arm circumference. When design collar loose style, one should consider collar height, the upper and lower collar differences. If it is a small chest, waist, with big butt man, except in the amount of time to be measured by the body to wear bras with breast care; the other hand, waistline should increase, in order to make it more fluid.Specific Cheongsam loose design is as follows :

Loose Parts	Chest	Back	Collar	Bust	Waist	Abdomen	Hip
Size	1~15	1~2	2~3	4~6	4~6	4~6	5~8
Directions	This table is for normal measurement, if measure of a finger of loose tape, one can adjust the size by reference the table.						

unit:cm

Through the above multi-site measurements, master the necessary data of human body and add loose design into cutting so as to reflect the sense of three-dimensional cheongsam.

B.   Cheongsam Structure Design

Cheongsam is with the characteristic of three-dimensional, focus on the shape of two-dimensional shape space. From the composition point of view is the curve cut, clothing, cuff released, according to body size reduction of the amount of multi-site human form, cutting according to the human structure, after sewing, cheongsam has become a totally decent shape. Take 160/84A shape, for example, measure its size according to above method, and add a reasonable amount for the production model loose.

CHEONGSAM MAIN PARTS STANDARD TABLE

Name	Size	Dlothes	Bust	Waist	Hip
Standad	160/84 A	120	90	72	94
Name	Fore waist	Back waist	Shoulder width	Collar	
Standad	41	38	39	36.5	

unit:cm

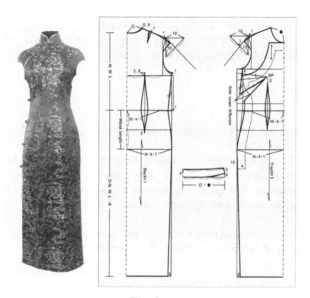

Fig. 4.

Bust loose is designed at 6cm, waist loose is 4cm, utt is2cm, collar is 2cm (figure 4)Main points of drawing the chart are as follows.

a)Shaping the bust, seam and back waist into both sides of the drum-shaped date pit, making it fit the change of the chest, waist and butt of human body, designing the beauty of the Trinity.

b)Shaping the side line into the round line according to human body, making it fit the chest, waist and butt.

c)Paring the fore shoulder line into slightly convex arc, the back shoulder line into slightly concave arc, making the shoulder more mellow and full.

d)Pay attention to adjust the anatomized relationship between collar height and outer edge.

e)The ultimate three-dimensional cheongsam is very tight, the shoulder is also attached to the body, which requires the performance of the scapula with the shoulder dart to the processes. When cutting fabric with color, the general recommendation would be transferred the shoulder to armhole, with the color fabric cutting, you can design directly in the shoulder around 1cm.

5   Conclusion

Cheongsam is the important remark of the change of Chinese women clothing style and structure. It is modified by the following and which eventually broke completely "flat decoration" of the continuation of the dress shapes for thousands of years, laid a foundation of "three-dimensional design" for Chinese women.

Comparison of human engineering point of view from the cheongsam clothing style and structure changes of dress design and style features, the application of analytical theory of human body measurements, the study of cheongsam three-dimensional design, not only help to draw the shape of contemporary designers and structural elements, apply to today's fashion design; also help explore the causes of cheongsam design, make further improvements to meet the cheongsam contemporary aesthetic.

References

1. Zheng, R., Zhang, H.: Robes of traditional and modern design, pp. 21–23. China Textile Industry Press, Beijing (2004)
2. Zhao, Y., Shan, L., Liu, J.: Body design clothing. Knitting Industry, 31–34 (May 2008)
3. Hang, H., Zheng, R., Xu, F., Wang, X.H.: Robes amount of space allocated provincial relations with the side seam shape. Beijing Institute of Clothing, 39–45 (March 2006)
4. Zhuang, L.: Shanghai cheongsam style and structure of the change. Silk, 50–52 (February 2008)
5. Hen, R., Chen, W.: Cheongsam design of the shape changes of the evolution and structure. Zhejiang University of Technology, 155–159 (February 2007)

A New Method of Pencil Drawing

Shuo Sun[1] and Chunbao Ge[2]

[1] School of Science, Tianjin Polytechnic University, Tianjin, China
[2] Shengshi Interactive Game, Beijing, China
`{sunshuo_0,hagcb}@163.com`

Abstract. This paper proposes an new method of pencil drawing generation technique based on Line Integral Convolution (LIC). The original LIC pencil filter utilizes image segmentation and texture direction detection techniques for defining outlines and stroke directions, and the quality of a resulting image depends largely on the result of the white noises and the texture directions. It may fail to generate a reasonable result when the white noises and the texture directions are not consistent with the texture structure of the input image. To solve this problem, we propose in this paper to improve the existed LIC-based method. First, a more accurate and rapid graph-based image segmentation method is introduced to divide the image into different regions. Second, we present a new region-based way to produce white noises and texture directions. We also demonstrate the enhanced LIC pencil drawing is closer to the real artistic style.

Keywords: NPR, pencil drawing, LIC, image segment.

1 Introduction

Recently, Non-Photo-Realistic rendering (NPR) has become one of the most important research topics of computer graphics. A number of techniques have been developed to simulate traditional artistic media and styles, such as pen and ink illustration[4-7], graphite and colored pencil drawing[8-10], impressionist styles[2], paintings of various materials including oil[1], water color[3] and so on. The existing researches on painterly image generation mainly take two different approaches. The first approach is to provide physical simulation to the materials and skills, and has been mainly combined with interactive painting systems or 3D non-photo-realistic rendering systems for generating realistic painterly images. The second approach is the painterly filtering, which involves taking an image and applying some kind of image processing or filtering techniques to convert it into an image of a painterly look. While many excellent painterly filtering techniques have been developed for generating brushstroke based paintings [11], relative few publications can be found on converting a source image into line stroke based drawings. In case of drawing, geometric information such as the outline of regions, the direction and shape of strokes becomes more critical, while it is usually difficult to extract such information from 2D raster images automatically. We propose in this paper to improve the existed LIC-based method. First, a more accurate and rapid graph-based image segmentation

J. Luo (Ed.): Soft Computing in Information Communication Technology, AISC 161, pp. 149–157.
springerlink.com © Springer-Verlag Berlin Heidelberg 2012

method is introduced to divide the image into different regions. Second, we present a new region-based way to produce white noises and texture directions.

2   Related Works

Pencil drawing has been an important topic since the beginning of painterly image generation research history. Related works focus on [3,7,8,9,10].The largest difference between our technique and all these existing techniques is that our technique can generate a pencil drawing from a source image in a completely automatic way while all these existing techniques rely, to certain extent, on user interventions, for specifying the attributes and directions of strokes. Several commercial packages provide some filters for creating pencil drawing effects. For example, Jasc Paint Shop Pro software supports a black pencil filter. However, to obtain a satisfactory result with those filters, a user usually needs to combine the effects of many other filters and explore the best generation process experimentally through trial and error for many times.

3   LIC Pencil Filter

3.1   LIC Algorithm

Line Integral Convolution (LIC) is a texture based vector field visualization technique [14]. It takes a 2D vector field and a white noise image as the input, and generates an image which has been smeared out in the direction of the vector field through the convolution of the white noise and the low-pass filter kernels defined on the local streamline of the vector filed.

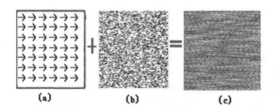

(a)   (b)   (c)

Fig. 1. Line Integral Convolution (LIC) (a) Input vector field; (b) Input white noise; (c) Output result

The images in figure 1 show the basic algorithm of the LIC. The inputs are the vector fields and white noises.

P is the output pixel, $\rho(\tau)$ is stream line $(-L \leq \tau \leq L)$, L is the half length of the stream line。 $T(\rho(\tau))$ is the noise texture value in the stream line, $K(\tau)$ is a convolution kernel。 So the pixel value in P is $T'(\rho(0))$:

$$T'(\rho(0)) = \frac{\int\limits_{-L}^{L} k(\tau) T(\rho(\tau)) d\tau}{\int\limits_{-L}^{L} k(\tau) d\tau}$$

The discrete form describes as follows: p_i is the discrete point in the stream line, W_i is the contribution of pi to P, namely the area that $K(\tau)$ cover between p_{i-1} with pi.

$$T'(P) = \frac{\sum\limits_{i=0}^{N} T(p_i) W_i}{\sum\limits_{i=0}^{N} W_i}$$

The idea of using LIC for pencil drawing generation was inspired by the visual similarity of LIC images and pencil drawings. As an LIC image is obtained by low-pass filtering a white noise along the local streamlines of a vector field, we can observe the traces of streamlines along which intensity varies randomly. Such traces have a similar appearance of pencil strokes where the variance of intensity is caused by the interaction of lead material and the roughness of paper surface.

3.2   The Existed LIC Pencil Drawing Method

In general, for producing a pencils drawing from a 2D source image, several steps are done.

1. Generate a white noise (Figure 2(b)) from the source image (Figure 2(a)).
2. Segment the input image (Figure 2(a)) into different regions (Figure 2(c)).
3. Extract region boundary (Figure 2(d)).
4. Generate the vector field (Figure 2(e)) representing the orientation of strokes.
5. Generate stroke image (Figure 2(f)) by applying LIC to the white noise (Figure 2(b)) and the vector field (Figure 2(e)).
6. Add the boundary (Figure 2(d)) to obtain the drawing with outlines (Figure 2(g)).
7. Composite the resulting image (Figure 2(g)) with the paper sample (Figure 2(h)) to obtain the finished pencil drawing (Figure 2(i)).

4   Enhanced Region-Based LIC Pencil Filter

4.1   Graph-Based Image Segmentation

A well used technique in pencil drawing for conveying the 3D shapes of objects and spatial relationship among different objects in a scene is to emphasize the boundary between two different regions by drawing outlines or changing the appearance of

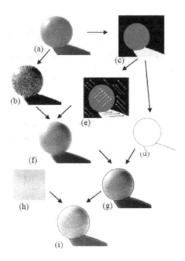

Fig. 2. The existed LIC algorithm

strokes in the two regions. To create such effect, we propose to divide the input image into different regions using existing image segmentation technique. In our current implementation, a Graph-Based image segmentation technique [14] is used for the region extraction. Contrasting to the method in [12], this method can dramatically promote the performance of our pencil filter.

Graph-based image segmentation techniques generally represent the problem in terms of a graph $G = (V;E)$ where each node vi $\in$ V corresponds to a pixel in the image, the edge set E is constructed by connecting pairs of pixels that are neighbors in an 8-connected sense (any other local neighborhood could be used). A weight is associated with each edge based on some property of the pixels that it connects, such as their image intensities. Neighbor edges are clustered into a forest, and each tree in the forest is related to a minimum spanning tree (MST). Finally each MST is a sub-area. To judge whether two trees can be merged into one tree, a predicate is defined. The predicate expression is showed as follow.

$$D(C_1,C_2)=\begin{cases}True\ if(Dif(C_1,C_2)>MinT(C_1,C_2))\\False\ otherwise\end{cases}$$

$D(C_1,C_2)$ means the merging predicate of the areas of C_1 and C_2, $Dif\ (C_1,\ C_2)$ means the difference between the area C1 and the area C2. $MinT\ (C_1,\ C_2)$ is the minimum internal difference.

The advantage of the method is that the accuracy of the image segmentation can be adjusted by users and some details of the certain regions can be ignored. This will improve on the effect of the pencil drawing. In addition, this results in a graph with O(n) edges for n image pixels, and an overall running time of the segmentation method of O(n log n) time.

4.2   Region-Based Noise Production

The white noise image is generated in a way that the probability a white value is set for a pixel is proportional to the intensity level of the corresponding pixel in the input image. The gray-scale tone of a resulting pencil drawing is mainly decided by the white noise image. To match the tone between the input image and the resulting pencil drawing, we use the tone of the input image to guide the distribution of noise.

An important characteristic of the pencil drawing is its ability to preserve detail in low-variability image regions while ignoring detail in high-variability regions. The input image is then divided into many small regions which have corresponding meanings. The method mentioned in [12] dealt with the noise according to uniform criterion. The result noises would be failure to distinguish the important elements and the unimportant elements, when the range of the intensity of the image is small. Our region-based method solves the question through dynamic adjusting the threshold value of different areas. It is very important to pencil drawing. We simply introduce the algorithm.

Because the processing of LIC is global, so some local regions with exquisite texture will be destroyed. In addition to, as a kind of sketch drawing, pencil drawing should keep some blank areas from the point of aesthetics. So we need to segment the original image into different region through above graph-based method, then we deal with the each area through giving certain threshold value. Commonly we select the average gray intensity of the certain regions as the threshold values.

Our method improves the producing way of white noise. We deal with the original gray image according to different gray intensity range in the respective regions. So the contrast between light and shade in the results is more eminent and more similar to the artistic style. The formulation of white noise is showed as follows.

Let Iinput be the intensity of a pixel in the input image, and P is a floating-point number generated with a pseudo-random function; Ri is the average intensity in the ith regions, and it is selected as the threshold to control the distribution of the noise. Then the intensity Inoise of the corresponding pixel in the noise image is decided in the following way:

$$I_{noise} = \begin{cases} I_{noise1} = \begin{cases} max1 : (if(P > T_1)) \\ min1 : (else) \end{cases} (if(Input \le R_i)) \\ I_{noise2} = \begin{cases} max2 : (if(P > T_2)) \\ min2 : (else) \end{cases} (if(Input > R_i)) \end{cases} \quad (P \in [0,1])$$

Here mxa1 and max2 mean the maximum gray value of the output pixels, commonly is 255; and min1 and min2 are the minimum gray value of the output pixels, commonly is Ri. However, we can adapt these values to fit for the whole tone of the pencil drawing.

$$T_1 = k_{i,1}(1 - \frac{I_{input}}{255}), T_2 = k_{i,2}(1 - \frac{I_{input}}{255})$$

$$k_{i,1} = \lambda_1 (1 - \frac{I_{input}}{255}), k_{i,2} = \lambda_2 (1 - \frac{I_{input}}{255})$$
$$\lambda_1, \lambda_2 \in [0,1]$$

According to our method, λ_1 and λ_2 are two experiential values. Our initial experiment result suggests that a default value of 0.7 and 0.3 produce a visually acceptable result for most scenes. We also allow users to interactively adjust the value of λ_1 and λ_2.

Through above method, the noise of each area can be controlled to match the features of the pencil drawing to a great extent. As shown in figure 3, the picture (a) show the noise produced by old way, and the picture (b) is produced by our method. Obviously, the result noise dealt with by our method can reflect the nature of the pencil drawing. Some areas need to be emphasized corresponding to more dense noises, and some areas need to be fade out corresponding to sparse noises.

(a)

(b)

Fig. 3. The existed noise algorithm (a), and Region-Based noise method (b)

4.3   Stroke Orientation

Here we propose a method to produce the direction of the stroke according to the features in the regions. First, the orientation vectors of the pixels are calculated by sobel operator. Then a criterion is presented to decide the direction of the stroke.

Let $V_{i,j}$ ($j = 0,1,2,...m_i$) be the field vector of the jth pixel in the ith area. It is computed by sobel operator. Here m_j is the numbers of the pixels in the jth area.

Then the mean of the field vectors in each area is calculated. At the same time, the CV (Coefficient of Variation) is computed to reflect the orientation's distribution.

$$V_i = (\sum_{j=0}^{m_i} V_{i,j}) / m_i$$

$$D_i = \sum_{j=0}^{m_i} (V_{i,j} - V_i)^2$$

The final field vector of the each pixel is determined through the following criterions:

$$FV_{i,j} = \begin{cases} V_{i,j} : if\,(D_i / m_i > T) \\ V_i : (else) \end{cases}$$

Here T is a given threshold. Experiments suggest that a default value of 0.7 produce a visually acceptable result for most scenes.

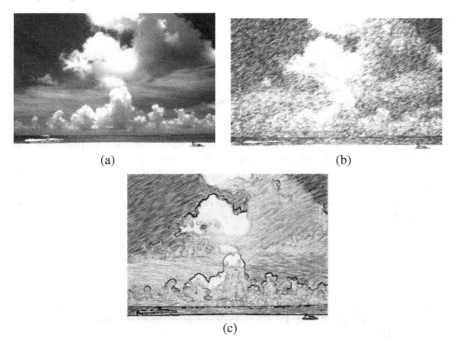

(a)   (b)

(c)

Fig. 4. The contrasts of the stroke directions

Figure 4 (a) is original image, and (b) is the result in [12]. The picture (c) is our result. Obviously, our result not only reflects the details in some areas but also embody the style of the pencil drawing.

5   Experiment Results

We have implemented all the techniques described above and built an automatic pencil drawing generating system on Windows environment. Basically, after the input image is specified, a pencil drawing style image will be generated automatically with a set of default value and other information derived by the system from the input image. Users are also allowed to interactively specify some parameters, such as the parameters for controlling the region tone of the output image, the threshold used for graph-based image segmentation and the length of convolution.

6   Conclusion

We propose in this paper to improve the existed LIC-based method. We believe the purpose of NPR systems is not to replace artists, but rather to provide a tool for users with no training in a particular medium. Potential application fields of our technique include producing posters from photo graphs, processing videos into pencil drawing style animations, obtaining the preparatory sketches for creating paintings of other styles, and so on.

References

1. Hertzmann: Painterly Rendering with Curved Brush Strokes of Multiple Sizes. In: SIGGRAPH 1998 Conference Proceedings, pp. 453–460 (1998)
2. Litwinowicz, P.: Processing Images and Video for An Impressionist Effect. In: SIGGRAPH 1997 Conference Proceedings, pp. 407–414 (1997)
3. Curtis, C.J., Anderson, S.E., Seims, J.E., Fleischer, K.W., Salesin, D.H.: Computer-Generated Watercolor. In: SIGGRAPH 1997 Conference Proceedings, pp. 421–430 (1997)
4. Winkerbach, G., Salesin, D.H.: Computer-Generated Pen-and-Ink Illustration. In: SIGGRAPH 1994 Conference Proceedings, pp. 91–100 (1994)
5. Salisbury, M.P., Anderson, S.E., Barzel, R., Salesin, D.H.: Interactive Pen–And–Ink Illustration. In: SIGGRAPH 1994 Conference Proceedings, pp. 101–108 (1994)
6. Winkenbach, G., Salesin, D.H.: Rendering Parametric Surfaces in Pen and Ink. In: SIGGRAPH 1996 Conference Proceedings, pp. 469–476 (1996)
7. Salisbury, M.P., Wong, M.T., Hughes, J.F., Salesin, D.H.: Orientable Textures for Image-Based Pen-and-Ink Illustration. In: SIGGRAPH 1997 Conference Proceedings, pp. 401–406 (1997)
8. Takagi, S., Fujishiro, I., Nakajima, M.: Volumetric modeling of colored pencil drawing. In: Pacific Graphics 1999 Conference Proceedings, pp. 250–258 (1999)
9. Sousa, M.C., Buchanan, J.W.: Observational Model of Blenders and Erasers in Computer-Generated Pencil Rendering. In: Graphics Interface 1999 Conference Proceedings, pp. 157–166 (1999)

10. Sousa, M.C., Buchanan, J.W.: Computer-Generated Graphite Pencil Rendering of 3D Polygonal Models
11. Gooch, Gooch: Non-Photorealistic Rendering. AK Peters Ltd. (2001)
12. Mao, X., Nagasaka, Y., Imamiya, A.: Automatic Generation of Pencil Drawing from 2D Images Using Line Integral Convolution. In: Proceedings of the Seventh International Conference on Computer Aided Design and Computer Graphics CAD/GRAPHICS 2001, pp. 240–248 (2001); Conference Proceedings, pp. 138–143 (1989)
13. Mao, X., et al.: Enhanced LIC Pencil Filter. In: CGIV 2004 (2004)
14. Cabral, B., Leedom, C.: Imaging Vector Field Using Line Integral Convolution. In: SIGGRAPH 1993 Conference Proceeding, pp. 263–270 (1993)
15. Vermeulen, H., Tanner, P.P.: PenchilSketch—A Pencil-Based Paint System. In: Graphics Interface 1989 (1989)
16. Tamura, H. (ed.): Introduction to Computer Image Processing. Soken Shuppan Publisher Inc. (1984) (in Japanese) ISBN 4-7952-6304-3

The Empirical Analysis on the Action of Binhai New Area FDI for Economic Development Based on the Two Visions of Capital Accumulation and Exports Level

Shenshen Wang and Chunhong Zhu

College of Economic, Tianjin Polytechnic University, Tianjin, China
{wangss_6676f,hong66096}@163.com

Abstract. This paper analyzed the role that Binhai New area FDI plays in the economic development from two visions: capital accumulation and exports level. Based on the analysis, this paper presented conclusions. The Binhai New Area should change the investment attracted model, and fully play the leading role of FDI technology transformation in the Binhai New Area innovation. And it also could adjust the foreign investment attracted policies and industry sectors, to pull the foreign investment to the industries which have higher prospects such as advanced processing and manufacturing and tertiary industry. At the same time, it should deepen the understanding of strengthening the foreign-enterprises-rooted.

Keywords: Binhai New Area, FDI, capital accumulation.

1 Introduction

View from the current study, there is a common view in academic world that FDI has promoting action to the regional economic growth and employment, and there are also many empirical analyses. However, the research on the capital accumulation and exports level is much less, especially the research on the relationship between the Binhai New Area FDI and the two visions. Based on this, this paper analyzed the Binhai New Area FDI from the two visions, and expects to get the measures to enhance the utilization rate of Binhai New Area FDI.

2 The Relationship between Binhai New Area FDI and Capital Accumulation

Early developmental economics theories emphasized the action of attracting FDI to the national capital accumulation. W.Rostow first proposed that the developing countries could improve the constraint of capital shortage by using FDI when the economy "takes off". H. Chenery and A. M. Strout, based on the Harrod-Domar model, put forward the "double gap" model. The model theoretically illustrated that attracting foreign investment in developing countries could solve the problem of insufficient domestic funds as well as relieving the pressure of foreign exchange

J. Luo (Ed.): Soft Computing in Information Communication Technology, AISC 161, pp. 159–164.
springerlink.com

shortage and fulfill the demands of domestic and imports investment. Therefore, FDI could promote the national economic growth. British economist Griffin thought that capital inflows may reduce the domestic savings and investment. From all above, it can be seen that different economists have different visions on this issue. And it also could conclude that the FDI has complex connections with the national investment.

Due to the lack of capital accumulation data in Binhai New Area, This paper chose the total social fixed assets investment (GUDING) of Binhai New Area as the only indicator to measure the regional capital accumulation. The PGUDING is the indicator that has excluding price changes of RMB. The unit is billion Yuan. Meanwhile specific to the two-way-effect of promotion and inhibition of FDI to the national capital accumulation, this paper chose PFII (the contracts amount of direct foreign investment in Binhai New Area) and PIINCOME (the total wage of Binhai New Area) as the explanatory variables of the PGUDING. Namely, the coefficient of the PFII illustrated the acceleration of FDI to Binhai New Area fixed capital accumulation, and the coefficient of the PIINCOME illustrated the acceleration of income of Binhai New Area to the fixed capital accumulation.

The graphs and scatter of the PFII and PIINCOME was shown as followed.

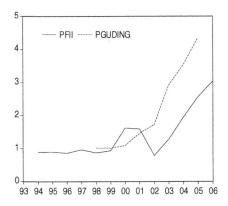

Fig. 1. The time curve of PFII and PGUDING

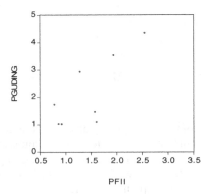

Fig. 2. The scatter of PFII and PGUDING

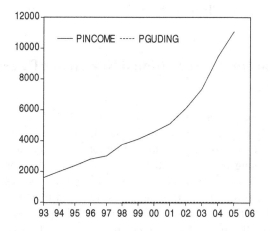

Fig. 3. The time curve of PIINCOME and PGUDING

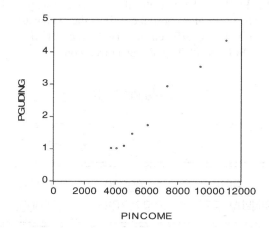

Fig. 4. The scatter of PIINCOME and PGUDING

By using the EViews software, doing the binary linear fitting to the data, and removing heteroscedasticity, the result could get like:

PGUDING   =   0.0004936271621*PIINCOME   -0.07435929591*PFII   - 0.9506855184

R-squared = 0.999983 S.E. of regression = 0.019128

The fitting coefficients of the model passed the T text, and the goodness of fit is 0.99. This means the model has good fitting results.

Economic significance of the model can be attributed as the following points. Firstly, in the case of other conditions remain unchanged, the PFII increased 1 billion, the PGUDING decreased 7,435,900 Yuan. Secondly, in the case of other conditions remain unchanged, the PIINCOME increased 10,000 Yuan, the PGUDING increased

49,400 Yuan. Thirdly, if the PIINCOME and PFII are all equal to 0, the PGUDING is 95,068,600 Yuan.

3   The Relationship between Binhai New Area FDI and Exports Level

Generally speaking, the effects of FDI to the national trade are to make the host country join the global division rapidly, and increase the exports markedly. Specifically, it contains two points: the impact on total trade and the impact on trade structure. As the missing of the data, the paper focuses on the impact on the total trade. In general, the foreign invest enterprises in less developed regions are always focus on the manufacturing and processing industries. And most of its products will use to fulfill the international market, this means high exports sales. Therefore, it could generally infer the dependence of the regional exports level to the foreign investments based on the impact of FDI.

This paper chose the Binhai New Area's exports as the measure, Unit: billion U.S. dollars. And the PEXPORT is the indicator that has excluding price changes of RMB. Unit: billion Yuan. The paper also chose the PFII as the explanatory variable.

Figure 5 is the correlation coefficient matrix between PEXPORT and PFII, and it shows that the PEXPORT and PFII have high correlation ship.

Correlation Matrix

	PFII	BINHAI_PEXP
PFII	1.000000	0.921706
BINHAI_PEXP	0.921706	1.000000

Fig. 5. Correlation coefficient matrix between PEXPORT and PFII

The graphs and scatter of the PEXPORT and PFII was shown as figure 6 and figure 7.

By using the WLS in EViews software, doing the linear fitting to the data, and removing heteroscedasticity, the result could get like:

PEXPORT = 4.51864586*PFII - 2.598529398

R-squared = 0.999890   S.E. of regression = 0.133512

The fitting coefficients of the model passed the T text, and the goodness of fit is 0.99. This means the model has good fitting results.

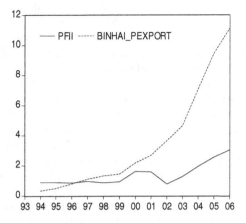

Fig. 6. The time curve of PFII and PEXPORT

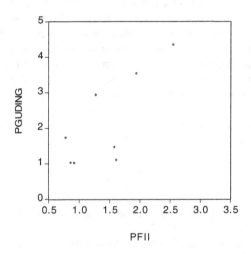

Fig. 7. The scatter of PFII and PEXPORT

Economic significance of the model can be attributed as the following points. Firstly, in the case of other conditions remain unchanged, the PFII increased 1 billion Yuan, the PEXPORT increased 4.52 billion Yuan. Secondly, if the PFII is equal to 0, the PEXPORT is 2.5985 billion Yuan.

4 Conclusions

The empirical analysis shows in the promotion of exports, the growth proportion between Binhai New Area FDI and exports is 1:4.5, it's close to the proportion between macro-economic growth and FDI. This shows that the export-oriented economy is still the main economic model at present in Binhai New Area. Manufacturing is still the main production model of the secondary industry in Binhai New Area. Thus, if the situation of that Binhai New Area is producing low value-added products in the international manufacturing industry chain need be changed, it should be encourage the foreign investment to establish the industry services center and R & D centers in order to enhance the industrial added value. It should fully play the leading role of FDI technology transfer to the Binhai New Area innovation, abandon the investment model that focuses on the scale and quantity, and put the innovation ability enhanced by FDI on the first place in order to transfer the "Attracting investment model on scale" to "Attracting investment model on innovation". The government should choose and control the entry of foreign investments, pay more attention on that where the foreign investments put in. It also should enhance the technical standards of the foreign investments entry, attach importance to the foreign-owned enterprises' ability on technology and organizational transfer. Based on this, the technology spillover effect of FDI could expand, and it can accord with the regional industrial upgrading and restructuring.

On the other hand, in terms of capital accumulation, it could be shown from the model, the major leading role of the Binhai New Area total social fixed assets investment is the total consumption of employees, and the entry of FDI has crowding-out effect. Thus the government should adjust the policy and industry sectors that attracting foreign investment, lead the foreign investment to the advanced process manufacturing and tertiary industry, such as automobile and equipment manufacturing, electronic information, new energy and new materials, biotechnology and modern medicine, petroleum marine chemical and so on. Meanwhile, the government should also improve the efficiency in the use of foreign capital by strengthening the foreign enterprises rooted.

References

1. Zheng, X., Yan, S.: Foreign direct investment and economic growth. Economic Theory and Economic Management 1, 11–16 (2002)
2. Huang, H.: The Empirical analysis on relationship between Foreign direct investment and economic growth. Nankai Economic Studies 5, 46–51 (2002)
3. Zhang, J.: Transnational corporations and direct investment, vol. 3, pp. 271–350. Fudan University Press, Shanghai (2004)

The Research on the Social Appraisal Index System of Reproduction of Media Politic to Sport Events

Xinyu Wu

Department of Physical Education, Hebei Polytechnic University, Tangshan, China
wuxinyu@heut.edu.cn

Abstract. We will conduct this project design and arrange investigation and research about the task by using questionnaire survey, interview and analytical hierarchy process at the point of the 16th game of college students in Hebei province. The questionnaire was given out in two stages, factors of each index will be confirmed at the first time. The ratio of relative importance among factors in each index will be confirmed in the second time in order to be used in the analytical hierarchy process. By researching the relation between the reproduction of media politic to provincial sports events and the social appraisal, the index system of that in order to reappear the initial mechanism of action of international media politic on social appraisal in Beijing Olympic Games. It will be an example of framework for the establishing social appraisal in the important international sport game in the future and promote the development of quantization research on international media and social appraisal on sports.

Keywords: The sport events, media politics, social appraisal, valuation index system.

1 Introduction

We will conduct this project design and arrange investigation and research about the task by using questionnaire survey, interview and analytical hierarchy process at the point of the 16th game of college students in Hebei province. It will be an example of framework for the establishing social appraisal in the important international sport game in the future and promote the development of quantization research on international media and social appraisal on sports.

2 The Object of Study

At the point of the 16th college student games in Hebei, We will arrange two students who got the training of questionnaire in charge of sending the questionnaires to the volunteers who serve for the leaders from schools and provincial and municipal administration from May 5th to 10th, 2010, the volunteers will distribute questionnaires to the units and leaders then collect them back point to point, they will distribute 74 questionnaires to team leader or general secretary from all the teams and collect 100%

J. Luo (Ed.): Soft Computing in Information Communication Technology, AISC 161, pp. 165–170.
springerlink.com © Springer-Verlag Berlin Heidelberg 2012

of them. Besides, all the 14 leaders from excluding counties, 24 coaches from all individual sports, 11 senior referees, 14 media workers of this games(from newspapers, radios, campus videos and so on) and 16 local experts. The total number of questionnaires is 153 and all of them are valid. The questionnaire was given out twice, factors in each index will be confirmed in the first time, the ratio of relative importance among factors in each index will be confirmed in the second time in order to be used in the analytical hierarchy process.

3   Construction of Index System on Sports Events to Social Appraisal

A.   Construction of Valuation Index System

Consulting a lot of related researches on sport politics, new media of sport, sport broadcasting, sport games and published thesis on analytical hierarchy process for references, I selected some representative materials such as the research findings on sport news media and sport games, the scholarship level of them are not equal. I take notes on the main findings. Excuse me for not taking others because of the limitation of page. I draw up some relative index on the reproduction of media politic on sport games of social appraisal and use the way of questionnaires to experts, I collect the judgments from experts on the importance of valuation index, classify them into extremely important, very important, important, average and not important, their marks are 5,4,3,2,1, the higher marks mean more importance, I determined the index of reproduction of media politic on sport games of social appraisal by comparing, analyzing and synthesizing.

B.   The Judgment the Influence of Each Index on the Reproduction of Media Politic to Sports Events Social Appraisal

This task is a quantitative analysis on influence degree among levels in the index system above. Analytical hierarchy process (AHP) is the multi-aim, multi-principle method suggested by American operational research experts Satty, it's a new simple way to make judgment on some complicated things. It was combined by qualitative analysis and quantitative analysis, it will decompose the question into some factors according the goal of the question in a general perspective and build a hierarchy model according the dominance relationship, determine the relative importance between factors of program decisions by using the comparison method in order to get a satisfactory result. The figure of relative importance among indexes reflect people's estimation on the importance of the index, for an example, the factor A in the first level has something to do with the factors B1, B2...Bn in the second level, the importance of comparison between B1, B2, Bn is expressed by the way of 1-9 and the reciprocal of them such as the Table 1. If Bij=1, it means factor Bi and Bj equally importance; Bij=3, means factor Bi is slightly important than Bj, and so on. There is a general rank sorting after the level rank, calculate the weight of relative importance of all the factors in the same level to the top level. This process is carried on from the top to the bottom like the followings.

Table 1. The proportion scale of relative importance

Scale markers	Meanings
1	Means the two factors are equally important
3	Means the latter is more important than the previous.
5	Means the previous one is more important than the latter, obviously.
7	Means the previous one is strongly important than the latter.
9	Means the previous one is extremely important than the latter.
2,4,6,8	The median between the two adjacent judgments
The reciprocal of the above	The ratio of importance between factor j and factor I is bij=1/bij if the ratio of importance between factor I and factor j bij is in the above.

If Bij=1, it means factor Bi and Bj equally importance; Bij=3, means factor Bi is slightly important than Bj, and so on. There is a general rank sorting after the level rank, calculate the weight of relative importance of all the factors in the same level to the top level. This process is carried on from the top to the bottom like the followings. Make the weight vector of A-B matrix W= (0.1018, 0.4781, 0.1440, 0.2760) rank in the second line in the table 2, then formulate the weight of B1-C matrix by sequence. So the calculation formula of general rank sorting of level C is:

0.103553×0.224288+0×0.4619828+0×0.143308+0×0.170422=0.023226

0.439340×0.224288+0×0.4619828+0×0.143308+0×0.170422=0.098539

…

Table 2. 0×0.224288+0×0.4619828+0×0.143308+0.423147×0.170422=0.072114 The general rank sorting of level C

Level B / Level C	B1 The factor of game	B2 The factor of media	B3 The factor of broadcasting	B4 The e factors of audience	The weight of general ranking	The ranking
	0.224288	0.461982	0.143308	0.170422		
C11 The brands in games	0.103553				0.023226	18
C12 The athletic stars	0.439340				0.098539	3
C12 The promotion and service of game	0.146447				0.032846	11
C14 The strength and international influence of host country	0.310660				0.069677	5
C21 The agenda setting		0.250984			0.115950	1
C22 The reported content		0.054883			0.025355	16
C23 The media service		0.063565			0.029366	12
C24 The assistant funds		0.057412			0.026523	14
C25 The quality of media workers		0.062738			0.028984	13

Table 2. (*continued*)

Factor					
C26 The media classifying	0.230401			0.106441	2
C27 The influence of media	0.134740			0.062247	6
C28 The development of new medias	0.093423			0.043160	8
C29 The run system of media	0.051854			0.023956	17
C31 The guarantee system		0.030541		0.004377	25
C32 The participation of government		0.326605		0.046805	7
C33 The participation of NGOs		0.070303		0.010075	20
C34 The participation of the sport industry		0.068091		0.009758	21
C35 The participation of the celebrities		0.056456		0.008091	23
C36 The spread in schools		0.16y0529		0.023005	18
C37 The theme activities		0.183991		0.026367	15
C38 The manifestation of political content		0.058290		0.008353	22
C39 The options of broadcasting ways		0.045194	0.227351	0.006477	24
C41 The educational background				0.038746	9
C42 The knowledge about sport issues			0.227351	0.038746	9
C43 The social stratification and income			0.122152	0.020817	19
C44 The recognition of athletic stars			0.423147	0.072114	4

The general rank sorting must be under conformance exam from the top to the bottom. Assume the conformance index of a factor in level C to Bi, the random conformance index is RI, so the conformance ratio of general rank sorting in level C is,

$$\frac{\sum\limits_{i=1}^{n} a_i CI_i}{\sum\limits_{i=1}^{n} a_i RI_i}$$

Where a_i is the ith component of W, when CR<0.1, the judgment that upward sloping has the conformance in a whole. The conformance exam of general rank sorting of this example is:

$$\sum\limits_{i=1}^{4} a_i CI_i = W \times CI + W \times CI + W \times CI + W \times CI =$$

0.2242888×0.040440+0.461982×0.116205+0.143308×0.086903+0.170422×0.003452
=0.075797

$$\sum_{i=1}^{4} a_i RI_i \quad = W \quad \times RI \quad + \quad W \quad \times RI \quad + \quad W \quad \times RI \quad + \quad W \quad \times RI$$

$$=0.2242888\times0.94+0.461982\times1.45+0.143308\times1.45+0.170422\times0.94=1.248700$$

$$CR= \frac{\sum_{i=1}^{n} a_i CI_i}{\sum_{i=1}^{n} a_i RI_i} = \frac{0.075797}{1.248700} = 0.060701 < 0.1,$$ so it's passed the examination in conformance.

Table 3. The ranking of all factors in the index system

Level C					Weight	The ranking
C21 The agenda setting		0.250984			0.115950	1
C26 The media classifying		0.230401			0.106441	2
C12 The athletic stars	0.439340				0.098539	3
C44 The cognition of athletic stars				0.423147	0.072114	4
C14 The strength and international influence of host country	0.310660				0.069677	5
C27 The influence of media		0.134740			0.062247	6
C32 The participation of government			0.326605		0.046805	7
C28 The development of new medias		0.093423			0.043160	8
C41 The educational background				0.227351	0.038746	9
C42 The knowledge about sport issues				0.227351	0.038746	9

4   Conclusion

The factors of the valuation index system are showed in Table 3, the first factor is agenda setting, it shows that there is a initial regulation in sport politic broadcasting, and it's not randomly compose of sport information. It concerns about broadcast strategy, media idea, concordance ability on information. In the important positions of western media politic, only the experts who get the ability of controlling the strategy of macroscopic sport information can do this. These experts contribute to the intelligence reference about the agenda setting in the reproduction of media politic on sports events of social appraisal, they make the agenda setting challenging by using their intelligence, knowledge and the influence of goal keepers of media. They get the status and high salary in this industry. They are the examples for all the people in this career. The fifth and tenth factor are strength of the host country, international influence, media influence, government engagement, new media development, education background, sport knowledge, etc. We can make the early four factors the coral ones and the latter six are half-coral ones, in the coral ones, we can make the media classifying the content of politic broadcast media, the agenda setting the soft strength of politic broadcast content, the star athletes the athletic subject of sports events (the subject of watching content), the stars cognition the watching subject of watching sports events. It can construct a framework of four-dimensional politic broadcast value and the four are the main facets of reproduction of media politic to sports events of social appraisal value, other six half-coral factors are the expansion of coral influence factors, the rest factors are the references to coral and half-coral ones.

References

1. Brookes, R.: Representing Sport. Arnold, London (2002)
2. Meisenheimer, M.: No Pants Required: A Behind-the-Scenes Look at Television Sports Broadcasting. Wheatmark, Tucson (2008)
3. Sandvoss, C.: A Game of Two Halves Football Fandom, Television and Globalisation. Routledge, New York (2003)
4. Hyatt, W.: Kicking Off the Week: A History of Monday Night Football on ABC Television, 1970-2005. McFarland, Jefferson (2007)
5. Boyle, D., Haynes, P.: New Media Sport. Sport in Society 5, 96–114 (2002)
6. Boyle, D.: Mobile Communication and the Sports Industry: The Case of 3G. Trends in Communication 12, 96–114 (2004)

The Development and Application of Valuation System on Competitive Sports Events

Xinyu Wu

Department of Physical Education, Hebei Polytechnic University, Tangshan, China
wuxinyu@heut.edu.cn

Abstract. The some related researching materials and fixed the index factors of the valuation system on sport events after comparing, analyzing and synthesizing. I calculated the weight of each factor in valuation system and used AHP method to construct a judgment matrix according to the scored data. All the factors have been examined on conformance by calculating. The sport events have become an important part of people's daily life with the coming of information society and popularity of sport. The development and application of valuation system on sport events will contribute a lot to construct a comprehensive feedback system on sport events, promote sports development and improve the quality of people's life. The value of this study and the social appraisal framework based on it is used generally to testify some practical problems and be worthwhile to be popularized.

Keywords: sport events, valuation system, development and application.

1 Introduction

We will conduct the program and arrange related research on this task at the point of hosting the 16th collage games in our school. The university game in Hebei which is held every two years is the pageant for college students in Hebei. This is the 16th of it. It's not only an examination of reform in colleges and popularization of sport events but also a chance for those colleges to show their athletic level. There are 76 colleges, 322 teams and more than 4900 people in individual sport of track and field, basketball, volleyball, Orienteering, aerobics and table tennis in this game. 30 teams got cups; 492 gold, silver and bronze medals were created; 25 groups and 53 teams got the moral awards; 7 people and 3 teams broke 8 records of college games, 12 people met the standard of first grade national sportsman; 171 people met the second grade. This game was hosted by provincial education ministry in Hebei, collaborated by the college sport association in Heibei, organized by Heibei University of Technology. There are 3 levels in the games, students in team one are from regular universities. Students in team two are from regular junior colleges. The team three concludes special-enrolled senior athletes and students in PE major. The period is from May 5th to 10th, 2010.

J. Luo (Ed.): Soft Computing in Information Communication Technology, AISC 161, pp. 171–176.
springerlink.com   © Springer-Verlag Berlin Heidelberg 2012

2   Construction of Valuation System on Sport Events

I confirmed the index factors of valuation system on sport events after looking up related materials [1-6] and making comparison, analysis and synthesis. The calculation of weight of all factors in the system is to construct a judgment matrix by the scored number. The judgment matrix of level A is demonstrated in Table 2. Get the judgment matrix of sub-criteria level in analog, the process is omitted. The calculation of judging the feature vector and characteristic root of the matrix (the method of calculating root).Calculate the product of factors in each line of the matrix A

$$Mi = \prod aij, i = 1,2,\wedge, n$$

Calculate the nth root of Mi

$$\overline{Wi} = \sqrt[n]{Mi}$$

Standardize Wi as

$$Wi = \frac{\overline{Wi}}{\sum_{j=1}^{n} \overline{Wj}}$$

So Wi is the feature vector.
Calculate the biggest characteristic value.

$$\lambda_{max} \approx \sum_{i=1}^{n} \frac{(AW)_i}{nW_i}$$

Calculate the process of W by referencing the weight calculation of index level B1-B4 corresponding to the criteria level A.

$$\begin{bmatrix} 1 & 1/2 & 2 & 1 \\ 2 & 1 & 3 & 3 \\ 1/2 & 1/3 & 1 & 1 \\ 1 & 1/3 & 1 & 1 \end{bmatrix} \xrightarrow{\text{每行之乘积}}$$

$$\begin{bmatrix} 1\times1/2\times2\times1 = 1 \\ 2\times1\times3\times3 = 18 \\ 1/2\times1/3\times1\times1 = 0.1667 \\ 1\times1/3\times1\times1 = 0.3333 \end{bmatrix} \xrightarrow{\text{求}M_i\text{的四次方根}} \begin{bmatrix} 1 \\ 2.0598 \\ 0.6389 \\ 0.7598 \end{bmatrix} \xrightarrow{\text{标准化}} \begin{bmatrix} 0.2243 \\ 0.4620 \\ 0.1433 \\ 0.1704 \end{bmatrix}$$

$$W \approx \begin{bmatrix} 0.2243 \\ 0.4620 \\ 0.1433 \\ 0.1704 \end{bmatrix}$$

$$\lambda_{max} = \sum_{i=1}^{n} \frac{(AW)_i}{nW_i} = \frac{0.9123}{4 \times 0.2243} + \frac{1.8517}{4 \times 0.4620} + \frac{0.5799}{4 \times 0.1433} + \frac{0.6920}{4 \times 0.1704} =$$

4.0457

Then solve the approximation of the biggest characteristic value.

Calculate the sequencing weight vector and examine them in conformance.

Table 1. The valuation index system of reproduction of media politics to sport events

The destination level A	Criteria level B	Index level C
The valuation on sport events	The influencing factors	The influencing index(measuring index)
	B1 The factors of game	The brands of games
		The athletic stars
		The promotion and service of games
		The strength and international influence of host country
	B2 The factors of media	The agenda setting
		The reported content
		The media service
		The assistant funds
		The quality of media workers
		The media classifying
		The influence of media
		The development of new medias
		The run system of media
	B3 The factors of broadcasting	The guarantee system
		The participation of government
		The participation of NGOs
		The participation of sport industry
		The participation of celebrities

Table 1. (*continued*)

		The broadcasting in schools
		The theme activities
		The manifestation of political content
		The options of broadcasting ways
	B4 The factors of audience	The educational background
		The knowledge about sport
		The social stratification and income
		The cognition of athletic stars

The conformance index is to judge the inconformity of the matrix and the sign is CI, defined as:

$$CI = \frac{\lambda_{max} - n}{n - 1}.$$

Obviously, the matrix judgment $CI = 0 \Leftrightarrow \lambda_{max} = n \Leftrightarrow$ is a matrix. The bigger CI is, the more unconformable the matrix is. How to use analytical hierarchy process to judge the unconformable range? Stochastic conformance index was introduced to solve that.

For a steady n, select a number as a_{ij} from $\frac{1}{9}$, $\wedge$, $\frac{1}{2}$, 1, 1, 2, $\wedge$, 9 and constitute

$$\hat{A} = (a_{ij})_{n \times n}, \quad (i < j)$$

The $\hat{A}$ as above is the most unconformable, assume the biggest characteristic value of it is $\hat{\lambda}_{max}$, then the definition is:

$$RI = \frac{\hat{\lambda}_{max} - n}{n - 1}.$$

The stochastic conformance index, when the stochastic conformance $CR = \frac{CI}{RI} < 0.1$, the discordance of A could be acceptable. Otherwise, the matrix judgment must be adjusted.

Besides, the data of conformance index RI to each matrix is given in Table 4 to make the examination easy. For the matrix of A,

$$\lambda_{max} = 4.0457, \text{CI} = 0.0152, \quad \text{RI} = 0.94, \text{CR} = 0.0162 < 0.1$$

Table 2. The judgment matrix of level A

A	B1	B2	B3	B4	W
B1	1	1/2	2	1	0.224288
B2	2	1	3	3	0.461982
B3	1/2	1/3	1	1	0.143308
B4	1	1/3	1	1	0.170422

So the checkout of conformance is passed, W could be the weight vector. The rest parts could pass the checkout by calculating, the process is omitted here.

Table 3. The index of stochastic conformance

N	1	2	3	4	5	6	7	8	9	10	11
RI	0	0	0.58	0.94	1.12	1.24	1.32	1.41	1.45	1.49	1.51

3   Conclusion and Limitation

With the coming of information society and the rise of new media, the traditional medias are threatened and need to make strategic adjustment. It's forming a situation in which there is a fight between the traditional media with the new ones. It's been an inevitable trend for media classifying. Common medias include newspapers, journals, books, radios, TV, movies and new medias, the initial principle among them are different, we can fulfill the goal of politic broadcasting and promote the broadcasting of political content better only by studying every single type of media and build up a run system of classifying medias. The third factor is the athletic stars. That's reasonable. The current problem is how to attract stars in some individual sport in the global games to participate fully because it reflects the influence of the game itself and the position it has got in sport industry. It is also an evidence that the game is an international game even the top international game. The crew of the task plan to make the rate to more than 80% of global star athletes in individual sport. Some influential games like the four open games in tennis, stars make their options according to their athletic conditions and environmental factors. The participation rate of stars in some games will be low because of that. The dispersion of the stars will decrease the influence of these games.

The existence of limitations and shortages in the research is unavoidable. It's reflected in a number of areas. One of the many reasons is that the researching object is a provincial sport events, it's very different from the global great games like modern Olympics, World Cup in the level no matter from the event level, media quality or the political factors, which confine the meaning of the research, the credibility and applicability of the index system. The second reason is that most of the respondents are leaders in schools and experts, few of them media workers, especially senior or experienced government circles and media workers. There is a deviation of school awareness in the construction of the index system, factor of broadcasting in schools highlighted in examining factors of broadcasting process can prove that., but it won't affect the value of this study and the popularity of the construction framework of social appraisal based on it. It could be used to testify some practical problems and is worthwhile to be popularized.

References

1. Kirsten, L.: Sport and New Media: A Profile of Internet Sport Journalists in Australia. Melbourne: Master's dissertation of Victoria University (2002)
2. Raymond, B.: Mobile Communication and the Sports Industry: The Case of 3G. Trends in Communication 12, 73–82 (2004)
3. Pedersen, Kimberly, S., Laucella, C.: Strategic Sport Communication. Human Kinetics Publishers, Champaign (2007)
4. Schultz, B.: Sports Media: Reporting, Producing, and Planning. Focal Press, Boston (2005)
5. Haynes, R.: Footballers' Image Rights in the New Media Age. European Sport Management Quarterly 7, 361–374 (2007)
6. Leonard, D.: New Media and Global Sporting Cultures: Moving Beyond the Clichés and Binaries. Sociology of Sport Journal 26, 1–16 (2009)

The Olympic Myth Macro Effect Mechanism Established by Internet Media Politics

Lu Huang

Department of Physical Education, Hebei Polytechnic University, Tangshan, China
huanglu0797@heut.edu.cn

Abstract. Check America's public opinion strategy on Beijing Olympic, with two basic dissemination theoretical system, news agenda-setting (timeline) and overall promotion strategy (spatial axis), as the research lead. It is perceived that the behavior of American Internet Media Politics was steady. Recognition of a great America was shaped domestic and abroad. And the American style Olympic legend and myth tradition was continued and maintained. Review China's public opinion strategy on Beijing Olympic, and examine the situation of Sino-America Internet Media Politics. It is perceived that the behavior of China's Internet Media Politics fell short of expectations. China's Internet Media Politics is, on a great deal of degree, influenced by western news review pattern and pre-assumed concept of value, insufficient of actively guiding the public opinion. Unveil the macro effect mechanism of Internet Media Politics' construction of Olympic legends. And provide reference for the China Internet Media Politics practice on 2012 London Olympic Games.

Keywords: Beijing Olympic Games, 2012 London Olympic Games, Internet Media Politics, Olympic Myth, the Impact of Public Opinion.

1 Introduction

Beijing Olympic has become the past and London Olympic is coming. How to summarize experiences, gains and losses of media development in Beijing Olympic between dialectics connection of present and past and build political route chart for London Olympic has become an immediate issue for post-Olympic process theory and practice. Olympic media politic is also called politicization of Olympic media, collusion of media and politic has become irreversible social trend. Politic express its desire and archive planned object by help of media, politics is more publicized and media is more publicized. 4 years circles of Olympic are actually a short period. Sports theory and practice researchers in China has no reason to despise and neglect but shall indicates to follow up and take beneficial theoretical position in order to prompt fast and healthy development of Olympic activities in China.

2 Olympic Tale Produced by Internet Media Politics: Reviewing USA

Beijing Olympic offers talkative opportunity for USA to reestablish image of world dominator and re-encourage its people. Pursing for golden medals also becomes focus

J. Luo (Ed.): Soft Computing in Information Communication Technology, AISC 161, pp. 177–183.
springerlink.com © Springer-Verlag Berlin Heidelberg 2012

area of USA's Olympic territory. USA lost past brilliance in golden metal competition and decent to position that relying on silver and bronze metals. Tradition and ambition to win golden metal was dismissed, MSNBC made vain calling: "China era is coming! [1]"

USA urgently needs to use former ways to launch media attack and change world structure by increasingly stable information monopoly that established in last century (media power that dominates the world). USA uses media to make agreement is not a kind of new art, it has long history and has unlimited importance compare to any other economic power. Making agreement will change every political planning and every political presupposition [2]. In major global events such as publicity confrontation between USA and former Soviet Union, Gulf war, Kosovo War, Afghan war and Iraq war after "9-11" incident and Beijing Olympic all sufficiently reflects impact of media attract / advantages of USA media to event result presupposition, this is also main expression method of USA national public diplomatic strategy.

Functions of publicity theory is unelectable in regard of analysis, guidance and decision implementing in Olympic marketing of USA medias, while two basic publicity theory systems "news agenda setting" and" general publicity strategy" played important roles. The two major theory systems have long history and passed many tests while developed together with global expansion of modern capitalism and reached perfect level at present. In regard of theory structure, it explains regular pattern between time axis and space axis; in regard of practice process, its unique and convenient practicality offers condition of implementing political willing.

USA will not give up to use Olympic as an opportunity to create American tale, its agenda including establishing powerful national image, expressing outstanding national identity, creating common community memory, fulfilling American style patriotic image and well planning American style Olympic Internet Media Politics events [3]. It's presuppose news agenda has close connections with time axle of Beijing Olympic (shown in Fig.1), therefore to create powerful recognizable image of USA. It carries historical heritage of American individualism and advocates multi-national capitalism spirit of team cooperation. Phillps and Dream Team of men basketball team is granted with spokesman image of USA is supported by iconic symbol system of USA national flag.. Medal ranking is also shown. There's highly motivated atmosphere without fear of challenges, more prompting events and celebrations after Olympic creates an Internet Media Politics Olympic recognition chart through with American will.

On space axis of Beijing Olympic, overall publicity strategy is placed on significant position to guide practice. Not only official statements, news report of commercial media, comment and following-up reports of entertainment media is required on gold medal competition that is the most core issue of Olympic game, but interactive reflections of space, Blog and Podcast is also needed to create a kind of successful gold medal recognition concept that covers all fields.

In historical memory fragments of two major military groups between East and West during cold war, as political agent and media marketing instrument, sports built connections between theoretical conflicting and military competition. US conducted active and effective over all publicity strategy and sports political marketing activity to risk level for former Soviet Union collapse [4-5]. At present, each part in overall publicity strategy has created high level tacit understanding and stick to their own character in order to complete each other's working task in process system, while also create partial publicity concept - unity of recognition to construct and maintain Olympic legend and tale tradition of US.

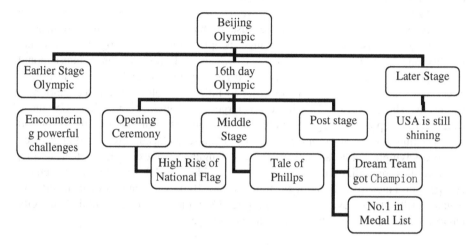

Fig. 1. Agenda setting chart of USA Internet Media Politics to create American style Olympic tale

3   Olympic Tale Produced by Internet Media Politics: Reviewing Beijing

International media highly rated Beijing Olympic including opening ceremony, contests, organization, investment revenue and other aspects. Even captious Western media also tended to overall neutrality and partial negation evaluation, symbol of disclosure including: "honors and surprising cost-effective spending plan that can be a reference for future Olympic host nations. Whether you agree with China's diplomatic policy and political theory or not, no one can deny this is a perfect Olympic. Congratulations, Beijing! [6]"

At meantime, Beijing Olympic showed the world with great success since opening and reform for 30 years, deep Chinese culture and tendency of national recovery that made old capitalism nations feel nervous and devalue that have a devoted and polite appearance. Former primer minister of UK- Tony Blair lamented that: "Giant nations including UK and other powerful nations in Europe and US in 20th century all in West since centuries ago. We have to accept a world that sharing right with Far East [7]". Such anxious and sense of loss roots in national dimension directly cause old capitalism nations feel strong jealous because of rapid emerging of China, symbol of disclosure including:" Number of Golden medals is 51 for China, the world is watching China with jealous. [8]" As mistress of world, US has particular same feeling and more symbolic feelings are included: China-West comparison theory, good or bed for capitalism political system or socialism political system, post-modern period emerging of China and fading of world super-power, restructuring of world power layout....

Compare to traditional media that strictly controlled by states, network media has more expansive power to deliver information. Developed nations in West plans to launch network publicity strike in order to distract, influence and even create public publicity environment of Olympic because of backward network management in developing nations, therefore to change people's Olympic recognition, national identity and post-Olympic policy design. US chose Beijing Olympic international media political activity with well organized and targets plan and careful selection to create news agenda setting chart in order to influence golden medal recognition in China (as shown in Fig.2) with clear target and obvious intension. US made exaggeration reports on negative public events such as withdraw of Liu Xiang and Lip-synching at opening ceremony, etc to create a reproduced, reformed and crooked Chinese Olympic public activity in political language context of US. At meantime, US told lies regarding human right issue, golden medal-oriented theory, ethnic conflicts in Tibet to make political traps around Beijing Olympic circle in order to negate brilliant achievements of Beijing Olympic, influence public publicity environment and slander China's international image.

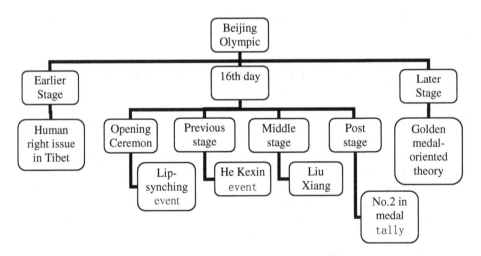

Fig. 2. Agenda setting chart of Internet Media Politics events specifically created by USA

On time axle of Olympic, content and intention of Internet Media Politics events agenda of US can be described as: vigorously create human right issue and Tibet issue in earlier stage of Olympic, fabricate China as a centralized autocratic empire without human right and freedom for ethnics; exaggeratedly reported lip synching and age issue of He Kexin to create a notorious image of world factory that all Chinese are fake products makers; excessively mis-interpreted culture symbolic significant for withdraw of Liu Xiang in order to describe China as a coward in East Asia that shrink back from difficulties, weary and couldn't hold position at crucial point; USA vigorously agitated importance of medal tally method in later stage of Olympic but not to talk about medal tally issue in order to reduce recognition level of historical achievements for public. US embroiled China's golden medal strategy with whole nation's power in order to create boycotting atmosphere for China's golden medal-oriented theory.

On space axle of Olympic, US use all controllable and manoeuvrable media powers such as newspaper, TV, radio and Internet to change subject identity of different speakers while demonize overall publicity strategy of Beijing Olympic to conduct American style implantation from all dimensions, levels and roles. On demonize Liu Xiang's withdraw, agent of politics, government, business and career successively emerged just like an unique auction that including both of serious salesman, capper whom bids at key time point and seeming honest watcher.

The audience has no time to make rational judgment and necessary thinking in front of all dimensional media strikes, by indulging in uncountable information created by Internet Media Politics, visual memory fragments are gradually produced to create recognition simulacrum of real happened events, but recognition frame delivered by Olympic information was just counterfeit selectively made by American Internet Media Politics which means Phillps representing American history, Bolt representing limitation of human body and Liu Xiang means fade of a nation. American Olympic contest is cost-effective event for whole nation but Olympic in China is poor efficient and expensive event.

4 Strengthen Training and Supervision of the Network Comments Agents

The theory and practice of New Media have great prospects. With the follow up of the sports fields, New Media is widely used in the Interdisciplinary research of sports media studies and physical education. Scholars make different understanding of the concept of New Media. With the renewal and development of new technologies, the connotation of New Media is constantly changing. Compared with the maturing sports new media technology, policy design, operation system and public opinion strategies of developed countries, the developing countries are still in the state of the laissez-faire slow development. There are many public safety problems and information vacuum. It is the main path of the developed countries' public opinion colonialism. The development of sports media is a complex and diversely dynamic development, including Media technology, system design, and practitioners (sports professionals, experienced Internet users, hackers, etc.), virtual social network, public ethics and other fields of multi-dimensional construct. It needs the lead of sports organization and the coordination of all forces. The New Media Political Practice focusing on the Olympics will be a construction of the field of ideology with difficult, complicated and intensified competition. In order to establish the right to speak at home and broad, sports official statement is not enough. It needs the coordination of sports media system innovation. We should attach importance to developing the field of sports new media and increase the construction of sports and new media marketing efforts.

As the axes of network media impact, web reviews has the very important role and significance of public opinion construction. Web review agents are the groups, which in accordance with the wishes of the employer, serve for some manufacturing sectors of public opinion. There is a analysis on the 3000 online sports reviews published on the three main new websites (Ynet, nrg and Walla!) of Israel. And it thinks the web reviews are in the new media expansion of business areas and becomes the new origin of access to sports information and debate with manufacture of sports media public opinion [9].

Public opinion influence is always the profit-making network-marketing tool that is web review agent in form. It is a Neo-liberal type of network economy. Common marketing models are BBS or paid blog reviews, such as ReviewMe 、 SponsoredReview 、 FeedSky 、 PayPerPost and other paid review service. Paid review services receive the advertising agency or client orders and employ agents in the salary payment form of bidding system for the task, reviewing words quotas and salary commission system. It is usually formed by social idle workers, work-study students or part-time composition. They focus on certain topic and give comment in the will of the payer. They always go to two extremes (Positive comments and negative comments) and change the established public opinion of events.

5   Conclusion

Based on observation of mediocrity game layout, historical great result to be on top of golden medal tally was not reorganized by pubic but even became target of social critics. The established gold medal cognition of Chinese civilian has been transferred / reformed by western media, and induced serious social influence / aftermath. By common sense, China should be reorganized as a powerful physical culture nation since it obtained the most many golden medals, but 4 years of Olympic contest works even didn't reach estimated result and the facts developed by opposite way, its mechanism of action shall be brought to the forefront. Hoping to strengthen researches in sports media theory and construction work of international speaking right, promote Chinese competitive sports to develop better and faster.

References

1. Boswell, T.: China setting a new gold standard. In: MSNBC, August 15 (2009), http://www.msnbc.msn.com/id/26211295/
2. Lippmann, W.: Public Opinion, April 20 (2009), http://xroads.virginia.edu/~Hyper/Lippman/contents.html
3. Dyreson, M.: Crafting Patriotism for Global Dominance: America at the Olympics, pp. 1–7. Routledge Press, London (2009)
4. Edelman, R.: Serious Fun: A History of Spectator Sports in the USSR. Oxford University Press, New York (1993)

5. Wagg, S., David, A.L.: East plays West: Sport and the Cold War. Routledge Press, London (2007)
6. Rogers, M.: Beijing trumps Athens ...and then some. In: YAHOO, August 24 (2008), http://sports.yahoo.com/olympics/news?slug=ro-beijinglegacy082408&prov=yhoo&type=lgns
7. Blair, T.: We Can Help China Embrace the Future. WSJ Online, August 26 (2008), http://online.wsj.com/article/SB121970878870671131.html?mod=opinion_main_commentaries
8. Weiwei, A.: Gold is not the real measure of a nation. In: GUARDIAN online, August 25 (2008), http://www.guardian.co.uk/commentisfree/2008/aug/25/olympics2008.china
9. Galily, Y.: The (Re)Shaping of the Israeli Sport Media: The Case of Talk-Back. International Journal of Sport Communication 1, 273–285 (2008)

Identifying Core Competence of Sports Journals in China from the Perspective of Author

Lu Huang

Department of Physical Education, Hebei Polytechnic University, Tangshan, China
huanglu0797@heut.edu.cn

Abstract. This paper is to identify the core competence of sports journals in China from the perspective of author and reader, combined with theories on core competence and analysis on macro-development of journals, and to discuss the cases of representative journals with core competence advantages, for example, manuscript evaluation process advantage of Journal of Capital Institute of Physical Education, network impact advantage of Journal of Physical Education, review period advantage of Journal of Tianjin University of Sport, popularity advantage of Journal of Sports and Science, institutional identity advantage of Journal of sports science, and etc. The times when identity is considered to determine fate has gone and now journals are faced with new circumstances externally and internally, which has promoted the innovation of journals.

Keywords: sports communication, academic journals of sports, core competence, online publishing.

1 Introduction

"Core competence" is a knowledge practicing process to coordinate different production skills and integrate various technical schools, and one or more production factors reach an irreplaceable professional level within the system or industry [1]. Core competence is an effective tool in identifying industrial competitive advantage; meanwhile, as a strategy and performance evaluation system, it plays a role of guider in selecting a strategy and practicing behavior for competitors. Here I expressed my views from the perspective of author and readers, combined with core competence and macro analysis on academic journals of physical education, to understand the core competitive advantages of academic journals. Hope editor teachers make comments.

2 Journal with Manuscript Evaluation Process Advantage – Journal of Capital Institute of Physical Education

With the development of computer software technology and journal networking, international famous journals are characterized with short publishing cycle, intense competition of manuscript source, collectivization of publishing scale and

J. Luo (Ed.): Soft Computing in Information Communication Technology, AISC 161, pp. 185–191.

specialization of content [2], which forces physical education journals to think about the surviving way and the editing work of physical education journals is faced with unprecedented challenge [3-6]. Although severe external environment of networking has been demonstrated early [7-8], there is always a distance between ideal and reality. *Journal of Physical Education* took the leading position in network practical exploration, but it is *journal of capital institute of physical education* that actually realized online gathering and editing, who applied strategic thinking and developing expectation into practice and accumulated rich experience , with a clear division of labor and detailed review procedure, including " manuscript reception", "preliminary review," "director review" "external review", "retreating for modification ", "editing", "layout", "proof" and even set up a "publishing" stage. Such an open review process with full tracking gives full respect to the right of know of authors in a humane way, which deepened the understanding between editors and authors on mutual trust, and enhanced the trust of journal. During the trials and hardships of online gathering and editing, *Journal of Capital Institute on Physical Education* has won a number of faithful audiences who recommend manuscripts, thus has formed a steady manuscript source, especially after it was promoted into the core journal of Beijing University. Thanks to the efforts of young and old authors, the number of manuscript received has exceeded ten thousands per year, and the enhanced popularity has improved its influential power and professional identification within the industry, which has provided basic condition for selecting excellent manuscripts, and spread the effect of journal for increasing the "support" and "trust" of authors [9]. Surely the plenty of manuscript received has increased the workload, and the speed on processing manuscripts will be affected when editors discriminate such a number of academic information with a strict attitude, especially the detailed division of manuscript review has added more review levels, where the detention of any procedure will affect the process of next procedure. Management, editing and service are important indicators for physical education journals to control quality, and *Journal of Capital Institute on Physical Education,* with advance thinking to guide the practice, to explore novelty, to be the pioneer, to break through tradition rules and to seize the opportunities, which has strengthened the pace of progress. The practice of *Journal of Capital Institute on Physical Education* has told us to make every effort to realize our dream.

3   Journal with Network Impact Advantage – Journal of Physical Education

After the construction of portal and the online manuscript gathering system, *Journal of Physical Education* is now focusing on the development strategy of "combine three websites into one" to explore the development model of physical education journals online publishing in the aspects of development concept, multi-media interaction, talent cultivation, financial guarantee and international development. It adheres to provide better online service to audience, grasps initiative of development in the information age, and has a great advantage in enhancing the network capability and constructing the overall validity of academic influence. *Journal of Physical Education* is rooted in Guangzhou, a first-tier city in China, with an open and active thinking, aiming at the development frontiers of network, and established three interactive systems, i.e. portal, sports online academic forum and the "Online Journal of physical education", which is conducted to extend the development space and enhance the brand recognition and academic influence. Network development strategy of *Journal of Physical Education* is endowed with modern marketing concepts, like the children marketing strategy of KFC, which is dedicated to providing free products with high added value and actually is to promote the increment and growth of journal around the identification system established by journal brand. Three interactive systems of *Journal of Physical Education* are in leading position in the industry. The portal is characterized with large amount of information, high stability and interaction, while "Online Journal of Physical Education" is to publicize the academic pure value, spread the achievement influence in daily life, and advocate an academic atmosphere of utilitarianism with a slogan of "article issued online can be published in journal with the recommendation of experts" and practical action. It is the first academic journal under network operation. Sports online academic forum takes root in the surviving rule of network community, who advocates academic controversy, free debate, resource sharing and achievement promotion. It has an accurate target population location, promotes academic controversy and has a good interaction between editorial department and e-friends, which can be regarded as three features of the forum and it provides an academic communication platform for physical education theory and practice operators. Certainly, there are more difficulties than expected, and because of limited human resources, the portal update slowly. Affected by the construction of online intellectual property rights, "Online Journal of Physical Education" failed to further improve quality and increase its influence. Sports online academic forum was utilized as a word attack platform by premeditators, for example, it was once accused of academic misconduct and academic fraud of some people, regardless of academic controversy logic other nasty things, which imposed a pressure to forum and even forced the forum to close for adjustment. What is more critical, because the influence of online journal can not be quantified and the network agrees to such bilateral relations, and unable to provide a clear return prospects for investors, there is insufficient motivation for high level to prepare and implement the sustainable development strategy, and the strategic core is still put on strengthening traditional journal. Because the researchers hold to the feelings and loyalty of online journal, they will consciously or unconsciously have bias to references when selecting references of the same subject and the same research level published in different journals, which makes it possible to achieve journal influence and article quotation rate, and which can not be explained by the web indicator quantitative analysis of database (CNKI, wanfang data, etc.). The practice of *Journal of Physical Education* has told us to persist on something correct with enough courage.

4   Journal with Review Period Advantage – Journal of Tianjin University of Sport

Journal of Tianjin University of Sport promises the review period is 15 days, which is in the leading position in PE journal industry at home and abroad, since generally the review period in China is 2 to 3 months and for other countries, it is generally 8 to 12 months. This promise has established a real time target for editor job performance and expected actual result, which is a process for editors to pursue the ideal state of journal editing. *Journal of Tianjin University of Sport* established the network editing system at the beginning of 2009, which is, although not the forerunner, the expression of advancing with times of the journal. It is rational and objective for Journal of Tianjin University of Sport to change the working direction to network editing, which is consistent with the philosophy of the journal. The editing work of Journal of Tianjin University of Sport pursues innovation but not prematurely, who pays close attention to the progress of fellow journals in the exploratory stage of journal network and actively quests for advanced concepts [16], to pursue other journals with the advantage of late-development on the successful operation of network editing system of Journal of Capital Institute on Physical Education and Journal of *Physical Education. Journal of Tianjin University of Sport* insists on the principle of article first, and emphasizes the quality of articles, diluting the influence of external factors, such as title, professional post, and subject. Such a practical philosophy runs throughout the work of journal consistently, highlights two core elements in the process of editing network one of which is to enhance the communication between editors and authors. The division of network editing process is simple and practical, including "preliminary review", "chief editor review", "external review", "retreat for modification" and "final review". The other core element is to shorten the review period. The preliminary period is generally 3 days, to effectively collect the first submission right of authors, that is to say, two or three days' preliminary review period can be ignored in terms of the time cost. Some people submit manuscript with a hope of getting review, since even if their manuscripts can not pass the preliminary review, they can get some comments which is helpful for them to enhance their academic ability, which is the reason of increased manuscript received, with a broader source of high-quality manuscripts. Meanwhile, the reduction of time cost has lower the academic moral hazard, because there is no need for the contributor to bear moral condemnation for "one paper submitted to several journals" to save two or three day, so the system is designed to effectively curb that behavior. Length of time in editorial work can not be simply equated with the quality of journals, and professional ethics and efficiency of journal editors can not be embodied by cumbersome process and lengthy time. The practice of *Journal of Tianjin University of Sport* has told us it is possible to race against time.

5   Journal with Popularity Advantage – Journal of Sports and Science

Debate and comment is the fine tradition of academic circles. International physical education fellow journals generally have retrospect's or review columns （Reviews, Review Essays, Book Review, etc.）, compared with which, we see little remarkable academic controversy with in physical education industry. Instead, it is depressing, which is caused by lack of encouraging innovation, preparation of journal layout, opened macro academic environment, and lack of thinker, academic school, characteristic and academic viewpoints in physical education [17]. Totally different and Idyllic Academic life of Journal of Sport History and Culture (marked with "sea of books " column) has been buried by today's standardization, industrialization and one-dimensional production, and now only *Journal of Physical Education* keeps review column ("Sports Information" column). There are few journals permanently set with controversy and comment, except for the "Open Forum" column of *Journal of Tianjin University of Sport*, "New Forum" column of Journal of sports and Science, "New vision, new view" column of Journal of Xi'an Institute of Physical Education, and " Exploration and Free Views" column of *Journal of Physical Education*, *Journal of Nanjing institute of Physical Education* of social science edition and *Guizhou Sports Science and Technology*, many of which are more in name than reality with a even worse situation. *Journal of Sports and Science* is the only provincial sports journal that was selected as core journal of Beijing University for five times. The indicators of *Journal of Sports and Science* rank among the best displayed by major journal evaluation systems (core of Beijing University, core of Wuhan University, CSSCI, Photocopied Data of People's University, etc.), which has reflected a unique style of the journal. Only with characteristics, it is possible to reflect the purpose of journal and obtain utility effect; to reflect the unique features and attract readers to reach a consensus effect; to reflect the exclusive advantages and get brand effect; to reflect the object consciousness, and form a stable readership, resulting in a group effect [18]. From the column "Physical Culture Research " and "Olympic Cultural Study" years ago, to current "Nanjing Youth Olympic Games" and permanent "New Forum" column, waving the banner of academic contention, resolutely getting out of the academic barriers that " Dare not totally repudiate, speak out the name or sharply criticize ", with an attitude to probe into the bottom of the matter, Professor Han Dan put forward the question "What is Physical Education?". The researcher Mao Peng insisted in questioning and correcting the main training ideology, while Professor Zhang Hongtan maneuvered and point out problems, all of which shows that journals stick to the feature and value positioned. Journals are trying to link closely with the sports practice, not only higher level, not only books, but only the fact speaks, giving theoretical thinking to the practice, shaping group identification and theoretical atmosphere for the practitioner of physical education to maximally break through the traditional relation ethics and institutional barriers, i.e. to avoid a kind of ethic constraint that "not want to say because of pride, dare not say even though actually know, speak nonsense because of feeling hard ". The person who speaks rationally will take advantage of voice and the literature cited. Journal of Sports and Science Practice is willing to listening to the suffering of front-line workers; dare to respond to main theory of physical education, so that those front-line physical culture workers who suffer with insulation are moved deeply. Interest subject can adjust the mutual quotation rate of high-end journals (such as CSSCI), but can not change the appeal of broader area (e.g. core of Beijing University). We should believe that the masses have sharp eyes and great power. The practice of *Journal of Sports and Science* has told us to firm the direction forward with characteristics as prerequisite.

6   Journal with Institutional and Status Advantage – Journal of Sports Science

Journal of Sports Science, superintended by State Sports General Administration and hosted by China Sports Science Society, has a integrated advantage in the links of operation, like institutional identity, academic quality, editing and binding, who continuously ranks in the top of various journals, with professional, authoritative and symbolic significance and is regarded as a platform yearned by those theoretical and practical workers of physical education to display their strength. Affected by the mechanism "one team, two names" during the social transition, the journal is highly specialized and in-system narrative, which is also the result of the formation of core competence based on comparative advantage. Too much authority may lead another developing direction. In traditional scholars' opinion, one high-level academic paper is convincing than many common ones, so Journal of Sports Science has become the only choice, where the published article is a symbol of high-grade identity and honor, and which is the goal of a traditional, standard, in-system scholar. More often, in-system narrative is completed in the form of introduction, and then one can get an "identity permit". In such an atmosphere, even without auxiliary system for journal advantage, like a network editing system, online marketing forum, 15-day review period promise, academic contention and book review column, the authority will never be affected. Surely, *Journal of Sports and Science, Journal of Physical Education* and *Journal of Tianjin University of Sport* once shortened the distance with *Journal of Sports Science*, effectively making use of innovative advantages, and brought the system vibration, inspiration and reflection. In recent years, *Journal of Sports Science* does not seek to lead the trend, but not to drop far behind, and has made some revolutionary adjustment in the aspects of network editing, review period, article structure, which let people see a series of encouraging changes. The application of network editing system has increased the transparency of copy edit, enhanced the communication between authors and editors, and accelerated the speed of copy edit. Article structure is becoming more equalized, and some achievements of humanistic sociology of sport system get much attention. The first author group is tending to be much younger, from which young scholars get more confidence and encouragement, especially after the publication of a paper contributed by a 25-year-old young man. After a short burst of loss, *Journal of Sports Science* is now running quickly on the way of innovation, leading the development of journal industry. The practice of *Journal of Sports Science* has told us that identity is certainly important, but also need to be perfectly justifiable.

"Is fate determined by identity or by struggle? [19]This enlightening whoop on *People's Daily* symbolized and reflected the developing characteristics of a new era. With the accelerated pace of reform and opening up in new century, the liberation of productive forces and the established relations of production of academic journals were restructured, and the external environment (internationalization and network) as well as the internal factors of innovation for journal production and operation are playing a more influential role. According to the traditions of physical education academic journals, the conventional ranking of journals undoubtedly is *Journal of Sports Science, China Sports Science and Technology, Journal of Beijing Sport University, Journal of Shanghai University of Sport,* and others. It is the internal and external new environment that promoted the innovation of journals and brought the traditions vibration and deconstruction. Now we are at a crossroads, and the era when identity determined the fate is gradually getting out. Innovation will make life more wonderful!

Acknowledgment. Thanks to the Chinese sports journal editors efforts.

References

1. Kornack, D., Rakic, P.: The Core Competence of the Corporation. Harvard Business Review 45, 79–91 (1990)
2. Feng, W., Huang, X.C.: Enlightenment of International Prestigious Journals to the Running of Chinese Sports Scientific Journals. Journal of Guangzhou Sport University 30, 120–123 (2010) (in Chinese)
3. Wang, Z.P., Wang, X.H.: The Networking Condition of Sports Science Journal and Measurements in China. China Sport Science and Technology 43, 31–36 (2007) (in Chinese)
4. Song, L., Lu, C.Y., Zhong, L.: Development and countermeasure of sports academic journals under the new situation. Journal of Shandong Institute of Physical Education and Sports 21, 127–128 (2005) (in Chinese)
5. Li, X.X., Qiu, J.R., Li, Q.H.: Research on Development of Sport Academic Journal of New China. China Sport Science 29, 3–23 (2009) (in Chinese)
6. Wang, H.J.: The Status Quo and Development Measures of Sports Journals E-Publication. Journal of Chengdu Sport University 34, 46–48 (2008) (in Chinese)
7. Zhao, Y., Wang, X.F.: The Status Quo and Countermeasures of the Networking of Acdemic Sports Journals. Journal of Beijing University of Physical Education 26, 610–612 (2003) (in Chinese)
8. Zhu, K., Li, S.X.: Effects of Data Base and Internet on the Sports Science and Technology Journal. Journal of Xi'an Institute of Physical Education 18, 29–31 (2001) (in Chinese)
9. Liu, L.X.: Investigation and analysis of communicative effects of sports journals in China. Acta Editologica 19, 391–394 (2007) (in Chinese)
10. Zhang, X.Y., Li, Y.Q.: On Necessity of Investigation toward Satisfaction Extent of Editing Work of Sports Academic Periodicals. Journal of Capital Institute of Physical Education 21, 295–298 (2009) (in Chinese)
11. Li, M.: On Innovative Consciousness for Editors of Journals of Sports Science. Journal of Shandong Physical Education Institute 20, 95–96 (2004) (in Chinese)
12. Ran, Q.H.: Development on quality control index for Chinas academic P. E. journals. Acta Editologica 20, 183–185 (2008) (in Chinese)
13. Zheng, Z.Y., Tan, G.X., Wang, W.L.: Practical exploration of the development of network publication of Journal of Physical Education. Journal of Physical Education 15, 14–19 (2008) (in Chinese)
14. Zheng, Z.Y., Qin, L., Tan, G.X.: Review on Value of "TIYUOL" Academic Internet Forum and Its Development. China Sport Science and Technology 45, 136–139 (2009) (in Chinese)
15. Wang, X.F., Li, A.Q.: Journal of Chinese core sports online academic influence and influencing factors of the investigation. China Sport Science and Technology, 70–75 (January 2009) (in Chinese)
16. Li, J., Wu, H.T.: Based on Opening Deposit and Withdrawal our Country Sports Science and Technology Periodical Publication Pattern Research. Journal of Capital Institute of Physical Education 21, 517–520 (2009)
17. Ma, X.J.: Status of Sports and Development of Academic Journals. Journal of Shanghai University of Sport 34, 34–36 (2010) (in Chinese)
18. Cheng, Z.L.: Characteristics and Development of the Journal of Sports Science. Sports Science and Technology 24, 95–98 (2003) (in Chinese)
19. Column Group of Current Observations, "Column "Current Affairs". Is fate determined by identity or struggle?" People's Daily, November 11 (2010) (in Chinese)

The Value of the Basketball and Hip-Hop Dancing's Invention and Its Feasibility

Chao Guo

Department of Physical Education, Hebei Polytechnic University, Tangshan, China
xiaochao654789@sina.com

Abstract. Basketball hip-hop is a new body-building form which combines hip-hop with basketball. Currently, the sport only appears in some parts of our country, and the media hasn't reported or propagandized it. After referring to large amount of materials, there is no research related to the sport. The paper makes preliminary study of the value of creation of basketball hip-hop, significance of popularization and the existing difficulties and expects to promote the development of this sport.

Keywords: basketball-and-hip-hop dancing, the value of its invention, popularization.

1 Introduction

Basketball hip-hop is a new body-building form which combines hip-hop with basketball. Currently, the sport only appears in some parts of our country, and the media hasn't reported or propagandized it. After referring to large amount of materials, I found that there is no research related to the sport. This paper makes preliminary study of the value of creation of basketball hip-hop, significance of popularization and the existing difficulties and expects to promote the development of this sport.

2 Research Object and Method

A. Object

77 members of dance class and 70 ordinary students from East China Jiaotong University; 21 members of hip-hop team from the East China Institute Of Technology (Nanchang Campus); 11 members of men's basketball team of high school from No.1 Middle School of Yichun and 11 junior school students from No.1 Middle School of Yichun.

B. Method

- Questionnaire Survey Method
200 questionnaires in total were given to the students, 192 were retrieved, among which 180 were effective questionnaires and the effective rate was 90%. Besides, 43

J. Luo (Ed.): Soft Computing in Information Communication Technology, AISC 161, pp. 193–198.

questionnaires in total were given to the students of experimental group, 43 were retrieved, among which 43 were effective questionnaires and the effective rate was 100%.

- Experimental Method

Take 43 students of the inquirers as the experimental objects; create a basketball hip-hop movement of 4 beats; teach them for 4 weeks and the total class hours are 32. Design an experimental method aiming at the basketball skill; compare the related data before the experiment with that after the experiment; and examine the heart rates and other physical function indexes of the participants before and after learning.

- Statistical Method

Conduct classified statistic and comparative analysis of the acquired materials; and examine and make statistical treatment of all the tables by turns.

3   Result and Analysis

A.   The Status of Development of Basketball Hip-Hop in China

With the continuous development of the economy and steady increase of people's living standard in China, people are increasingly conscious of the importance of keeping physical health and body contouring. They walk into the health clubs and training centers of amateur dance, which definitely promotes the development of fitness and entertainment methods in our country——from aerobics to Latin calisthenics, from tae kwon do to Kwando aerobics, etc, different kinds of creative projects for fitness and entertainment emerge in succession. However, the basketball hip-hop has not yet appeared in the domestic health clubs. Here, let's understand it better: it is a special presentation form of dance, which adds the actions in basketball sport such as dribbling and passing to the hip-hop movements and masterly integrates the flexible characteristics of the latter into the former. It combines the basketball sport with dance and makes it a sport outside the basket. However, in our country, this form of entertainment and fitness is in the embryonic stage, and only during the interval of the basketball game that we can see the similar performance. The public knows little about it; therefore, we shall focus on the value of creating the movements and promotion at the present stage.

B.   Creating Value of Basketball Hip-Hop

The basketball hip-hop has novelty, creativity and interest. Its creation is aimed at the fitness and entertainment, and it can assist and enhance the skills of basketball sport and hip-hop.

Table 1. Statistical Table of Questionnaire Survey on the Creating Value of Basketball Hip-hop (N=180)

	Characteristics (Multiple Choice)				Effects (Single Choice)			
	Novelty	Creativity	Interest	Other	Fitness	Basketball	Hip-hop	Other
Number of People (People)	180	174	62	24	74	18	56	32
Proportion (%)	100	96.7	34.4	13.3	41.1	10	31.1	17.8

Table 2. Measurement of physical function indexes before and after experiment (N=43)

		HeartRate (T/min)	Squat-and-straighten-up Index of 30s30times	StepIndex
Boy	Before Experiment	76.13±3.47	7.85±0.36	48.01±1.3
	After Experiment	136.40±2.97	7.09±0.24	51.79±2.03
Girl	Before Experiment	77.67±2.86	8.73±0.28	44.00±1.82
	After Experiment	138.72±3.68	7.71±0.22	48.31±1.70

- Effect on National Physical Fitness

The analysis of the change of heart rate before and after the experiment shows that the sport meets the standard of public aerobics exercise, (220-age of the exerciser)×[0.6, 0.8]. Take 20 years old as the average age of the participants, the optimal heart rate of the aerobics exercise is between 120 T/min and 160 T/min, which indicates that the sport meets the standard of public aerobics exercise.

The indexes of Skubic step test and 30s30 times squat-and-straighten-up are important indicators for evaluating cardiovascular function and effective indicators for assessing the aerobic capacity. High index of Skubic step test indicates good aerobic capacity; while the index of 30s30 times squat-and-straighten-up presents synchronous decrease and has obvious difference (P<0.05), which shows that the basketball hip-hop exercise promotes the shape, function and adjustment capacity of the cardiovascular system of the organism and generates good adaptation, thus enhance the aerobic capacity. Therefore, the above mentioned data indicates that this sport event can be applied to public fitness and it is a good method for fitness.

The survey result shows that 75.7% of the students would like to learn the dance, which indicates that the dance has wide acceptability and can be learned by the public. According to the experience of promoting tae kwon do and Kwando aerobics in China, it attracts three groups of people of physical fitness, namely tae kwon do sport, Sanda sport and body building exercise. This shows that the compound-type fitness project has large base of people, and it has more learning people compared with that of the single-type fitness project. Try to imagine that how many basketball and hip-hop fans will be attracted and people will try this unprecedented fitness project when this project enters the health club. Therefore, in terms of public fitness, this project has huge developmental potential.

- Effect on Increasing the Level of Basketball Skills.

The swerve around sticks exercise of dribbling is as follows: the test distance is 10 meters, 3 sticks are placed in the middle as the barriers (the position of the sticks are respectively 2.5 meters, 5 meters and 7.5 meters from the starting line), the distance between the sticks (width) is 5 meters, the tested students dribble and run around the sticks in Z shape and calculate the time of finishing the full distance.

The 30s ball intercepting experiment is as follows: 30 seconds in total, take the center circle scope of the standard basketball field as the test site, conduct the competing ball intercepting exercise of two people in dribbling, and calculate by the times of mistakes in dribbling.

The standing pass experiment is as follows: 1 minute in total, the distance is 5 meters, calculate the times of passing ball between two people in one group (2 times per one pass).

Footwork combination experiment is as follows: the test distance is 20 meters, 6 sticks are placed in straight line in the middle as the barriers, the tested students run to the stick and conduct the cross step and up-and-under dribble action after rapid stop and backspin crossing action (repeat 3 cycles of 2 methods), calculate the time of finishing the full distance.

Table 3. Statistics of Effects of Designing Experiment on Different Skills of Basketball
(N=43 35 boys<including 9 members of basketball team>, 8 girls)

		① swerve around sticks exercise of dribbling	② 30s ball intercepting exercise (time)	③ standing pass exercise (time)	④ footwork combination exercise (time)
Experimental Group	Boy	9.48	2.17	---	---
	Girl	10.72	3.51	---	---
	Total	9.72	2.44	48.33	12.15
Control Group	Boy	9.81	2.64	---	---
	Girl	10.76	3.78	---	---
	Total	10.00	2.87	48.38	12.17

From table 3, we can figure out that the basketball hip-hop has distinctive effect on increasing the basketball control capacity, while indistinctive effects on the skills such as basketball pass and footwork. The increasing radian of the ball control data achievement of boys is higher than that of the girls, which shows that the sport has obvious effect on improving the level of basketball control skills. And the increasing rate of the students of basketball team is lower than that of ordinary students, which shows that the sport has obvious effect on the beginners of basketball. Thus it can be inferred that aiming at the uninteresting development of basketball control capacity of the basketball players in our country, developing the ball control capacity of the students by this sport can not only realize "happy teaching", but also improving the level of basketball skills. Implementing this method on the beginners of basketball may improve their ball control capacity in a short period, and make them be more interested in basketball sport according to the correlation between the increase of ball control capacity and that of other basketball skills.

C.   Promotion of Basketball Hip-hop

- As the learning of this sport requires certain dance background and sense of basketball, the survey shows that the middle school students have difficulty in accepting it due to their physical quality and other aspects; the high school students have less difficulty in accepting it; while the college students may accept it easily. Therefore, the sport shall face people above 16 years old and give priority to the youth. It is advised that before formal study, class of basic basketball control skills and hip-hop class shall be opened to decrease the difficulty so as to arouse the interest of students.

- As the creation of movements of basketball hip-hop is flexible, the individual may create movements by himself (herself) according to his (her) own characteristics, goal and other factors. This paper is aimed at providing a creative form of entertainment and fitness and basis of creation value and promotion, which may bring certain inspiration to the fitness enthusiasts and physical culture workers.
- As the sport is in embryonic stage, therefore, creating movements shall take the principle of being simple and easy to learn; arousing the interest of students and giving proper attention to the aesthetic feeling of movements.
- We shall mainly use rubber basketball in learning the sport, as it is not only beneficial for the beginners to practice sense of ball, but also has cheap price.

4   Conclusion

Basketball hip-hop is a new dance and fitness form which combines the hip-hop with basketball sport. This form of entertainment and fitness has not appeared yet in the health clubs or training centers of amateur dance. At present, we shall give priority to the value of creating movements and the promotion. The sport is novel, creative and interesting. In terms of public fitness, this sport meets the standard of public fitness and exercise and it is a relatively good method for the public to keep fit. This sport has obvious effect on improving the basketball control capacity.

5   Suggestions

A.   Give Priority to Promotion of Health Club

The basketball hip-hop has similar structure of the tae kwon do and Kwando aerobics, and both of them are new forms of entertainment and fitness which integrates another kind of sport into one sport. One of the important reasons for the smooth promotion of tae kwon do and Kwando is that it gives priority to "first rate" teaching; fully plays the media role of health club and builds bridge between teaching and learning. Therefore, the basketball hip-hop shall also give priority to the promotion of health club.

B.   Give Priority to TV Media Promotion

In the late 1980s, the introduction of fitness exercises to our country took the TV promotion as the main method, and the representative column was the "5 minutes every day" on the sports channel of China Central Television. Afterwards, it has gained great favor from the sports exercisers and been widely spread in the whole country. The basketball hip-hop may also let people know it elementarily with the help of modern media methods, and then the people may change from spectators to exercisers.

C.   Targeted Promotion

As the sport has effect on assisting and improving the basketball control capacity, we may take it as the breakthrough point and apply it to the daily training from part to the whole. This is beneficial for the wide promotion of this sport.

References

1. Coetzee, M., Spamer, M.: Assessing the Determinants which May Influence the Creating of a Learning Climate in Physical Education. Journal of Human Movement Studies 38, 1–22 (2004)
2. Hastie, P., Trost, S.: Student Physical Activity Levels During a Season of Sport Education. Pediatric Exercise Science 14, 64–74 (2002)
3. MacPhail, A., Kirk, D., Kinchin, G.: Sport Education: Promotion Team Affiliation Through Physical Education. Pediatric Exercise Science 23, 106–122 (2004)
4. Wallhead, T., Ntoumanis, N.: Effects of a Sport Education Intervention on Students' Motivation Responses in Physical Education. Journal of Teaching in Physical Education 23, 4–18 (2004)

Graphic Design of Poetic Thought

Sisi Lv and Weiwei Suo

Institute of Art & Fashion, Tianjin Polytechnic University, Tianjin, China
{lss_0401,shirley3977}@126.com

Abstract. Poetic thought in modern graphic design plays a unique role, it will visual elements complete refine and expression comes out, with the most concise and effective visual form show more design theme profound connotations. Poetry imagery as design, graphic design vision language manifestation of rhythm and cadence with poetry imagery exist the subtle relation.

Keywords: Graphic design, Poetic thought, Visual language, Poetry artistic conception.

1 Introduction

With the development and evolution of graphic design more towards the expression of the theme in the style of the outstanding creative visual language, visual language has the appeal, impact and convey information accuracy, objectivity, emotions become a graphic design of the most important means.

Graphic Design in China, the development of modern graphic design process of the gradual integration of Eastern and Western ideas, a new development, designers in the concept, form, artistic expression has a new outlook. As for the present, how to innovate and inherit the question becomes the graphic design field a very important topic. A conception of the design is good design, and as a cultural and ideological positions shown poetic thinking, it is the way to poetry to graphic design to create a mood, trying to give a three-dimensional graphic design, open thinking Inheritance of new ideas, good modern design ideas of the designer.

2 Summarize

A. The Relationship between Poetic Thought and Graphic Design

Rational thinking is people who know about the advanced consciousness, thinking, results used in modern graphic design school, graphic design thinking from the dependence of perceptual knowledge to rational knowledge level, the study of graphic design tends to be more theoretical and systematization. Poetic thinking is rational thinking, logical thinking it as a guide to the concept of the basic unit of thinking, in the image on the basis of thinking, through the use of analysis, synthesis, abstraction, generalization and other methods, the various attributes of things to think processing

J. Luo (Ed.): Soft Computing in Information Communication Technology, AISC 161, pp. 199–203.
springerlink.com   © Springer-Verlag Berlin Heidelberg 2012

extracted from the concept reflects the essential nature of things, the nature of things more complicated to judge, to determine the linkages between things, reasoning out complex matters and the internal relationship between a law of thought. Poetic thinking in graphic design, to free, eclectic approach, there are. Poetic thinking that graphic design from the infinite wealth of things and forms to screen some of its special things - simple concise visual elements. By melting and differentiation, symbol of the gradual loss of three-dimensional, gravity and material, they become flat and purely visual things. Therefore, with the concise expression language, the work will be revealed pure and clear. With poetry, the design is successfully express a concept of artistic and business philosophy, graphic works more smoothly smart.

"Poetry" in ancient Greek meaning is produced, produced or created so that both can be called poetry out of something Whether in literature, drawings or other art. Elegant poetry, full of content, it comes from life, from the public, the use of the community, infection with the community, the development of the society, it describes a profound life and feelings, to seek the true meaning of life. Mood in the Chinese classical aesthetics than in the "image" more popular aesthetic category, that is, the mood outside the pursuit of images. The mood and spirit of poetry is in line with contemporary aesthetic taste. Art from the concentrated observation of life, the design is refined from carefully on the screen. Design is a concept, the purpose is the pursuit of the perfect design, vivid, and return to nature. Whether writing poetry or graphic design art, their creative material, inspiration, language, construction, artistic beauty are interlinked. The aura, simple design and poetry is the meeting point, which implies that a simple and elegant, rich content of the aura, the United States. Single lines, vivid graphics, aura, symbol, these are the backbone and soul. Line drawing is the poetry of the most commonly used methods, a few strokes with a little hook out of it an elegant realm. If the form of poetry, charm used in graphic design, showing a deeper meaning, a rich picture, the picture is more content, meaning, and charm, and can achieve a state of profound mood. Therefore, graphic designers are mostly poets, who in the design process based on a given material, into the boundless realm of creativity and imagination, create new ideas and business practices by attracting the attention of the viewer, to create another a world. Graphic design work of thinking through the poetic expression of the designer not only emotional, but also to arouse the viewer's emotions. This is designed to create a world more full of human charm, but also full of rich poetry, more in line with modern people's aesthetic taste. As a designer, poetry and design should be sorted and stored and fully applied to the creation of graphic design.

B.   Poetic Thinking in the Use of Graphic Design

Graphic design is a graphic language, it is like as a literary language vocabulary, grammar, syntax, rhetoric, and with the poetic language of thinking contains contrast, rhythm, balance, unity, firing, rhythm, space and other principles and methods, this is not only the concept of performance, and accurately convey information useful, you can make graph concise easy to remember, harmony and unity, correct and clear. Graphic design is the designer of visual images combined with the visual elements and write a poem composition techniques. Can be seen from the data, the European "visual poet" who for the grasp of graphic design graphic element is very flexible. The meaning of the formation contains a rich picture, a visual construction of coordination, enabling viewers to get a strong visual impact and aesthetic pleasure to Gunter Ram bow is one of Europe's visual poet, early works the body in a concise and organically integrated graphics and the environment, trying to build a visual coordination, visual works created by conflict and internal coordination, so that viewers get a strong visual impact and aesthetic pleasure. The author's work is undoubtedly a decent style of poetry. Most of his works are no written posters, Poster design, but does not affect the visual symbols to convey results. His use of photography, collage and other technology, scattered elements of reconstruction, the image of reality into abstract images with graphics, concise text with layout, form a powerful visual impact of graphic design work, the rhythm layout and form of rhythmic poetry of art and very close to nature.

The work, 《 Flying Dutchman 》 " drama posters, graphics and text hawk, background, texture, and many other elements in place, the focus, contrast and other aspects of integrated applications, so that the area of partition layout to differ, but the sense of weight but the area is similar to form a balanced picture effect.

Poetry in the plane, the emphasis is on charm, and graphic design if the link with the poetry, the poems in the form of interest applied to the graphic design, and then after the designer of the processing, sublimation, with point, line, surface Organic interludes and echoing. The meaning of the formation contains a rich picture, a visual construction of coordination, enabling viewers to get a strong visual impact and aesthetic pleasure. Poetry in Graphic Design to change the text to be opened, and then reunited to form the image of a similar meaning, but meaning the same, the same as the rich poetry through the art of processed, resulting in more extensive form of graphic design. There are deeply interested in poetry shallow design in tastes, and poetry writing comes from life, graphic design, visual elements extracted but also in social life. Designers have to go deep into life, creative life, beautify life and enhance life, life-changing, so that works with real easy to understand, strong visual shock, the real design to achieve a high state. Through designers enhance the visual language, the sublimated into another image of the object, forming a strong realism and visual impact, which is finally realized impressive, to enlighten the viewer, these are, and poetry have the same purpose. Poetry can be used as the visual design elements can also be designed in the advertisement, so as to achieve a very high Mood of Poetry. The use of poetic thinking design should be followed: to extract from life, from the local culture to enhance, through the fusion of the designer, after sublimation, so that the audience fully accepts the results and achieve enlightenment.

Graphic Design Master Kan Tai-Keung has been explored in his works the language of Chinese ink painting and poetry blend of language, the pursuit of simple composition and the integration of the richness of communication. Focus on national culture in the discovery and establishment of local consciousness, Kan Tai-Keung also attaches great importance to this classic affair and the fusion of modern design, he absorbing, widely take Europe and America, the experience of Japan advanced to the attitude of experimental positive development and evolution of their own design style. Late 80's to 90's, he perfected his own typical style: solid composition, elegant picture, color and sometimes joy, sometimes elegant, timeless and fills the Zen conception. Works can be seen from time to time the wisdom of traditional Chinese literati harmony with modern commercial civilization dialogue. His design for the Beijing 2008 Olympic Games posters, was in 2006 China Taiwan International Poster Design

Award Gold Class Special theme topics, the content reflects the "Sport, Sunshine, and Health ", still adhere to the elements of his Chinese ink. Produced posters with the whole lot of artistic expression of traditional Chinese ink painting, let the mayonnaise and the words charm, injects a torrent for the local posters, the list is undoubtedly reflected in work style. Most of the images seem simple Su Jie, a large blank area of the design method, reminiscent of the oriental concept of space. As the design work is widely spread, designers have to make contact with overseas foreign land feel the subtleties of Chinese culture, in promoting the culture of their motherland, contributed.

C.   China's Poetic Thinking in Graphic Design

Today, the aesthetic characteristics of graphic design is the "performance" rather than "reproduction" Simple screen image has been unable to support all life from a work must proceed from a focus on internal and personality, highlighting artistic spirit and personal style.

Poetry is the art of design and to the community in an effort to extract the general public, designers need adequate nutrients and fresh air. While graphic design is definitely not the pursuit of Chinese ink painting and poetry of eternity, but the effectiveness of visual communication of information, but they are required to have artistry. Advocated by traditional Chinese aesthetic beauty of the image, the same can be reflected in the design, its design has a guide on how to poetry, but also makes the design more depth of Chinese culture. Some people might consider graphic design as a practical art, the recipient must take into account the aesthetic needs of information and aesthetic standards, in other words graphic design to fully consider the popularity. China's traditional culture is a national treasure, but its advocates than merely copying the traditional elements, while the nation should be more on the psychological, philosophical, aesthetic habits of analysis, the national spirit and The visual language of modern combined to form the style of contemporary Chinese graphic design. As a designer, in the context of China's diverse culture, should continue to learn all aspects of nutrition, must meet the standards of the local culture, certainly Chinese poetry culture, and fashionable, be better applied to the graphic design order to get to speed up the pace of development of Chinese graphic design.

3   Conclusion

Overall, poetic thinking in contemporary graphic design plays a unique role, which will refine the visual element of the full manifested in the most simple and effective design of visual elements to show more profound meaning of the theme, poetry can be used as a basis for the design elements most directly to the use of poetry can also be said that the designer inexhaustible source of creativity. Chinese culture is profound, no matter what kind of art is diverse, and tends to harmony. In this soil has a long and glorious cultural history, we as a modern graphic designer, is constantly subject to external shocks, to learn Chinese and Western combined to the main, the formation of national character, designers to Chinese traditional culture with modern design concept, creating a poetic flavor with Chinese culture, style of thinking, so that our national cultural heritage can be continuously. We need to learn graphic design to promote poetry in thought, not opposed to diverse ideas, diverse and rich language of the means to carry graphic design, but the design is the inclusion of poetic thinking in particular aesthetic poetry, Chinese-style, is certainly a field applied to the poetic words rhyme creative graphic design applications.

References

1. Wang, S.: World History of Graphic Design. China Youth Press, Beijing (2005)
2. Zhu, G.: Poetics. Shanghai Ancient Books Publishing House, Shanghai (2005)
3. Chen, F.: World Masters design ideas. Heilongjiang Fine Arts Publishing House, Heilongjiang (2000)
4. Zhao, E.: Cultural Studies of Modern Advertising. Shandong People's Publishing House, Jinan (1998)
5. Tan, P.: Design, In the name of heritage. Sichuan Fine Arts Publishing House, Sichuan (2005)

Analysis of the Needs of College Media Sports Information and Its Accessing Channels

Xinyu Qiu[1] and Jian Cao[2]

[1] Department of Physical Education, Hebei United University, Tangshan, China
qxy6801@sina.com
[2] Library, Hebei United University, Tangshan, China
caoj1967@163.com

Abstract. With the advance of information-based society, mass media information will be important resource for college students. By investigating 2216 students from 25 colleges of Hebei province, we analysis the needs of college media sports information and its accessing channels in this paper. The result shows that college students have several motivations of contacting media sports information, different media have the different motivations. For college students, the major motivations of contacting media sports information are acquiring knowledge, learning and entertainment, in order to meet their reality and entertainment needs. Mass media is the important accessing channel. Among them, electronic media (include Internet, television, radio) are the major approach and internet has become the preferred way to sports information.

Keywords: college students, mass media, sports information, needs, approach.

1 Introduction

With the advance of information-based society, times characteristics like "knowledge-based society, internet-based information and socialized study" bring new challenges, new knowledge and new opportunities to college students. For these students, mass media information will be their important resource. Modern sports and mass media are inextricably linked. Impact of mass media is becoming widely concerned, especially on the sports life. Therefore, studying the mass media's role on college students and understanding the needs of college media sports information and its accessing channels have a positive and relevance effect on promote effective sports education policies, creating good social environment for healthy growth, promoting students form correct sports awareness and values and improving initiative to participate in sports.

2 Research Methods

A. Literature Material Method

More than 20 thesis and related books are the reference of this research.

J. Luo (Ed.): Soft Computing in Information Communication Technology, AISC 161, pp. 205–209.
springerlink.com   © Springer-Verlag Berlin Heidelberg 2012

B.   Questionnaire Survey

We took out 2216 students who have different genders, majors and grades from 25 colleges of Hebei province. Through this survey, we broadly understand university students' motivation and access to sports information media contact. In this survey, we put out 2500 questionnaires and withdraw 2216 efficacious questionnaires.

C.   Mathematical Statistics

Use Excel to analyze and process the acquired data.

D.   Interviews

We make face-to-face communication with college students to know the relevant situation.

3   Results and Analysis

A.   Motivation of Contacting Media Sports Information

In psychology, human needs are dependence and demands of human survival and development of their own objective conditions. It is an inception trend of human. Need is the foundation of motivation; motivation is the expression of need and direct cause and internal power of behavior. When need points to certain target, expresses the possibility of achieving target, and possesses the approach of need satisfaction, it can form the motivation and drive of activity. Therefore, motivation expresses in the process of target's need satisfaction.

There are many human needs. As Marx said, "the infinity and extensiveness of human needs differ human from other animals." The information need of college students is the reflection of their information awareness on sports information. It is affected by information environment, personal knowledge structure, cultural level, individual abilities, interests and intelligence [1]. While media need is an important spiritual need which includes the needs of reality, relaxation, escapism, emotional stimulation, communication demands, happiness demands and peace demands. The broadcasting behavior of audience submits to needs and gets satisfaction from spreading activities. Studying on the media needs of audience must start from their motivation of participating activities [2].

According to these needs, we designed this questionnaire on motivation of contacting media for college students, modified on the base of questionnaire of Huilin Wang and Wei Yan [3], fifteen "motives" on college students' contact media were investigated, include: "to increase their sports knowledge", "for sports news", "to seek solutions to the problem", "for amusement", "personal interest", "kill time", "Shopping Guide" , "interpersonal relationships", "sports stars", "author, host, narrator's style", "referee's style", "long habit", "narcissism", "passive participation" and "other".

Through conducting statistics and summary on these motivations of contacting media for college students, draw the outcome as follows: (See table 1).

Table 1. Motivation of contacting media sports information

Motivation	Frequency of Media Sports Information (%)					
	Internet	Television	Radio	Newspaper	Magazine	Book
to increase their sports knowledge	43.82	23.56	13.65	25.53	26.32	21.58
for sports news	52.41	58.12	34.21	31.67	21.59	3.56
to seek solutions to the problem	26.85	13.85	9.63	11.65	8.32	20.54
for amusement	31.25	26.79	25.96	25.55	43.67	6.94
personal interest	29.61	16.75	20.46	12.46	13.45	12.88
kill time	22.32	16.54	6.56	16.99	26.12	9.47
Shopping Guide	36.52	11.33	7.89	3.25	8.65	3.55
interpersonal relationships	13.89	10.25	3.45	3.45	9.17	2.58
sports stars	33.56	38.74	9.65	21.36	31.45	7.64
author, host, narrator's style	5.23	16.87	3.36	7.25	7.87	8.68
referee's style	2.56	8.98	1.25	1.54	3.89	2.54
long habit	21.58	11.23	8.62	12.39	6.98	5.67
narcissism	17.63	12.65	4.63	7.68	12.34	3.58
passive participation	3.56	5.21	3.97	3.32	1.58	5.21
other	6.57	3.69	4.54	5.68	5.64	3.63

The result shows that for internet media, the shares of "to increase their sports knowledge", "for sports news", "Shopping Guide", "like sports stars" and "to seek solutions to the problem" are 43.82%, 52.41%, 36.52%, 33.56% and 26.85%. It shows the shares of "style of the author, host, narrator" and "the elegance of referees" are all lower than 10%, which means internet media can hardly satisfy those needs of college students.

For television, the choice of "for sports news" occupies 58.12%, and "like sports stars" occupies 38.74%. It shows TV media can satisfy those needs of college students.

For radio, the major choices are "for sports news", "for amusement", "personal interest" and so on. Other motivations are less chosen. It means those motivations can hardly satisfy those needs of college students.

For newspaper, the major choices are "for sports news", "to increase their sports knowledge", "for amusement", "personal interest" and "sports stars". The frequency shows that in the 15 motivations, the shares of "author, host, narrator's style" and "interpersonal relationships" are all lower than 10%, which means newspaper can hardly satisfy those needs of college students.

For magazine, the choices are "for sports news", "to increase their sports knowledge", "for amusement", "personal interest", "kill time", and "sports stars".

For books, the major choices are "to increase their sports knowledge" and "to seek solutions to the problem". Other motivations are less chosen. It means those motivations can hardly satisfy those needs of college students.

There are various needs of students in their growth process. When those needs points to media, they form some media expectations, which advocate college students to approach some media or some mass media educational information and guide them to process those information to some degree of satisfaction. [4]The survey shows that students contact the media actively.

To sum up, the main motivations of contacting media sports information are "for sports news", "to increase their sports knowledge", "for amusement", and "to seek solutions to the problem". The college students use those motivations to satisfy their needs of reality and entertainment. The survey shows that college students mainly use internet, television to satisfy their needs of acquiring information and knowledge and entertainment, use magazine and books to satisfy their needs of reality and entertainment, use newspaper to satisfy their needs of reality, and use radio to satisfy their needs of reality. It further illustrates that acquiring knowledge, learning and entertainment are the major motivation of contacting media sports information to the college students.

Table 2. Channels of College Students Accessing sports information

Media	Select Frequency	Percentage	Order
Internet	1609	72.61	1
television	1260	56.87	2
radio	1010	45.62	3
newspaper	832	37.56	4
magazines	583	26.31	5
teachers	480	21.68	6
books	351	15.87	7
friend	165	7.45	8
parents	129	5.82	9
other	102	4.60	10

B.   The Accessing Channels

In order to investigate the processing channels of acquiring information, besides of "TV", "Internet", "radio", "newspapers", "books", "magazines" and other mass media, we add four options such as "teachers", "friends", "parents" and "other" in the survey, in order to confirm the importance of mass media in the channels of acquiring information from the side. The result shows the channels are in this order: Internet, television, newspapers, radio, magazines, teachers, books, friends, parents and other. Before the rise of the network, the researcher had survey on the channels in the different periods. The result is that TV, newspapers and radio are the main source

(Boning Zhu, 1990). In our research, the frequency of internet is 1609, occupying 72.61%. It shows the network has gradually become the first choice for college students to obtain sports information after television, newspapers and radio. At the same time, the frequencies of teachers, friends, parents, other channels are lower, which proves mass media is the most important channel of acquiring sports information and knowledge. Among them, the electronic media (include Internet, television, radio) are more important (See table 2).

4   Conclusion and Suggestion

Mass media has important influence on college students. There are several motivations of contacting media sports information. Different media have the different motivations. Acquiring knowledge, learning and entertainment are the major motivation of contacting media sports information. And meet their reality and entertainment needs in the maximum extent.

Mass media is the most important channel of acquiring sports information and knowledge, among which the electronic media are more important. Internet has become the preferred way to sports information.

We suggest government and college strengthen the functions of guidance and education and develop appropriate measures for student's media.

References

1. Wang, L.: Investigation and research on undergraduates' information demand and information-used action in network environment. Information Science 20, 217–220 (2002)
2. Chang, C.: Mass communication: impact model, pp. 61–62. China Social Sciences Press, Beijing (2000)
3. Wang, H., Yan, W.: Analysis on the demand of sports information through mass media of college students. Journal of Chengdu Sport University 31, 40–43 (2005)
4. Lu, H.: Psychology of schooling, pp. 126–128. Northeastern University Press, Shenyang (2000)
5. Hong, Y.: Investigation and analysis on the present situation of media - sports information receive of the students in colleges of Jiangxi province. Journal of Yichun College 130, 150–153 (2008)

Study on the Fatigue Coefficient of World Elite Male Athletes in Swimming

Xinyu Qiu

Department of Physical Education
Hebei United University
Tangshan, China
qxy6801@sina.com

Abstract. By conducting statistic analysis of competition results of the top 20 elite male swimmers of each year, we have figured out the fatigue coefficient of different events with the same swimming stroke of the world elite male swimmers. Fatigue coefficients of 100m, 200m, 400m freestyle are 4.53s, 9.77s, 12.83s; when 200m butterfly is 11.75s, and the 200m backstroke is 9.51s; when 200m breaststroke is 10.02s, 400m individual medley is 15.23s. In addition, data analyses of typical cases are undertaken to provide reasonable quantitative indicators for the formulation of training, competition and goals of the national excellent male swimmers. Fatigue coefficient is universal. It applies not only to elite athletes in the world, but also to elite athletes in China.

Keywords: male, swimming, fatigue coefficient, goal, Prediction.

1 Introduction

Plenty of facts show that many world elite swimmers can gain outstanding performances when participating in various events of the same competition, and there exist a close relationship among results of different events with the same swimming stroke. By contrast, the national athletes are obviously far behind. By taking Fatigue coefficient of swimmers in different events with the same swimming stroke as entry point, the article conducts data statistics on scores of the world top 20 excellent male swimmers from 2004 to 2008 and then figures out fatigue coefficient of different events with the same swimming stroke. The aim is that provides a practical method and quantitative reference data for the coaches and athletes.

2 Research Object

Elite male swimmers whose competition scores rank top 20 each year from 2004 to 2008 in formal events of 50m, 100m, 200m, 400m freestyle, 100 m, 200 m butterfly, 100m, 200 m backstroke, 100 m, 200m breaststroke, and 200m, 400m individual medley. In the World Swimming Championship in 2009, competition results of

J. Luo (Ed.): Soft Computing in Information Communication Technology, AISC 161, pp. 211–215.

swimmers increase significantly due to extensive use of high-tech swimsuits, resulting in low comparability with previous data. In addition, FINA starts to restrict usage of high-tech swimsuits this year, therefore, the research does not put data of 2009 into application.

3   Research Methods

A.   Literature Material Method

Check each individual performance as well as related information of the top 20 world elite male swimmers from 2004 to 2008 on the official website of FINA; Look up 6 literatures regarding fatigue coefficient research to provide theoretical support for the article.

B.   Mathematical Statistics

Conduct statistic and summary on each individual performance of the world top 20 elite male swimmers from 2004 to 2008, calculate average score and general average score of them in every event each year and draw the average fatigue coefficient as well as the general one of different events with the same swimming stroke.

C.   Prediction

Predict scores and objectives of others events with the same swimming stroke by putting fatigue coefficient principle into application and according to performance of athlete so as to provide reference of quantitative indicators for regulating training actions of the athletes and instructors.

4   Results and Discussions

A.   Introduction and Analysis of Fatigue Coefficient

The conception of fatigue coefficient of swimmers in competitions, which is researched in the article, is introduced from that of runners in the running competition. Time difference exists between a runner's speed of running 100m and 200m, and more time difference exists between a runner's speed of running 200m and 400m. Time difference or delay between different events can be called as fatigue coefficient [1]. We can draw forth with similar theory that time difference exists between a swimmer's speed of swimming 100m and 200m, and time difference exists between a swimmer's speed of swimming 200m and 400m. As a result, time difference or delay of two different events with the same swimming stroke can be named as fatigue coefficient. Therefore, calculating equations of fatigue coefficient of swimmers in the competition are (in these equations, make "χ_1" stand for "100m swimming fatigue coefficient", "χ_2" stand for "200m swimming fatigue coefficient", "χ_3" stand for "400m swimming fatigue coefficient", "τ_1" stand for "50m time", "τ_2 stand for "100m time", "τ_3" stand for "200m time", "τ_4" stand for "400m time".):

$$\chi_1 = \tau_2 - \tau_1 \cdot 2 \tag{1}$$

$$\chi_2 = \tau_3 - \tau_2 \cdot 2 \tag{2}$$

$$\chi_3 = \tau_4 - \tau_3 \cdot 2 \tag{3}$$

In theory, the lower the fatigue coefficient, the better the performance. However, fatigue coefficient can only decease gradually with the increase of athletes' skills and there is no possibility that it will decrease to zero.

B.   Fatigue Coefficient of Some Events of the Elite Male Swimmers in the World

Through conducting statistics and summary on performance of athletes, draw the outcome as follows: the world elite male athletes' fatigue coefficients of 100m, 200m, 400m freestyle are 4.53 seconds, 9.77 seconds, 12.83 seconds; the fatigue coefficient of 200m butterfly is 11.75 seconds, and the 200m backstroke is 9.51 seconds; when 200m breaststroke is 10.02 seconds, 400m medley is 15.23 seconds (See table 1, table 2 and table 3).

C.   Validation and Forecast

Next, let's analyze and validate it with examples. American athlete Phelps won eight gold medals in Beijing in 2008. Among them, he got the gold medal in a time of 1:54.23 in the man's 200m Individual medley. If calculated by fatigue coefficient, his performance of 400m Individual medley should be 4:03.69. In fact, he won the champions in a time of 4:03.84. Two results are almost the same. It indicated that he brought the 200m butterfly races into play to an extreme level, and in fact, he won the championship overwhelmingly with a leading of half length of the body.

Table 1. Top 20 Athletes Average Performance and Fatigue Coefficient of Freestyle

	Time (Second)				Fatigue Coefficient (Second)		
	50m Freestyle	*100m Freestyle*	*200m Freestyle*	*400m Freestyle*	*100m Freestyle*	*200m Freestyle*	*400m Freestyle*
2004	22.17	48.93	107.52	228.02	4.59	9.66	12.98
2005	22.16	48.94	107.64	227.78	4.62	9.76	12.50
2006	22.21	48.95	107.31	227.65	4.53	9.41	13.03
2007	22.10	48.66	107.18	227.13	4.46	9.86	12.77
2008	21.69	47.86	105.91	224.67	4.48	10.19	12.85
average	22.07	48.67	107.11	227.05	4.53	9.77	12.83

Table 2. Top 20 Athletes Average Performance and Fatigue Coefficient of Butterfly and Backstroke

	Time (Second)				Fatigue Coefficient (Second)	
	100m Butterfly	*200m Butterfly*	*100m Backstroke*	*200m Backstroke*	*200m Butterfly*	*200m Backstroke*
2004	52.34	116.62	54.57	118.43	11.94	9.29
2005	52.55	117.09	54.55	118.67	11.99	9.57
2006	52.76	116.67	54.27	118.14	11.15	9.60
2007	52.02	115.83	54.06	117.45	11.79	9.33
2008	51.44	114.71	53.38	116.51	11.83	9.75
average	52.22	116.19	54.17	117.84	11.75	9.51

Table 3. Top 20 athletes average performance and fatigue coefficient of breaststroke and medley

	Time (Second)				Fatigue Coefficient (Second)	
	100m Breaststroke	*200m Breaststroke*	*200m Medley*	*400m Medley*	*200m Breaststroke*	*400m Medley*
2004	61.12	132.13	120.12	255.33	9.89	15.09
2005	60.94	132.56	120.58	256.27	10.68	15.11
2006	60.78	132.02	120.22	255.51	10.46	15.07
2007	60.85	131.57	119.42	254.43	9.87	15.59
2008	60.21	129.62	118.39	252.12	9.20	15.34
average	60.78	131.58	119.75	254.73	10.02	15.23

Kosuke Kitajima, a Japanese athlete, defended the honor of the Asian male athlete by winning the two gold medals of 100m and 200m breaststroke in Beijing Olympic Games. Among them, his 100m breaststroke achievements is 58.91 seconds. If calculated by fatigue coefficient, his score of 200m breaststroke should be 2:07.84, and his actual score when he won the championship in a time of 2:07.64, they are differing only about 0.2 seconds.

Bernard got the third place in the men's 50m freestyle performance with a score of 21.49 seconds in the 29th Olympic Games, and if calculated by fatigue coefficient, his score of 100m freestyle should be 47.51 seconds, but actually he won the gold medal in 47.20, it showed that he did not perform perfectly in his 50m freestyle races in some cases. Park Tae-hwan, a Korea athlete, won the men's 400m freestyle gold medal and the 200m freestyle silver medal in Beijing Olympic Games. His score of 200m freestyle was 1:44.85, if calculated by fatigue coefficient, his score of 400m freestyle should be 3:42.53, but he got the gold medal in 3:41.86 in the final of 400m freestyle, 0.67 seconds faster than the prediction, and basically in line with expectations. Zhang Lin, a rising star of China, won the silver medal of 400m freestyle in 3:42.44, it's a "breakthrough" because it's the first medal of Chinese men swimmers in the Olympic Games, and Zhang Lin himself become one of the excellent athletes in the world. If calculated by fatigue coefficient, his score of 200m freestyle can be 1:44.82, but his best score was only 1:46.13, and this proves that there is still great space for his improvement.

Peirsol, a American athlete, won the gold medal of 100m backstroke and the silver medal of 200m backstroke. His score of 100m backstroke was 52.54 seconds, calculated by fatigue coefficient, his score of 200m backstroke should be 1:54.59, but his actual score was 1:54.33, 0.26 seconds faster than the calculation. Arkady Vyatchanin, the winner of the two bronze medals of 100m backstroke and 200m backstroke in the 29th Olympic Games, got a score of 59.18 seconds in the 100m final, predicted by fatigue coefficient, which his score of 200m should be 2:05.87, and his actual score was 2:04.93. It shows that he played even better in the 200m backstroke.

Then we make a prediction on the men's swimming world record of some similar events by using the principles of the fatigue coefficient. The current world record of 50m freestyle is 20.91 seconds, calculating according to the formula above, the world record of 100m freestyle can be 46.35 seconds (20.91×2+4.53), 0.56s faster than the current world record of 46.91s. It showed that the world record of 100m freestyle also had a great room for growth. The current world record of 200m freestyle is 1:42.00, and 400m freestyle's is 3:40.83, making use of the principle of fatigue coefficient calculated from the 200m world record to the 400m, it is 3:36.83. It also shows that 400m world record had greatly improved significantly.

Through the study, we also found that Phelps and Kosuke Kitajima, as the representatives of the world excellent athletes mostly can exert normal level in the competition, even beyond the normal level. That a number of world records are rewritten is a clear example in the 29th Olympic Games.

5   Conclusion

The world elite male athletes' fatigue coefficients of 100m, 200m, 400m freestyle are 4.53 seconds, 9.77 seconds, 12.83 seconds; the fatigue coefficient of 200m butterfly is 11.75 seconds, and the 200m backstroke is 9.51 seconds; when 200m breaststroke is 10.02 seconds, 400m medley is 15.23 seconds.

Fatigue coefficient is universal. It applies not only to elite athletes in the world, but also to athletes in China.

Through the research of the fatigue coefficient in the competition of the world elite male swimming athletes, it has a practical value for the excellent male athletes in China of formulating their training target and improving their performance.

References

1. Cai, Z.-X.: Research on the fatigue coefficient in running and application. Track and Field, 7–9 (April 1994)
2. Shi, D.-M., Tong, Y.-H.: Research on the Fatigue Coefficient of World Excellent Male Athletes in Running. Journal of Xi'an Institute of Physical Education 19, 68–70 (2002)
3. http://www.fina.org/H2O/
   index.php?option=com_wrapper&view=wrapper&Itemid=634/

To Improve College Teachers Multimedia Software Design Capability Theory of Basic Research

Wangdengcai and Hankun

Mathematics and the Information Technology Institute,
Hebei Normal University of Science & Technology, Qinhuangdao, China
wangdengcai@163.com, hankunkun@126.com

Abstract. Informatization teaching methods are widely applied in colleges and universities, especially with the growing popularity of multimedia technology of software. Not only can active teaching atmosphere, expand the student's thinking, improve teaching efficiency, but also strengthens the teaching effect and improve the teaching efficiency. Now because of many multimedia software designers, resulting in lack of theoretical support, Multimedia information design does not reach the Ideal target, the side-effect is the deviation of teaching theory and practice. All kinds of micro teaching and learning strategies are not fully effective in multi-software also. Therefore, the role of theory on practice is obvious.

Keywords: Multimedia software, design capability, universities, theory, research.

1 Introduction

Multimedia teaching software is mainly in the direction of the theory, the design that meets the needs of teaching program, including the teaching content and structure design, the navigation strategy and interface design work, and it continuously revised in the teaching practice, can make the development of multimedia teaching software Accord with teaching needs and obtain satisfactory teaching effect.

From the teaching software development process point (see chart 1), theoretical support has always been there from beginning to end.

1.1 To Improve College Teachers' Teaching Software Development in Topic Ability

When college teachers select the topic in making multimedia teaching software, it must be reflected the multimedia technology advantage, maximum full play advantage of multimedia technology. Through questionnaires of college and university teachers, result see table 1:

Table 1. The topic about teaching effect whether have influence or not

options	yes	no
proportion	66%	34%

J. Luo (Ed.): Soft Computing in Information Communication Technology, AISC 161, pp. 217–221.
springerlink.com
© Springer-Verlag Berlin Heidelberg 2012

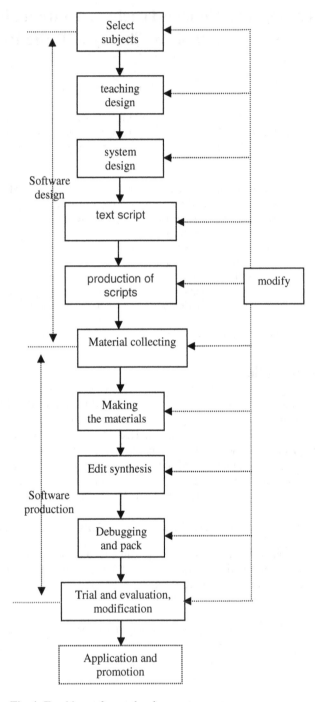

Fig. 1. Teaching software development process

Many people believe there are many effects In the choice.But, a few problems that not allow to ignore also discovered in investigation, have certain catholicity because of the lack of theoretical basis.

1.1.1 Technological Advantages
Traditional way can not reach the teaching aim, and the latest way give full play to the advantages of multimedia technology, Through the technical support so that students understand and apply their knowledge in a short time.

1.1.2 Individualized Teaching
Lesson in school is limited, the teaching contents cannot fully be extended, according to the characteristics of multimedia software,makeing a teaching software that can provide individualized learning and feedback, which can act as tutor role

1.1.3 To Improve Efficiency, the Effect and Benefit
Based on a long working time, or experimental operation in the high cost of teaching contents, make full use of multimedia technology advantage, build a virtual real environment, reduce in the cost, improve efficiency and effect.

1.2 To Improve College Teachers' Teaching Software the Development of Teaching Design Capabilities

1.2.1 System Method
The system of basic method based on understanding system, the application system science, systematic thinking, system theory, system engineering and system analysis and other methods, instructs people to deal with scientific and technological problems of a scientific method.

1.2.2 Psychology
Psychology is human and animal psychological phenomenon occurrence, development and for the activities of the law a science. Through the research of learning objects psychology, can guarantee design colleg, their aptitude.

1.2.3 Pedagogy
Pedagogy based on education phenomena, education problem is summarized as research object, the education activity of human science theories and practice, exploring solutions education activity generation, development encountered in the process of practical education problem, thus revealing general education law of the social science. Through the education theory guidance, can maximize to solve the problem of the integration about information technology with curriculum.

1.3 To Improve College Teachers' Teaching Software Development on College Teachers" System Design Ability

System design, mainly includes structure design, cover design, interface design, interaction design and navigation strategy design, etc.

It conforms to content need In structural design, the cover, the title page, the content page and the Last Page Present appropriate information

In interactive design, according to the need of expressing knowledge, using different control modes.

1.4   The Writing of Script

1.4.1   Text Script

Text script order based on teaching content, According to the content of the need to apply the proper media support, And for the production of scripts to lay the foundation. Text script is generally witten by the Teaching master-hand.

1.4.2   Production of Scripts

Production of scripts can display software production structure, including the corresponding teaching function, and give software developers theoretical basis. production of scripts are usually written by Computer professionals design, Production of scripts can meet the need of school teachers.

1.5   Media Material Choice

Media choices, to be in teaching contents as the center, improve the teaching effect and efficiency as the purpose, the biggest may improve teaching efficiency. About media choices survey distribution in table 2:

Table 2. Media choices distribution

Media choices	effect	efficiency	benefit
proportion	85.2%	95.1%	60.5%

From the results of the survey, teachers can choose media with content as the center, improve teaching efficiency and efficiency, but do not pay enough attention to improve benefit.

1.6   Collection, Editors and Synthesis

Theoretical basis can be used in software, you can search through search engines, or repository. take appropriate tools to get, modify and produce, and use better tools in synthesis. Thus, for material collecting tools , machining tools and the characteristics of integration tools must be familiar with, so that it can be satisfy with the content demand.

1.7   Debugging, Evaluation, Modification and Extension

The software integrates later, does not represent you have just finished the work. If you find that software still cannot meet the requirements in a certain field, must be debugged to continuously. Commissioning is finished, let students and teachers use through trial, if error is discovered according to the feedback, modify perfect and achieves the aim of promotion.

2   Epilogue

To sum up, the software design theory of multimedia teaching is a more categories of knowledge in system science. To a breakthrough in theory, not only in the teaching theory, learning theory, psychology and pedagogy,but also in the application software principle thorough digging deep enough, only in this way, can take the teaching goal based on technique principle can be realized, achieve all-round improvement.

References

1. Wangqian, Zhangbaolong: Talk about the multimedia teaching the requirements for teachers. Science & Technology Information (25) (2010)
2. Chenhua: The multimedia teaching effect and the insufficiency. Journal of Guangxi University (Natural Science Edition) (s1) (2007)
3. Zhengxunan, Hemingge: University of multimedia teaching empirical research. China University Teaching (11) (2008)
4. Wangjuan: University of factor analysis and suggestions about multimedia teaching effect influence. E-education Research (5) (2009)
5. Jinyan: Multimedia courseware for teaching quality and teaching effect factors probing. E-education Research (5) (2007)
6. Niexiaoyi, Zhangjun: Multimedia courseware 5/10/20/30/40 laws. China Education Info. (22) (2008)

The Study on the Application of Sports Dance in Golf Technical Teaching

Li-Jiquan and Qin Li

The Department of Physical Education, Hunan International Economics College,
Changsha, China
403399321@qq.com, qinli19820117@163.com

Abstract. As a new rising sports program, the study on methods and means of golf technical teaching is insufficient. In order to make beginners master the key points of golf technique, this paper points out the concept of introducing sports dance into golf technical teaching. It is believed that permeating sports dance into every link in the process of golf technical teaching can help students establishing the concept of core strength, perceive proper shift of center of gravity, and learn to control the rhythm of movement, so as to stimulate students' interest in learning. The results of teaching experiment that using this teaching method demonstrate that the average score of special skill assessment for students in experimental group is obviously superior to the matched group, and this difference is highly distinct ($P < 0.01$), so it indicates that this teaching methodology has higher value of application and popularity.

Keywords: golf, sports dance, technical teaching, teaching experiment.

1 Introduction

At present, because schools which have opened the project of the golf technical teaching are less and the study on teaching methods for golf techniques is seriously sufficient. In the process of teaching, the method of teaching and explaining-demonstrating, afterwards with a lot of repetitive practice is generally adopted, which is not only helpless to build up the concept of movement for students as soon as possible, but also frustrates the study interest for them, easy to make students come into being bored. We have tried to introduce sports dance into the golf teaching as the teaching method of auxiliary practice, achieving relatively good effect. Sports dance is a unique form of movement, with the integration of sound, body and beauty as a whole. The penetration of sports dance not only has changed traditionally bored teaching and aroused students' interest by ways of graceful movement, vivid rhythm and joyful music, but it also has trained students' ability of body coordination and control, deepened students' understanding and application for golf technical movements and promoted the improvement of the effect on technical teaching.

J. Luo (Ed.): Soft Computing in Information Communication Technology, AISC 161, pp. 223–227.
springerlink.com   © Springer-Verlag Berlin Heidelberg 2012

2   Teaching Experiment

In order to test and verify the effect of sports dance on golf technical teaching as an auxiliary practice, it has been designed and implemented teaching experiment with the design method, implementation process and experimental result as following.

A.   Research Object

Select at random two classes with all boys of 2009 golf social sports (golf-oriented) in sports department of Hunan International Economics University, every class with 16 boys respectively as the experimental group and matched group. Before experimental teaching, the boys for study have no contact with golf sports without any experience of golf skills training.

B.   The Design of Teaching Experiment and Implementation Control

Experiment time: 2010 March—June, totally 72 class hours. Adopt the method of teaching experiment contrast and implement teaching experiment of golf swing techniques for the students in the two groups according to the teaching process. The students in the two groups are taught and evaluated by the same teacher. The experimental group adopts the teaching pattern with the combination of sports dance teaching and normal teaching, namely giving sports dance teaching before every technical class. Matched group adopts purely normal teaching pattern. After the experiment, the students in the two groups will be given NO.7 iron club swing driving skill assessment (short for NO.7 iron skill assessment). There are three standards of movement, distance and degree of accuracy in the skill assessment. The teacher gives the score of the skill assessment in hundred-mark system according to the comprehensive performance of the experimental object on the basis of these three standards. Then the comparison is given between the two groups for the data result after the experiment. The class hours for the experimental group with sports dance teaching account for 30% in the whole class hours.

C.   Result and Analysis

1)   Experimental Result

Table 1. Chart 1 Mark table for air shot swing skill assessment after experiment

project	group	number	Score assessment		homogeneity test of variances		T	P
			$\bar{X}$	S	F	P		
NO.7 iron skill assessment	Experimental group	16	8.27	0.67	0.004	0.950	4.92	<0.01
	Matched group	16	7.08	0.70				

It can be judged from the chart 1: (1) the homogeneity of variance between the experimental group and matched group in the NO.7 iron skill assessment. (2) after experiment, the average score of the NO.7 iron skill assessment for the students in the experimental group is obviously higher than that of the matched group, and the difference between them is distinct (P<0.01).

2) The Result Analysis

The experimental result shows that introducing sports dance into golf skill training as auxiliary practice is useful to improve students' ability in the air shot swing skill. This method can be considered into popularization.

After deep analysis, it has been found that sports dance as auxiliary practice can play the following roles for the students during the practice teaching in the process of golf technical teaching.

a) Build Up the Concept of Core Strength

In the 1990s, the core strength training was widely used in foreign countries. In the early years in the 21st century, this concept has been gradually accepted, recognized and highly praised by competitive sports training field in our country. "From the perspective of anthropotomy, the core for human being refers to the location that spinal column, hip joint, pelvis place, and they rightly place in the binding site of upper and lower limbs of human body, playing connecting pivotal role in sports." [1] Core stable strength plays significant role nearly in all the competitive sports items, also including golf sport.

From the perspective of technical decomposition, firstly, when the time is for golf swing to back swing, after finishing the shift of left shoulder, "the center of gravity should be shifted into the right hip joint as the axis for the purpose of turning body later." [2] Direct viewing speaking, the technical movement at this moment is that pelvis moves around vertical axis with clockwise rotation. In the second place, the "whiplash" effect formed in the process of "downswing-release-driving" series of movements needs the participation of core and stable strength. It is to form the near-end fix and improve the strength of the tail-end muscle in this process by the body "beyond instrument", accurately and quickly transfer the strength of the lower limbs and trunk muscle into the upper limbs, combining the whole strength in the body to form the whiplash effect. The technical movement at this time is that pelvis turns around vertical axis with counterclockwise rotation.

Besides, on one hand, reasonably applying core strength can build excellent supporting point for agonistic muscle to send strength, improve the strength from different muscle and link for orderly sports, achieving the purpose of maintaining the stability and balance for the center of the gravity; on the other hand, it can reduce the burden on irrelevant muscle and joint, and prevent the occurrence of acute scathing, achieving the purpose of taking precaution of scathing.

Similarly, the accomplishment of many movements in sports dance rightly needs the participation of core parts. A typical movement in Latin dance is the reverse of the word of "∞", namely the transverse circular movement that hip joint does the circumduction of flexion and extension, adduction and abduction, turning inward and outward three groups movement. In Latin dance, "the movement of waist and abdomen plays leading role in the law of motion for crotch." [3] In addition, "rumba,

ChaCha, samba and other dance in Latin dance require two feet to seize the ground, the performance of body language is transmitted to feet, ankle, knee, waist, shoulder and arm through the floor, forming a chained dynamic system such as core stability requires." [4] While this is consistent with the requirement of golf technical movement.

b)   Comprehending Proper Shift of the Center of Gravity

The shift of the center of the gravity is the exchange and change of supporting point. In the golf technical teaching, the center of the gravity is an extremely significant technical link, and its shift is convenient for the movement of hip, attaining the max club head speed and hitting distance. The shift process of the center of the gravity is as follows: from the warming-up movement to the top of swing, 80% gravity for body should fall over the right hip joint and right leg. From downswing and golf swing to swing control, the center of gravity should be shifted into the left side of body. Golf experts once pointed out "the shift of the center of gravity in golf is also a subtle process, similar to the shift during dance." [2][54] Moreover, "excellent golf swing must go with the movement of hip twist." [2][41]

The most important link in sports dance is hip twist and the shift of center of gravity. "No matter it is ballroom dance or Latin dance, it is the key to hold the center of gravity for dance." [5] Only the proper shift of center of gravity in place can show graceful dancing posture. In the process of golf technical teaching, teachers can teach students to slowly study, experience and master the shift of the center of gravity, making them feel nature and steady shift process of center of gravity through the teaching of individual movement, or the combination of movements, and the organization of practice.

c)   Comprehending the Control of Movement Rhythm

In sports, rhythm is a kind of regular and complete sports form which should go ahead continuously. The rhythm of golf swing in golf sports namely is the time scale for different part of movement in the process of golf swing. Some scholars point out "the correct rhythm of golf swing should be like this: back swing in no hurry, with transient stop in the summit of backswing, smoothly transfer to level of downswing and smoothly accelerate during down swing, then swing and exert the strength at the bottom of the curve. The club head speed and strength must come to the summit at the instant moment of driving by swing and hit in such rhythm." [6] The rhythm of swing is the key to play good balls continuously. As far as the beginners are concerned, it is easy to form wrong movement patterns in the process of the study of technical due to lack of understanding about technical knowledge or the transfer disturbance of other sports technical. Most practicers blindly accelerate the speed of rotation from pelvis to speed up the velocity of club head, completely without any control to the rhythm. Therefore, it is hard to teach the requirement on rhythm by normal teaching and let the beginners quickly master in a short time.

Experts point out: "The rhythm for golf swing is controlled by hip. The rotation speed of hip decides the movement speed of arms and club head during swing time." [2][43] In Latin dance, the rhythm is mainly expressed by the twist of waist an d hip. Rhythmicity is not only the basic element for dance art, but also the basic art characteristic for dance. The rhythm of sports dance is extraordinarily outstanding, and every dance has its own style and characteristic. Consequently, in the golf technical teaching, properly adding sports dance with music accompaniment as auxiliary exercise can improve students perceiving and understanding about the rhythm of the movement of waist, pelvis and hip joint, which can be directly perceived by students' visual sense, also directly perceived by their auditory sense, so as to better control the rotation speed of core parts, and better control the rhythm of swing.

d) Stimulating Students' Interest in Study

Vivacious, brisk and full of times dynamic sports dance is a civilized, elegant and new rising sports event program. Sports dance also have strong participation and it is easy to mobilize the initiative and activity for students. It can transfer dull, tense and boring traditional class into active, vivid and relaxed class, not only mobilizing the learning initiative for students and enliven the class atmosphere, but also enrich and replenish the teaching content, better promoting the task accomplishment of sports teaching. In this aspect, it has obvious effect on the new rising golf technical teaching with single technical movement, lagged teaching method and study.

3   Conclusion

A. The golf technical teaching in experimental group effectively makes good use of sports dance as auxiliary teaching. Students can quickly master technical movement with vivid rhythm, fluent shift of center of gravity and excellent effect.

B. Permeating sports dance into every link in the process of golf technical teaching can help students to build up the concept of core strength, perceive proper shift of center of gravity, learn control the rhythm of movement, and also stimulate students' interest in study.

Acknowledgment. Hunan Province Sport Science Society Youth Excellent Research Project in 2010, No: KT10-57.

References

1. Huang, J., Zhao, S.: The History of the Application of the Core Power Training in the Practice of Competitive Sports in China. Journal of Physical Education (5), 74–76 (2010)
2. Mc Adams, Thomassy, T.J. (eds.), Li, C., Zhang, J., Huo, J., Gao, Z. (trans.) The encyclopedia of Golf, vol. 44. Beijing Sport University Press, Beijing (2008)
3. Tian, L.: On Latin Dance Teaching. Journal of Hebei Institute of Physical Education 17(3), 33–34 (2003)
4. Gao, L.: The Meaning of Core Strength Training to Latin Dance Training. China Education Innovation Herald (31), 200–202 (2009)
5. Lili, K.: Discussion on the Basic Teaching of Latin Dance in Sports Dance. Guizhou Sports Science and Technology (3), 77–78 (2004)
6. Geng, Y.: The Theory of Golf Swing, vol. 24. Beijing Sport University Press, Beijing (2009)

Garments Enterprise Pattern Analysis Used B2B2C

Zhengjia Rui

Institute of Art and Fashion, Tianjin Polytechnic University, Tianjin, China
oojiajia_2008@126.com

Abstract. In the global economy under the pressure of the environment since 2009, stagnated the overall apparel industry, and commerce have emerged satisfactory trend, the growth rate is gratifying. In this case, the traditional garment enterprises began to explore new intensive business model. In fact, many companies have plans to enter the traditional clothing of fine details of e-commerce are among the manipulator. B2B2C model of e-commerce industry's concerns caused, many experts believe that this is China's economic system mode, more in line with national conditions, are more easily profitable model. Experts predict that by 2011 many of the traditional garment enterprises will complete their e-commerce services, the number of burst into the network marketing year, China's apparel online shopping is expected to deal size will be more than 70 billion Yuan.

Keywords: B2B2C, Garment Operations, e-commerce.

1 Introduction

There is the high cost of the apparel industry channels, channel-level and more disadvantages of gross margin. And rely on the Internet e-commerce, because of its simplification of trade procedures, can transcend time and space, low cost and profit advantages, as the new century, new growth economies the trump card in trade form. Promote the use of e-commerce enterprises to upgrade traditional clothing, is not satisfied with simply more and more clothing retail business, asked the question raised on their own. Business system learn from the mature platform, relying on the Chinese garment industry background, give full play to industry influence or the brand power of influence and integration of resources for its brand to provide a second sales space, but also give consumers unlimited shopping fun.

However, over the past few years, the development of e-commerce, such sales space is not perfect, for example, sales of single-brand, quality and after sales has been criticized, and even some consumers have abandoned the market return to the traditional network market. Forerunner of a large number of e-commerce market through various channels to spare no effort to network the knowledge of the concept of consumer culture, user experience, service and consumption habits of the guide, as well as their perfect after-sales service, selling new products to promote and reduce online shopping The difference with the field materials and other methods to attract customers' attention. Compared to the past few years, and now has rapidly growing area of electronic commerce, began tipping by Internet users of all ages from

J. Luo (Ed.): Soft Computing in Information Communication Technology, AISC 161, pp. 229–233.

favorable. In the end what mode of these pioneers created a network of transactions that resulted in garment enterprises have stepped into these field giants Nuggets?

B2B2C model of e-commerce industry's concerns caused, many experts believe that this is China's economic system mode, more in line with national conditions, are more easily profitable model. Especially in terms of apparel e-commerce services.

2   Interpretation of the Meaning of B2B2C Model

B2B2C model concept: The first B refers to the raw material providers and product manufacturers; C refers to the end customers are also consumers; second B refers to the vendors. B2B2C aim is to build a clothing business e-commerce platform with the payment mechanism so that enterprises can sell their products through the network out of the store. Before then there had been B to C, C to C, but in fact, become increasingly blurred boundaries between the two, and is being or has been turned into a B2B2C model. B2B2C integration occurred before the B to B, B to C model works as follows: to B to C-based, B to B as the focus, these two individuals effectively link up business processes. With the market mechanism because this model: 1, retail and distribution businesses determine the characteristic of work is very heavy; 2, end-user is not willing to pay a lower amount of goods relatively high distribution costs. Faced with these practical obstacles, B to C mode to introduce B to B model of natural, that is downstream distributors as a sales channel to B to B, B to C integrated into B to B to C e-commerce model, that B2B2C. This pattern of distribution to reduce pressure on the one hand, on the other hand can also change the burden of inventory and reduce costs. Online shopping reflects the rapid, cheap and other advantages.

B2B2C is the core of the force: The second B vendor revenue by providing quality service for customers is convenient to buy in terms of their satisfaction and purchase of goods brought about by the process of shopping, look for sellers that Is the rapid sale of goods. In particular, vendors can create distinctive B clothing virtual community, for many customers to provide a platform for exchange and communication, through the popularity gather to form a large customer base, which is B2B2C that an important condition for sustainable development Moreover, Apparel buyers, especially women have a high desire to share the experience of wearing, people willing to explore in this virtual community experience buy clothes, to share views on a certain brand. On this basis, the middle of the B to the customer will have a full understanding of, and know how to improve their viscosity to attract more customers, the producer of such resources is needed clothing, so the middle of the B This information is provided to the first B, and for their product design, production, marketing and other areas to provide services, to build a clothing manufacturer and a bridge between the end customer, it is for manufacturers and sales growth Increased profits for the customer this means in terms of convenient shopping and fun. Of course, this bridge can play a role in B asked to provide a special service, through the establishment of a powerful information system, the front end and back-end clothing manufacturer organically connected customers to achieve effective integration purposes.

3   The Funtions and Characteristics of the B2B2C Model

A.   B2B2C Model and the Combination of Traditional Industries

Apparel producers can have the conditions, if the middlemen to sell their products, and will not affect their traditional way of selling. B2B2C mode in all participants will receive a huge harvest, the vested interests of all participants will not be damaged, so the business model of traditional business models will be praised hands.

B.   Middlemen B Does Not Need to Build Their Own Storage Centers and Distribution Channels

Not the same as in China and Europe, America and Europe the level of infrastructure than non-Chinese, and its transportation, postal, express, electronic settlement is not progress in a short time we can catch up. The development of 21st century trends in the apparel industry is "more variety, small batch, high quality, fast delivery ", while apparel products and fashion, and seasonal characteristics, specific to the enterprise, requires the reproduction process, sales process must be a high degree of automation and rapid response capability. B2B2C model fully into account China's current situation and the reality of e-commerce, set all the force, to play their respective advantages and garment enterprises and strength of his attention on improving their overall strength.

C.   Mode of Middlemen B Requires Focus on E-Commerce Platform Construction

A certain extent, determines the size of the customer base the size of middlemen profits, and customer base the size of the quality of services provided by them to decide. Therefore, brokers should focus on improving the quality of services provided.

D.   B2B2C Model Based on the Information Integration between Supply and Demand

It can make the garment immediately get B from the broker customer needs, timely guidance to garment manufacturers to adjust product structure to meet their needs, not only reduce inventory and greatly reduced costs. According to need and can follow the customer's own design and production of clothing unique personality, these are a complete information system has been implemented.

E.   B2B2C Model Integrates Upstream and Downstream Channels and Resources

Sales for the clothing producers a platform for customers to buy clothing to provide convenient conditions and benefits, and all have been organically integrated, sustained and rapid operation of the whole system, the formation of a pioneering market impact and meet the market Inclusive.

4   Apparel E-Commerce Business Condition of the Realization of B2B2C

All aspects of corporate and social soft and hard conditions are ripe, the need to find professionals to create websites. In order to improve network station visibility, the company issued in addition to the publicity pictures prominently published Web site, is also a larger issue than the well-known newspapers and magazines for print ads; stationed in the famous search engine YAHOO and GOOGLE. According to the website to collect the users a message, suggestions, ideas, improve marketing. Clothing enterprises to start business: B to C and retailers expand business. Website shortly after the normal operation, the companies then will the development and cooperation among retailers B to C business channel, between the traditional business and retailers moved to the Internet, but there will be a lot of the burden of inventory and consumers B to C to start the business. Because direct access to customer needs, and quickly carry out related business, but may compete with the retailing of friction.

The B2B2C model allows end users from the corporate Web site or order through retail stores, website summary orders received information passed to the corporate headquarters, headquarters with the producers and logistics and distribution management system for online between agencies of information processing, By the logistics and distribution sector from the consumer goods distribution to the most recent or consumer specific stores or home delivery to the consumer payment delivery.

5   B2B2C Business Process Analysis

A.   The Operation of the Core E-Commerce B2B2C Clothing - Information Flow

In the 21 century, customer resource has become the most important resource. How to communicate effectively with customers to obtain the customer information has become one of the important tasks. E-marketing activities with respect to the maximum advantage of traditional marketing activities in e-commerce environment, companies by means of modern information network technology, making the flow of information flow has become more open. B2B2C model only in the smooth flow of information to be able to under the front of garment manufacturers and customers effectively connected back up, as their communication.

B.   The Operation of E-Commerce Security B2B2C Clothing - Logistics

B2B2C model greatly simplifies the business processes, reduce operating costs of enterprises. The advantage of this cost to establish and maintain reliable and efficient logistics operation to ensure that the competition of modern enterprises in the key to victory. When consumers order online, the only complete ownership of the delivery of goods, trading activity was not completed, only through the logistics and distribution, will be the real goods to consumers, trading activity before it was to end.

C. Apparel E-Commerce Operations B2B2C Means - Cash Flow

As one of the three streams flow of funds is to achieve the integration of e-commerce an indispensable means of marketing channels. In network marketing, the bank is connected to the production enterprises, commercial enterprises and consumers' link plays a vital role, banks are able to effectively achieve B2B2C electronic payments have become an important part of model.

6   The Prospects of E-Commerce Model of B2B2C

With the era of network economy, network technology, computer technology and communication technology on the development of traditional enterprise enormous impact and influence, traditional companies seek to enter e-commerce, is becoming a worldwide phenomenon. Traditional enterprises is an important pillar of China's economic development, the future will remain so for a considerable period. As a production and management entities, our traditional business in its long-term production and management, with a wide range of partners, including the wide range of suppliers, agents, transport operators, service departments, the use of network resources and e-commerce enterprises to restructure way can make the traditional business and the formation of social network partners, sharing of resources. Moreover, in this network, co-operation among partners, mutual benefit and common development market, expanding the content and the outer edge of individual enterprises, the formation of networked virtual enterprises, and business flow, logistics, capital flow, information flow integration, Completed the individual alone can not complete their own business, so that all enterprises can concentrate on strengthening its core competencies, improve management standards and social prestige, to enhance the overall value of traditional enterprise and competitiveness, but also a tradition to the future development of enterprises Not take initiative, and build competitive advantages.

However, for a particular enterprise, we must actively explore business models suited to their characteristics, so as to achieve the purpose of profit. But anyway, must be adhered to the "customer-centric" principle, in order to grasp market trends and make positive response to the greatest extent possible to meet customer needs and improve the competitiveness of enterprises so that enterprises in the fierce competition in the market Always invincible.

References

1. Zhou, J.: B2B2C - e-commerce businesses to make a profit. Taiyuan City Vocational College of Technology (1) (2006)
2. Duan, Y.P.: B2B2C platform for future models. E-commerce World (5) (2005)

A Brief Analysis of Present Status and Development of Digital Corporate Image Design

Luping Lu

Institute of Art and Fashion
Tianjin Polytechnic University
Tianjin, China
lulup1234@126.com

Abstract. This paper puts emphasis on discussing the expansion of corporate image design in the background of the rapid development of digital technology, and looks ahead to the future development of diversification through analysing its important role in the development of modern enterprises.

Keywords: corporate image design, digitalization, network communication.

1 The Concept of Digital Corporate Image

Digital corporate image takes advantage of the basis, mode and technologies of digitalization to conduct the establishment and the management of creating, publicizing and popularizing corporate image. It came into being on the basis of popularization and application of computer and the Internet.

Digital corporate image follows the fundamental principles of traditional corporate image and plays an important part in serving enterprise development. It differs from traditional corporate image in terms of application fields, communication modes, design methods and etc. Traditional CI lacks a whole set of methods for such jobs as issuing programs, utilizing training and popularizing and communicating. Yet the handbook of CI is nothing but a set of standard template, which is lacking in the textbook function of combining learning and training together. Thus, this makes it difficult to put CI programs into implementation that enterprises need to spend a huge number of manpower, material resources and time in image educating.

Traditional concepts and means of CI popularization have became completely outdated and inefficient under circumstances of constantly changing new technologies. Digital technologies bring about the diversifying expansion of corporate image design language, make corporate image design more humanized, promote the digital process of corporate image and enrich corporate image design in the digital era.

2 Characteristics of Digital Corporate Image

2.1 Properties of Interaction and Participation

During its design process, digital corporate image pays attention to interactions and participations of audiences. The integrity of communication theme relies on experiences of audiences.

J. Luo (Ed.): Soft Computing in Information Communication Technology, AISC 161, pp. 235–239.

The Internet communication is a kind of communication based on interactions with a omnibearing space frame. Digital corporate image communicates through network. Thus, in fact, people's participation leads to their interactions with network media, interactions between participants and designers, as well as those between one participant and another... Interaction is not only an existing mode, but also a communication mode. Take the cooperation between Super Girl (a singing competition) and Baidu (a network search engine) for instance. Interactive marketing, being conducted between traditional media and on-line community, provides a practical way for the development of new media. They clear away the boundary lying between audiences and media. Now users are able to obtain media contents interactively instead of unidirectionally, or even able to create contents.

Digital corporate image exists everywhere during processes of design scheming, introducing and implementing of corporate image, as well as popularizing and appling. This kind of interaction provides guidance for people on participating. Besides, designers give partial controlling power of design to audiences, who can choose intentionally to see it or not, and think about consciously what to see and how to see. Their behaviours, like choices, participating modes and degrees, form a series of data which will be gathered and fed back through the network and then be processed and integrated by designers and related personnel, so as to shape a mode which is universally accepted. In this process, a participant's status of being an audience is changing. A number of participants, along with designers, accomplish the whole design process, so to speak.

2.2   Omnibearing Communications

Traditional corporate image presents audiences with industrial properties, cultures, quality foundations and value orientations of enterprises by means of visual communication, while digital corporate image realizes its communication in all respects through sight, sound, touch, taste and scent. This comprehensive image representation expresses and communicates more precisely, condenses the spiritual essence of enterprise by rendering atmosphere without any deformation, and reduces the number of misreading in communication.

The hyperlink of network breaks the linear receiving mode of traditional corporate image, so that static plane and dynamic display of plane are endowed with properties of space level and three-dimensions. The network browser will download the assigned content in a web page automatically, when a user clicks on a hyperlink. Characteristic of quick jump and conversion enables users to get their own designated theme easily.

Traditional corporate image is static, while digital corporate image focuses more upon dynamic representation and atmosphere rendering, into which it incorporates performance and plot. Owing to the technical support of application software, both dynamic and static images have the opportunity of coexisting and relying on each other. Therefore, digital corporate image is endowed with both dynamic and static beauty.

With the maturity and development of streaming media playing technology, the joining of sound element enables traditional corporate image to make a significant leap to digital corporate image. It is an upgrading function of digital corporate image that changing traditional corporate image from visual communication to audio-visual communication. This function breaks the the situation that visual sense rules the whole land of corporate image. Sound is not only a constituent element that can form an overall identification of corporate image together with other elements, but also an independently existed element. Because it's a symbolic identification in itself. This sort of sound can be a special piece of music or just a segment of particular sound effect (usually used in the network, such as the sound of firecrackers and impact sound).

2.3   Interdisciplinary Integration

Diversified visual ideas imply that new modes of visual communication will break through boundaries of traditional design categories and design will be a carrier which is able to integrate a variety of disciplines. Categories of knowledge required for the complexity and the comprehensiveness of digital corporate image design have far exceeded the controlling and managing range, to which a designer's capability can reach. Thus, it is necessary to have a team, an interdisciplinary new team of design.

Design team of traditional corporate image is composed mainly of graphic designers. Yet the team of digital corporate image consists of not only graphic designers, but also personnel of multimedia design, network maintenance and three-dimensional design. They should work together to accomplish corporate image design.

Interdisciplinary cooperation or integration of traditional design categories, firstly, realizes the multimedia image with audio-visual effect technically. It is possible to add sound elements to corporate image to produce sound effect by utilizing later composition softwares, multimedia design softwares and network application softwares. People's cognition degree will be increased and finally become audio recognition of corporate image. It's able to simulate various tactile sensations and effects by taking advantage of the function of filter in planar design software, as well as material editing and rendering in three-dimensional animation software. It is also able to create virtual three-dimensional objects by 3d animation software to form 360°panoramic displays of their images.

The final purpose of interdisciplinary cooperation of design is to form integration of not only thoughts and methods in the design field, but also spiritual cultures. Creative thinking aroused by contradictions and confrontations among different disciplines expands people's field of view and train of thought. New art forms are coming into being gradually in interdisciplinary conflicts and surges.

3   Thinking on Values of Digital Corporate Image Design

International development is now an irresistible trend for enterprises. Digital corporate image took shape in the infinitely connected network, not being confined to time and space. Undoubtedly, the application of digital corporate image is an effective measure to meet the requirement of global competition, as well as the declaration of an enterprise for marching toward internationalization. Digital corporate image helps enterprises in realizing their dreams of diversification and internationalization. Strategies of diversification are not only important manifestations of brand extension, but also active tactics of confronting problems like increasingly fierce market competition and the diversification of consumer demand. With the diversified development of business fields of enterprises, traditional corporate image becomes insufficient in containing existing businesses and future strategic extensions, and also in emboding the whole contents and values of a brand. However, digital corporate image is able to relieve this kind of contradiction and pressure.

The real intention of corporate image is to establish an unique identification that differs from other enterprises. It is no more embarrassing than being identical for corporate images, which calls the significance of corporate image into question. Digital corporate image is a kind of three-dimensional and multiple identification. On the one hand it retains the spiritual temperament of enterprises of being independent and complete, and on the other, it avoids the embarrassment of plagiarizing, misreading, and being identical.

The network is a medium with a relatively high performance to price ratio, on which properties of digital corporate image rely. Therefore, network advantages can be taken to the utmost extent, so as to get a larger revenue with less investment. The huge potential of network has been shown. The earlier to start to use image adapted to network communication, the more possible to take the lead in finding potential consumer group. In this respect, digital corporate image and advertisement play a similar role. It is inevitable that an enterprise will be confronted with setbacks. However, the network opens a window for enterprises which have difficulties in taking a step. Enterprises with their arms opening widely will get the whole world, yet those being conservative and complacent are tantamount to alienating themselves from the tide, as well as the market.

Science and technology are the primary productive forces. Digital corporate image comes into being by the catalysis of digital information technology, which is consistent with the law of historical development and is also the result of productive forces development acting on the realm of ideology. The development of digital corporate image will optimize enterprises in terms of efficiency, communication, organization and etc. This optimization and high efficiency that bring manpower into full play will definitely lead to the emancipation and the development of productive forces.

4   Development Trends of Digital Corporate Image Design

Currently, digital corporate image can be only carried out in certain enterprises which have realized digital existence. The majority of enterprises still focus on traditional corporate image. Although the Internet develops rapidly at present, there are still such hidden troubles as "Hacker" and "Trojan Horse Program", which make the network security a problem needed to be solved urgently. Only by intensifying the network construction and management, the harmonious coexistence of network society and real society can be achieved. Apparently, it requires not only the progress of technology, but also the improvement of the overall quality of mankind. However, it is predictable that the network will certainly mature as people become more dependent on digital devices.

Looking ahead, network media and mobile network media grounded on digital information technologies and network techniques possess higher communicating capacity and efficiency. They enable the range and the speed of information dissemination to develop by leaps and bounds. The strong power of transmission, communication and sharing informations makes people break through boundaries of time and space.

In order to further strengthen the corporate identification, new identifying systems, such as virtual reality and simulation, sensing techniques, display system and other technologies and devices, can be added to the network to provide users with a three-dimensional environment which reflects authentically changes of operational objectives and their interactions by means of simulating reality. Thus, a virtual world will be created and a three-dimensional interface which interacts with the virtual world will be offered to users through particular devices as well.

All kinds of design works provide consumers with lifelike visual effects and present clients with design programs through various approaches and manufacturing modes, in order to highly unify corporate image, make full use of corporate visual communication resources and finally achieve the optimum effect of brand communicating.

Changes brought by forces of science and technology are unattainable even for the richest imagination. Digitalization is the irreversible leading trend of the development of contemporary society, which brings about profound changes in terms of development pattern, economy and technology, the mode of life, etc, and also influences changes in spiritual and cultural fields. The progress of digitalization of corporate image is irreversible as well. Digital corporate image will definitely put on an extraordinarily brilliant performance on the stage of digital media communication.

References

1. Chen, Y., Wei, B., Yi, C.: Digitalized enterprises. Tsinghua University Press (September 2003) (in Chinese)
2. Guo, M.: The power of communication——Strategies for corporate media. Nanjing University Press (May 2006) (in Chinese)
3. Huang, M.: Digital arts. Shanghai Academia Press (December 2004) (in Chinese)
4. Negreponte: Being digital. Hainan Press (March 2000) (in Chinese)
5. Jian, M., Friedrich, S.: The era of picture——The theoretical analysis of visual culture communication. Fudan Press (October 2005) (in Chinese)

Thinking of Human Design of Furniture

Lan Ma

Institute of Art and Fashion
Tianjin Polytechnic University
Tianjin, China
hellokid1972@yahoo.com.cn

Abstract. The human design is a contemporary new design concept developing from human-orientation. This kind of design pays special attention to social vulnerable groups, and takes the human personality, emotional and cultural significance into account, viewing the earth as the source of resources and the ecological environment, pursuing the century theme of sustainable human development and humanization improvement.

Keywords: human-oriented, human design, function design, ecological design.

1 Introduction

Recently, the "human-oriented" design has become a hot topic in the design industry, which uses art and technology to show the humanistic spirit, and it is also the perfect combination of human and material, man and nature. As for furniture design, it could date back to the Paris Exposition in 1909, when the Scandinavian Furniture, with its smooth curves, friendly timber and natural style, won the favor of the public, creating a precedent for human design of furniture.

Firstly, let us see what human design is. The human design is a contemporary new design concept developing from human-orientation, distinguishing human from other creatures in a sense of "category" and clearly pointing to and implementing human's outlook of life, cosmology and social values in the designers' design concepts, then determining the design objectives and significance. This new design converts the "upper", "lower" concepts in the past into the public-oriented design, particularly showing great concern for social vulnerable groups, and taking it as the design issue, regarding the human personality, emotional and cultural significance as the important category, viewing the earth as the source of resources and the ecological environment, and pursuing the century theme of sustainable human development and humanization improvement.

The human design of furniture is human-oriented, emphasizing the inter-dependence and mutual promotion among people, furniture, environment and society. Besides, it attaches importance to humanistic spirit of the furniture design, so it is the combination of spiritual and emotional needs.

In this article, the author attempts to analyze the humane design from the following four aspects: individuation design (color, shape), function design, scale design (ergonomics), and eco-design of furniture.

J. Luo (Ed.): Soft Computing in Information Communication Technology, AISC 161, pp. 241–245.

2 Individuation Design

For furniture, the individuation mainly reflects in the color and shape, and through pursuing it, the contemporaries achieve to innovate, change, show their characters and enjoy simple but exquisite life .As for the color design, charming colors greatly reduce the life pressure of urban people and lead the fashion trend of the current world; while the new and unique shape design is also the consumers' negation of the brassbound external form of old products.

A.   Color Design

The Individuation design of colors should be based on color matching principle, using colors reasonably according to the psychological characteristics and physical characteristics of different groups, then achieving the requirement that the color serves human. The color design of furniture is often determined by the human's personality, age and occupation. For example, the boys' personality, to some extent, is offensive and restless, and they possess better abstract thinking and logical reasoning abilities, so they mostly tend to love cool colors such as blue and green; while girls are quiet and gentle, they own a better image thinking ability, so most girls like the warm colors such as red and yellow, especially pink. As for people of different ages, the elderly should use the furniture with harmonious and quiet colors, such as plain gray, serene blue and pure white; However, the color of the furniture chosen for children should be pure colors, such as red, yellow, blue, white, green, etc., thus creating the innocent and lovely "childish" atmosphere.

B.   Form Design

The individuation design of furniture forms mainly reflects in: taking the ergonomics as the guide, then designing beautiful and comfortable furniture of different forms. For example, when designing furniture for children, the designer can apply the animated cartoons and comic books, to attract the children's interest. Meanwhile, the form design of the furniture for the elderly should be given full consideration to safety; for example, furniture edges and corners have to be handled by the milling operation in order to avoid causing bumps to the old mobility-impaired people, and furniture handles must be thick enough for the elderly to grasp. Another example is the Bra chair designed by the Italian designer Zve Vanghn in the 80's. He used the traditional chair structure, but adopted the soft, full and shapely female body shape to design the back of the chair. Sitting in the chair, people feel comfortable and their thoughts are random, which is very interesting. Therefore, it can be seen that, different forms would lead to different emotional experience and psychological feelings, called "emotional designs". As the Danish designer Kay • Paul Jason 's famous remark: "Let the line smile," which explains the form design should be human and humanistic. Another example also can illustrate my view. A sofa with armrests, which was in general shape, became famous because of the designer's naming-- "Mom", meaning the sofa can provide a sense of protection, warmth and comfort, just liking lying in mom's arms. Practical besides, he made us think and dream to a certain extent. Therefore, the nature and characteristics of furniture must be clear, concrete and materialized through creating some shapes.

3 Function Design

The function design of furniture mainly meet people's all needs from the material point of view. For example, the designer should take the automatic lift function into consideration when designing the bed for the disabled and the elderly. While designing the furniture for the children, he ought to fully consider developing the game functions; for example, in the design process, the designer adopts the principle to design plug-in furniture for children, which combines the consolidation principle that has concentration and dispersion characteristics, and the replication principle that owns the characteristic of structural changes,. According to the learning level of the children at different ages, the designer designs the corresponding cartoon image card, and the children can choose their preferringprints, then they can combine the prints together to make various shapes; for instance, this kind of design changes the singularity of the fixed prints of traditional furniture for children from learning to read words with the help of the Chinese Phonetic Alphabet to writing poetry , from grouping the English words to grouping sentence puzzles. Besides, currently, the disassembly functions of children's furniture still focus on some simple functions, such as the increasing, shortening, widening functions of tables and chairs, to meet the children's basic growth needs and extend their working life. If the furniture for children is developed, from the perspective of mens et manus, to design the robot-toy random combination furniture, then it can not only reflect the practical function of furniture, but also expands its added educational function, which is conducive to children's practical ability and thinking innovation. Meanwhile, it will bring a revolution in children's furniture designation.

4 Scale Design

The human design of the furniture scales is an ergonomics-based comfortable design, while the ergonomics embodies the designer's caring and love for people, and achieving the goal that the design promotes human nature; for example, when designing a child chair, the designer should fully taken into account that children have equal status as adults in the right of speech, so he need to avoid the traditional way that children look up to adults to speak. Moreover, the height of the chair can be shortened according to the children's growth, which fully embodies the human design of furniture. What's more, the height of the bed for people in wheelchairs should be equivalent to the height of the wheelchair, that is, the space at the bed bottom should approximately be 200mm deep and 300mm high so that the wheelchair riders can move the wheelchair more closely to the bed, and this kind of design is also more convenient to move the disabled or elderly to the bed.

5 Eco-Design

The eco-design of furniture is also known as the green design. A seat design, called as a "solar leisure chairs ", is powered by the solar energy, automatically redirecting with the sun. Its innovative design interprets the concept of the green design of modern people. Sim Van der Ryn and Stuart Cownis define it: any kind of designs, which coordinate with the ecological process and minimize the environmental detriment as much as possible, are referred to the green design, life cycle design or environmental design. The core is "3R" principles: Reduce, Reuse and Recycle, which means to minimize material and energy consumption and reduce emissions of harmful substances, besides, making it convenient to do the separation, recovery, recycling or reuse of the products and components. The green design of furniture mainly includes three main areas:

A.   Furniture Material Selection and Use

For example, bamboo can be used to produce coffee tables, sofa, chairs, recliners, tables, etc. Besides, it is also used as furniture fittings to compose a set of furniture with wood, glass, metal and textile fabric.

B.   The Optimized Design of Furniture

The modular design is the main method of optimized design of furniture, and its most prominent advantage is interchangeability and demolition, which helps to achieve the reuse, replacement and recycling of furniture. For example, a set of furniture, made by scrap cardboards and recyclable polypropylene, is only composed of five hardboards when it goes out from the factory,   which is very convenient to transport and store and you can quickly put them together, like the self-assembly furniture of IKEA.

C.   Recyclable and Reusable Furniture Design

Use recycled components or materials: if the designer considers recovery and recycling when designing, it will greatly improve the renewable rate of waste products. For example, there were public seats designed with scrap hoses in the foreign furniture fair.

The rise of the green furniture design will bring a new industrial revolution. It is not only a technical innovation, but also a conceptual change. Besides, it requires the designer to create furniture in an environmentally responsible manner. The green furniture design will focus on ecological functions and environmental effects of the whole process of the product design, manufacturing, use and recycling, to realize the harmony among man - furniture - the ecological environment.

The human design of furniture is not static; instead, it is a dynamic system, changing along with the development of productive forces, improvement of science and technology, social consciousness, ideas and aesthetic levels. In the future of modernization, whether the furniture and environment are in harmony appears to be more important, the designer should make furniture become an integral part of people. As the Danish furniture design leader, Kay • Bojay •Kim said: "the thing we have produced is a living creature with heart beats ---- they should be human, alive and warm." When designing furniture, the designers should give furniture "humanized" characters, making it possess emotion, personality, fun and life through giving "human" factor into the form, function, scale, ecology of the furniture. The human expression of furniture designs is to use the visible "material state" to carry and reflect the invisible "spiritual state."

References

1. Lu, J.: On the Human Design. Wuxi Nanyang University (2006)
2. Zhou, S., Jiang, F.: Analysis of the Human Design of Furniture for Children. Ars Nova (2007)
3. Liu, Z., Liu, G.: Green Design. Engineering Industry Publishing House, Beijing (1999)
4. Sun, S.: Environmental Materials. Chemical Industry Press, Beijing (2002)

The Microscopic Construction of China's Government Functions in the Universities' Transformation of Sci-tech Achievements

Jianhua Zhang and Xiaomin Fan

Humanities and Laws Faculty, Tianjin Polytechnic University, Tianjin, China
zhongbin1979@yahoo.com.cn

Abstract. The low ratio of the universities' transformation of Sci-tech achievements is an important aspect of the problem of the transformation of our country's Sci-tech achievements. Whether or not can our government effectively promote the transformation of the universities' Sci-tech achievements, and improve the ratio of the universities' Sci-tech achievements, will directly affect the overall Sci-tech development level of China's future and sustainable economic development. From the five micro aspects of laws and regulations, investment system, Sci-tech management system, assessment mechanisms and building service system , the article describes our government functions' construction in the universities' transformation of Sci-tech achievements.

Keywords: University Transformation of Sci-tech achievements, Government function, Microscopic construction.

1   Introduction

In the process of the universities' transformation of Sci-tech achievements, the government should play its role of macroscopic regulation and guidance for the transformation of Sci-tech achievements to create a favorable policy environment and provide the institutional guarantee. But the macroscopic functions are implemented only at the micro level, to ensure that the government plays a real and effective function in the transformation process of Sci-tech achievements. In the process of promoting the transformation of government functions, it is important to note the following aspects of the micro-implementation.

2   Improve the Laws and Regulations of the Universities' Transformation of Sci-tech Achievements

With the gradual establishment of market economy and the rapid development of science and technology, the laws and regulations of the transformation of Sci-tech achievements must adapt to the new features, and amend timely. For example, China's "Law of promoting the transformation of Sci-tech achievements" (1996), although the law played a great role in promoting the transformation of Sci-tech achievements, but

J. Luo (Ed.): Soft Computing in Information Communication Technology, AISC 161, pp. 247–253.
springerlink.com   © Springer-Verlag Berlin Heidelberg 2012

there are a lot of articles no longer meet the actual needs of the transformation and lagging behind the times. So we must not only amend and improve the relevant provisions, especially in its operational aspects, but also strengthen supporting measures and interpretative legislation. To further speed up the pace of legislation in the universities' transformation of Sci-tech achievements and improve the legal system of technology market, you need to pay attention to the following four areas.

First, we must develop a normative law which supports the high-tech enterprise's development to increase demand for scientific and technological achievements and make the universities' transformation of Sci-tech achievements has a huge market space.

Second, through legislation we should further strengthen the support and regulation of all the Sci-tech agents, especially for technology assessment agencies, technology trade organizations, technology incubators, technology advisory bodies and so on. Through legislation we can regulate the management of Sci-tech agencies, and create an external environment which in favor of the universities' Sci-tech achievements' development, deployment, transfer and diffusion.

Third, we should further improve the intellectual property law and through intellectual property system support effective protection to intellectual labor achievements of university's scientific research workers. As the improvement of China's market economy system, we can further draw on U.S. intellectual property system which through legislation stipulates that universities obtain the ownership of the innovation achievements from the university research projects funded by the Government, but also introduces mandatory regulations that the university can prorate license fee from the transfer of ownership between the inventor and the inventor's college ,so that intellectual property rights are legally protected.

Fourth, we should further modify and improve the related tax and other policies. As the universities' transformation of Sci-tech achievements involves a very wide, the associated regulations and policies are more, we need to further strengthen the tax policy of Sci-tech achievements and in time make changes and adjustments in order to fully mobilize the various research subjects' enthusiasm for Sci-tech Achievements conversion to provide organizational and policy support. For example, in the implementation of diversification of the preferential tax policies, both has the traditional preferential policies and may take some tax-related ways. These methods include reducing tax rates, part of the tax return, technology funds, accelerated depreciation and other aspects.

3   Reform the Management System of Sci-tech

Although China adopted the market economy for years, but in talent, technology, universities and other management system, reform is not in place, there is a certain degree of program traces. Government's Sci-tech management system as the most important external environment in the university's transformation of Sci-tech achievements influences the universities' R & D and the transformation of Sci-tech achievements from multiple dimensions and aspects. Only the establishment of Sci-tech and technology management system can promote or facilitate the university's transformation of Sci-tech achievements. It requires a good job of macro design, and be reflected at the micro implementation of the management system to prevent backward on Sci-tech achievements delay or impediment.

On the one hand, we should rebuild the "triple helix" relationship among the government, universities and enterprises. China's universities, particularly research universities, and government research institutions the two scientific research institution aspect has "the two-track system", does things their own way. We must draw on foreign experience in advanced scientific and technological power and really establish a research system with research universities as the major force for basic research, government research institutions as the main force for applied research and enterprises as the main force research for development research. Through being clear about the research university and the government scientific research institution localization and respective characteristic and a clear division of labor, we can reduce duplication and waste of resources, increase exchanges and cooperation between the two, and re-appropriate resources to complement each other.

On the other hand, in the new situation, the Government should encourage and support the reform of micro-technology and human resources management system of universities according to various characteristics, in the work of universities' transformation of Sci-tech achievements. Firstly, universities should act, according to the national policy, the measures and systems of Sci-tech, Sci-tech achievements and the transformation of Sci-tech achievements and increase Sci-tech research and development market consciousness, lays the important foundation for the universities' transformation of Sci-tech achievements. Secondly, universities should establish and improve the system and mechanism of the human resources work, in particular, establish the type talent evaluation index system which takes the achievement as the key point, and is constituted by the character, knowledge, ability and other factors. Through the classified appraisal, universities can encourage university teachers according to their strengths in Sci-tech to make its due contribution to the transformation. Thirdly, around the transformation of Sci-tech achievements, universities should establish an open, competitive and merit-oriented selection and appointment mechanism which is conducive to outstanding talents of R & D and achievements transformation of Sci-tech come to the fore, and fully display their abilities, so as to create a large number of leading talents of R & D and achievements transformation of Sci-tech.

4　Establish the Multi-channel Investment System of the Sci-tech Achievements' Transformation

To further promote the universities' transformation of Sci-tech achievements, we must continue to improve universities' Sci-tech investment and financing system, and build a scientific Sci-tech investment and financing system, so as to effectively address capital investment issues in the process of industrialization of university's Sci-tech achievements. Under the conditions of the socialist market economy, in order to improve the universities' Sci-tech investment and financing system, we shouldn't just rely on government to increase financial investment in science and technology, but also to actively explore market-oriented financing mechanism to achieve diversification of investment and financing, market financing, and to build and improve a diverse, multi-channel, multi-level investment and financing system of universities' transformation of Sci-tech achievements which is led by the government and is combined with direct financing and indirect financing.

First, optimize the Sci-tech financial investment structure, increase efficiency in the use of funds invested. Through the appropriate expenditure, giving direct financial assistance to industrialization of university's Sci-tech achievements has become an important policy in general implemented by the world's governments. In the current context of limited government financial resources, we can not put too much emphasis on a substantial increase in university's Sci-tech expenditure and the field of Sci-tech financial investment should be adjusted to highlight the focus of investment in financial technology, and optimize financial technology investment structure in order to give full play to the guiding role of Sci-tech financial investment. Financial investment can be considered as a lever to attract funds of social capital, and we can set up science and technology venture capital funds, technology risk compensation fund, University Technology Enterprise Finance Guarantee Fund, the Youth Innovation Fund and so on. At the same time, we can consider joint ventures and cooperation with domestic and foreign investment institutions to set up Sci-tech fund management company which manage and operate the fund according to the market mechanism; on this basis, we should improve the Fund reporting system and introduce expert committees or professional advisory body to declare a rigorous assessment of research projects and optimization on university's Sci-tech workers, in order to achieve greater capital efficiency, and play the guiding role. Meanwhile, investment funds should strengthen management and improve efficiency in the use of funds invested. Properly handle the distribution of the various stages of financial investment, put the technology development stage as an important reference for the input of funds, and also put the transformation of achievements as an important part of the subject application. We should establish special funds for basic research to ensure the follow-up after the conversion exercise. This will also promote a greater focus on the interface between results of each study and development, and promote scientific research and market docking.

Second, we should sound risk investment system to promote university's transformation of Sci-tech achievements. We should comprehensively understand the role of risk investment in the process of university's transformation of Sci-tech achievements and be clear about government's role localization in the development of venture capital. We also should clearly recognize, in the different stages of development of venture investment, the manner and extent of the government in venture capital. In the process of the development of venture capital, the government shall promote China's venture capital industry's rapid development in accordance with the basic requirements of "formulate policies to create the environment, strengthen the supervision and control risk". Meanwhile, to speed up the establishment and improvement of multi-level capital market, guarantee the smooth exit of venture capital, the government must strengthen risk investment's management, and follow the basic principles which compliance management according to law, according to specifications, supervision according to law, in order to make it into the legal system and standardized track.

5 Establish a Fair and Reasonable Assessment Mechanism

The key of the establishment of a fair and reasonable evaluation mechanism is assessment of Sci-tech achievements and the government's position in the evaluation, it is science important prerequisite for defining responsibility of government.

First, innovate in the concept related to evaluations of Sci-tech achievements. The evaluation is often the basis which the encouragement draws support, therefore, to encourage scientific and technological personnel's main role in the transformation, we must eliminate the drawbacks of the existing evaluation system. For one thing, we should weaken the fact appraisal and strengthen the value appraisal. We should establish the concept based on the economic evaluation and change the practice that focuses only on the level of completion of the evaluation results, and ignore evaluating the maturity of technology or the degree of industrialization. In the process of setting the evaluation indicators, we should reduce the score of outcome's theoretical value, and increase the proportion of market demanding, economic efficiency, social efficiency proportions and other indicators. For the other thing, we should change a single way of peer review, and positively introduce the evaluation of external systems, in particular, the persons concerned should be involved in the business community. Simultaneously we should also clean the negative trend of peer review and develop a accountability system of peer review.

Second, reform the current assessment system. In the Sci-tech assessment system, the government's Sci-tech achievements' evaluation system would play an important role in guiding the university Sci-tech research and development and transformation of Sci-tech achievements. It displays that the government realizes the guidance or constraint on the direction or range of university R & D activities through the university's own impact assessment system. In the Sci-tech assessment system, in order to ensure the smooth transformation of Sci-tech achievements, the developed countries attach great importance to establish and improve Sci-tech achievements appraisal system. Such as the United States, Britain, Germany and other developed countries have significant numbers of institutions specializing in Sci-tech assessment. These institutions as a third party, mainly from the advanced nature, practicality and reliability of the university's Sci-tech achievements, the price of Sci-tech achievements, the business prospects and risks of Sci-tech achievements and several other aspects, carry on the appraisal in order to provide the support for transformation of the university's Sci-tech achievements.

6 Strengthen the Service System of the University's Transformation of Sci-tech Achievements Actively

The core of the service system of the university's transformation of Sci-tech achievements is Sci-tech agencies. As their lower service capability, the government needs to support them at various levels.

Firstly, establish the laws and regulations of the development of Sci-tech agencies. The laws and regulations contain some key points as following: the Sci-tech agencies should streamline the relationship with the government through legislation, clearing their legal status, organizational systems and development models; strengthen the mandatory qualifications and qualification identity of the legal form, improving market access standards; specify the service activities of the Sci-tech agencies and ensure equal competition.

Secondly, formulate and adjust the specific support policies of the development of the Sci-tech agencies. These policies include strengthening the financial capital support, tax incentive support, administrative support (it can be divided into the government procurement support and administrative delegate support), personnel policy support, brand support, strategic planning, technology policy support, and investment support and so on.

Thirdly, improve the construction of the information environment which benefits the rapid development of the Sci-tech agencies. Their business based on the information resources platform. In view of the current situations of the technology intermediaries information and the network building, we should strengthen the information platform step by step: First of all, build a national public information platform to ensure the authority and guidance of the information; the next, set up a regional public information platform which maintains a close cooperation of the national information platform; once more, create the search and service function of international information and form an international information network platform; finally, to build the above platforms ,we should supply necessary financial and technical supports and make a unified information services and charges.

Forth, improve the market environment of the Sci-tech agencies. The government departments should delegate powers to lower levels gradually and play the role of service, supervision and coordination. Their main function is formulating the "horse racing rules" instead of packaging administrative powers .They should ensure the t Sci-tech agencies engaged fair and neutrally. Cut off the connection between the government and the intermediary organizations in the economic, using the legal instruments to prohibit any government departments to divide the income of intermediary organizations. This will ensure the independence and autonomy in the respect of the personnel management, financial management and business management. Through the system innovation and transformation of Sci-tech agencies, they will know their market legal status clearly and ascertain the strong position in society.

Fifth, strengthen the incentive of the Sci-tech agencies. As for the government, they can set up a special "Science and Technology Promotion Award". With the establishment of this award, especially in the Sci-tech achievements transformation activities, it could give the related personnel some reward and publicity on physical and mental aspects. As for the colleges, they should also encourage the personnel and organizations which engaged in Sci-tech achievements transformation. Such as taking the funds, or extracting a certain proportion of the finance by transferring the Sci-tech achievements. As for the enterprise, they could set up special funds or pick up certain percentage money to award the related personnel and institutions. They still could give some institutions and staff some shares or involve them in the rank of decision or management. Thus, it will tie up and encourages them in a long term.

References

1. Yin, C.: Relations between research universities and government research agencies–International Comparative Study. Natural Dialectics Research (4), 78–81 (2006)
2. Peters: The Future of Governing: Four Emerging Models. University Press of Kansas, Kansas (1996)
3. Henig, J.R.: School Choice–limits of the Market metaphor. Princeton University Press, Princeton (1994)
4. Coleman, J.S., Hoffer, T.: Public and Private Schools: The Impact of Conununities. Basic Books, New York (1987)

References

The Overall Design of the Government Functions of the Higher Institute Tech Achievements Transformation

Jianhua Zhang and Wenbo Zhu

Humanities and Laws Faculty
Tianjin Polytechnic University
Tianjin, China
zhongbin1979@yahoo.com.cn

Abstract. Government function is one of the key factors of restricting the conversion rate of tech achievement transformation in higher institutions of China, thus the scientific define of the overall design of the government function become an urgent issue about the tech achievement transformation in higher institutions. This paper pointed out the roles and responsibilities of the government from the angle of role: the provider of the transformation of the Sci-tech achievements system, defenders of the market order and the public service providers. Meanwhile, from the macroscopic level, it also indicated five overall functions in the transformation of Sci-tech achievements in universities of China: regulations drive function, interest coordination function, and incentive guided function, effectively balance function, and intermediary service work function.

Keywords: higher institutions, tech achievement transformation, government function, overall design.

1 Introduction

There are some obstacles in the transform process from higher institutions tech achievement to productivity like technological gap, financing gap, and information technology gap, etc. Such gaps could not be conquered in their own colleges and enterprises as well as the intermediary institutions, and they formed a "death valley "in the leading of "market failure" during the process. Government must play a due role to cross the "death valley". From the angle of content, the government functions include government duties and functions. According the social development and demand, government duties and functions would have the corresponding changes. Correspondingly, the function change of government exactly means the changes, transformation and development of government duties and functions. To avoid the failure of the market and the government and solve the problems in the transform process, this paper made an overall design on the government function from the two aspects of duties and functions.

J. Luo (Ed.): Soft Computing in Information Communication Technology, AISC 161, pp. 255–261.
springerlink.com   © Springer-Verlag Berlin Heidelberg 2012

2 Government Functions in the Higher Institute Tech Transform Process

As mentioned above, Technology Transformation is not simply a technological behavior or economic behavior; it is not only related to the universities and enterprises, but also related to the relevant functional departments of the government. We need the coordination of industry, academia, research and government departments to increase the tech achievement transformation rate. Government should reduce the direct intervention on tech achievement transformation, and give full play on the basic role of the market during the transformation process. However, from the angle of government duties, the government can not withdraw completely from the market, and it should acts as the provider of the transformation of the tech achievements system, defenders of the market order and the public service providers.

A. The Providers' Duty of the Transformation of the Tech Achievements System

New Institutional Economics achievements have indicated that during the transformation process, the arrangement of government system is extremely important. Government should do something to encourage the legislation on scientific and technological innovation and the tech achievement transformation. And it should create a sound environment system from the establishment and completeness of the incentive mechanism, competition mechanism, and restraint mechanism and guarantee mechanism.

First, improve the related legal systems of technological achievements. To provide legal protection, and optimize the legal environment are the main methods to consummate the socialist market economic system. Government should provide available policy and regulatory environment to the higher institutes, and encourage university researchers to realize the transformation of the scientific and technological achievements into innovative. Encourage fair competition, venture capital and co-development.

Second, establish a competition mechanism of the transformation of scientific and technological achievements. To improve the conversion rate of university scientific and technological achievements further, the Government should strengthen scientific research organization and management innovation. And it should confirm the commitment unit of the qualified major research project and tech achievement transform project by the way of bidding. Encourage the higher institutes take apart in the biding competition actively.

Third, establish the incentive mechanism of tech achievement transformation. At present, one of the objective reasons of the low conversation rate isn't the scientific and technological achievements could not be transformed, but some scientific researchers do not want to do it. The reason lies in the inadequate distribution of earnings and surplus after the transformation, which affected the enthusiasm of the personnel. Therefore, the Government should apply itself on the following things: set up an conductive incentive mechanism for higher institutes' tech achievements transformation; clear the interests distribution of achievements transfer; reward the higher institute's personnel who made important contributions; lead the transformation by interest driven mechanism .The main incentive mechanisms of

Government as follows: set up an issue subsidize system; make concentrate use of the issue funding; adopt the awards methods for scientific research; strengthen the management of issue process; formulate the tax preferential policies in the transformation of higher institutes tech achievements.

Forth, establish restraint mechanism and guarantee mechanism of scientific and technological achievements. We should establish and improve scientific research and transform contract and project responsibility system and mobilize the positively of personnel to the full, in this way, the management and transformation of the achievements would proceed effectively.   Contemporary, reform the acceptance and certification schemes of scientific and technological achievements. Finally, we'll check whether the value of achievements in line with the needs of our country and if it get the good results and the market or not. Strengthen the metaphase project appraisal and tracking management of the scientific and technological transformation; enhance financial supervision and carry out the special responsibility system to manage the major projects. Establish related protection mechanisms of the scientific and technological achievements, such as the protection of the intellectual property from infringement, protection of national technology innovation; perfect the risk investment mechanism to ensure the supply of capital market of the achievement's development and transformation; complete the supply system of the bank and non-bank financial institutions; accelerate the development of various technology markets, and establish the market network hub of institutions of higher learning technology; improve the intermediary service system of scientific and technological achievements and offer the assurance mechanism to accelerate the conversation of the higher institutes tech achievements.

B.   The Defenders' Duty of the Market Order

As a defender of the market order and judgment of the market dispute, government should eliminate the potential disputes in the bud during the transform process by promulgating regulations and perfecting the market rules. And reduce transaction costs to the maximum extent, and ensure the technology transformed smoothly.

C.   The Providers' Duty of the Public Service

The higher institutes' research and achievements transformation need a good infrastructure and public services. While the quality of library and information services is poor; network communications facilities is lag; tenement facilities is insufficient; venture capital mechanism is not perfect; the test base of scientific achievements transformation is unsound. This entire have became the bottleneck of the achievements transformation. Therefore, the Government should try to improve efficiency and supply quality of the public goods and services, and lay a sound infrastructure for the achievements transformation.

In short, during the achievements transformation process, the intervention of the government is limited. The government should focus on constructing the environment of the achievements transformation and entrepreneurship rather than on conducting specific issues; it should provide policy guidance instead of excessive administrative interference; it should offer meticulous services rather than arbitrary meddling; it should formulate the scientific and transparent rules instead of playing the game itself; it should   provide appropriate temperature to incubate the   high technology enterprises rather than incubating them itself. During the transformation process, though the government plays an important impact on the financial expenditures, land use, taxation, planning and personnel, however, one fact cannot be overlooked that various independent intermediary agencies have also begun to play an increasingly important role on scientific and technological achievements transformation. Increase the transformation rate is not a product of levels of governments' policy, and it is not the result of inputting funds, either. During the transformation process, the role of the government only lies in strengthening the services, supplying the system and maintaining the order.

3   Government Functions in the Transformation of the Higher Institutes' Achievements

Because of the science and technology have become the economic reason, and now that the government wants to regulate the economic process, of course, it will control the achievements transformation process. Government Functions in the trans-formation of the higher institutes' achievements is the relationship between the government and the main actors of the technological achievements. Such as the Government and the higher institute as well as the enterprise. The regulation activities among the higher institutes and the enterprises and related organizes belong to the macroeconomic level of the government function and which displayed at the aspects of optimizing, encouraging and supervision, and coordination. From the point of administer, the essence of the above functions is straighten all kinds of interest relationship. China is in the transition period, the technology demands of the State's development are strong. The core of our Government's macrograph function of the transformation process is based on the equal and interact relationship of various interests, which formed effective governance.   According to the government function of "big government, big society" and "supervise-service" and the duties of the three roles, the overall function of the government in the transformation were manifested at the following five aspects:

A.   Regulations Drive Functions

"Regulation drive" is the substitution of the "administration drive", it is the fundamental requirement for the sustainable development of the achievements transformation. The transformation of the higher institutes is a commonality affair involved the government, university, enterprise and technology intermediary bodies. Practice has proved:" administration drive" often did not work effectively; it must achieve a good governance pattern in the legal fragment. To this end, the Government should bring the achievements transformation into the economic and social development plans, formulate and improve the systematic policy and management which good for the achievements transformation. This will make the transformation has laws to go by.   The government use the way of establishing the rights, responsibilities and duties of enterprise and higher institutes, which could improve their respective initiative and enthusiasm. Through the construct of the policies and the strictly enforce, the government would set up a must and willing situation for the technological achievements transformation.

B.   Interest Coordination Function

Interest coordination function is coordinating the interests between the colleges and universities, which enable them to achieve mutual development. In the higher institute's transformation process of china, in the guidance of the economic construction as the central focus, the higher institutes always paid more and affected the teaching and academic functioning of achievements transformation to a certain extents. On the condition of the market economic, interest coordination functions are the motive power and link of the healthy function. Based on the governance philosophy, the participators of the transformation will focus on the benefits from their own interest. The Government concerned about whether they got the deserved interests, or whether it would be able to produce good social benefits. The enterprises focused on whether the achievements accord with the need of production and development, and the technical expertise and technical margin. The university and scientific organization take care whether the technique input value for money; whether the school environment has improved; whether their management strength and the social reputation have been enhanced. Therefore, the government helps the higher institute obtained continuously achievements which coordinate y of ways.

C.   Incentive Guided Function

The aim of the incentive guided function is leading the higher institute's transformation into a virtuous circle, which consolidation and development the scientific and technological achievements. The content of the incentive contains four aspects: First, power incentive, for example, to appoint the principals of the enterprise as the members of the board of directors of the higher institutions or intermediary institutions. And let the related personnel of the relevant organizations hold the post of the assistants. Institutions or enterprises also serve as an intermediary deputies or relevant personnel of the relevant departments of the deputies; Second, interest incentive. For example, to engage the technical personnel to join in the higher institute scientific research and give a certain rights; third, reward incentive. To pay the personnel who made great contribution and the innovation achievement a fixed reward at the allowable framework of the national policy. Forth, honor incentive. For example, award the silk banner or certificate to the better enterprises and teachers, and report it through the media. According these incentives measures, the government would promote the higher institute tech achievement transformation to get a further development in depth and healthy.

D.   Effectively Balance Function

Norms contain two vital elements -- the balances and restrictions, which are good to the healthy development of higher institute tech achievements transform. According to the agency theory, the key point of governance is to establish an effective balance mechanism. On one hand, it can stimulate the agents to make efforts to realize the binding agents' goals; on the other hand, it can restrict the agents' actions effectively. During the building of the balance mechanism of higher institute tech achievements transforms, the government should pay more attention to the restraint and supervision of the power-- which also can be divided into internal supervision and external supervision. The external supervision includes administrative supervision and social public opinion supervision of the government and social institutions, which are implemented to all the parties relating to the higher institute tech achievements, transform; the internal supervision is implemented to the agent by establishing an internal supervision system between the universities and enterprises and relating supervisory measures. Governance sense of higher institute tech achievements transform is not like that one side shows the charity, help and support to the other, however, it should be that both sides can get benefits through taking the common responsibility and obligation. Therefore, the parties shall sign cooperation agreement with legal norms and stipulate the parties' rights, obligations and violations very clearly. At the same time, enterprises and higher institutes should improve the management system, strengthen the institutional constraints, enhance the moral education and strengthen the moral constraints in accordance with the principle of all-wins.

E.   Intermediary Service Work Function

Under the market economy condition, the "strong government" is a limited government, and we also need a "strong society". The government has to strengthen the establishment of the intermediary service system, to maintain a certain distance from the market, because the fact that the innovation ability of our universities and enterprises is relatively weak. The Government should gradually change their functions, rely on the intermediary organizations to improve the managements and services and entrust them to operate some affairs. Vigorously carry out "administrative policy advisory system", in making policy decisions more scientific, democratic and scientific, meanwhile, providing more room for development of the intermediary organizations, such to make them through practice gradually raise their own professional level and ability of service. Should build a number of science and technology intermediary organizations which the market badly needs based on the advantages of the colleges and universities. Colleges and universities should encourage scientific and technological workers to start some professional science and technology intermediary such as science and technology assessment, scientific and technological advice, technical contracts arbitration agencies and certificates of professional intermediaries science and technology, improving science and technology intermediary service system, to accelerate the technology transfer of colleges and universities and the transformation of the scientific and technological achievements.

At the same time, in order to transfer the scientific and technological advantages into real productive forces, universities should open the technology resources to the science and technology intermediary institutions, provide them with favorable conditions, and encourage teachers and students to cooperate actively with those institutions.

References

1. Mintrom, M.: Policy Entrepreneurs and School Choice. George town University Press, Washington, D.C. (2000)
2. Peters, G.: The Future of Governing: Four Models. University Press of Kansas, Kansas (1996)
3. Romer, P.M.: Eudogeuous Technological change. Journal Political Economy 98(5) (1990)

The Blend of Modern Graphic Design and the Culture and Art of Traditional Chinese Paper-Cut

Liu Hui

Institute of Art and Fashion, Tianjin Polytechnic University, Tianjin, China
llhhliuhui@126.com

Abstract. Chinese folk paper-cut is the most representative of artistic form in Chinese folk, which is the basis of folk art and the maternal art, providing good learning and reference for modern graphic design has played. This paper focuses on an in-depth analysis of how the traditional Chinese folk paper-cut art to make combination and fusion in modern graphic design.

Keywords: folk paper-cut, graphic design, aesthetic implication.

1 Introduction

Paper-cut art originated from China from the Han to the Northern and Southern Dynasties, which is a kind of Chinese folk tradition decoration art. Because of its simple material, low cost, wide adaptability, varied patterns, plain and vivid image, it has deep popularity of people. With a distinctive artistic features and living interests, Chinese folk paper-cut art works reflect people's pursuit of aesthetics and spiritual quality, and it provides good reference for modern graphic design, integrating traditional paper-cut art into modern graphic design, and enlighten the inspiration of modern graphic design and deepen the connotation and details of modern graphic design, so that the excellently traditional Chinese art and modern design can be in good fusion.

Today, in a social environment where economy is develop increasingly, the trend of global economic integration, gradually shaping strong culture to make erosion of weak culture. Regional, cross-country globalization process constantly changes all areas all over the world from all levels. How to combine modern design thinking with regional, ethnic culture traditions to build polybasic culture framework has become an important issue of modern graphic design. In a lot of good graphic design work, it is not difficult to find that their success is just often to use the relationship of traditional culture to the point, which not only reflects fashionable concept of modern graphic design, but also reflects different history and cultural characteristics and aesthetic orientation in all national regions.

For example: Tanaka, design works of a famous Japanese master of graphic design, his works not only reflect the charm of Japanese traditional culture, but also a lot of added modern element, combining traditional culture and modern ideas as one. In the "Japanese dance" poster designed by him, we can feel the impact of this tradition to show the very rich expression by a very simple form, called the model of traditionally geometric style in Japanese design. (Figure 1)

J. Luo (Ed.): Soft Computing in Information Communication Technology, AISC 161, pp. 263–268.
springerlink.com   © Springer-Verlag Berlin Heidelberg 2012

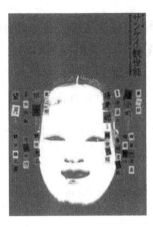

Fig. 1. The design work of Japanese prominent graphic designer Ikko Tanaka

There are well-known Chinese designers, Han Meilin, Mr. Chen Shaohua designed "logo of 2008 Olympic bid "by making use of the interactive form formed by propitiously traditional Chinese knot and Tai Chi. The shape of this work is the shape of a Chinese knot, and Chinese knot is a traditional Chinese folk art, known as the " Four circles are implemented 、Everything is well-illuminated ", and "Knot" take the meaning of "Continuous". Composed of the mutually surrounded and buckled five Olympic rings colors, pentagram is borrowed to symbolize the unity, cooperation, exchange and development of five continents in the world to create a new century hand in hand. The shape of this pentagram is similar in shape of human form doing Tai Chi, which represents the essence of Chinese traditional sports culture. (Figure 2)

Fig. 2. "Logo of 2008 Olympic bid"

The above two examples are living example for designers properly applying traditional culture art into modern graphic designs. In fact, traditional Chinese folk paper-cut art is the body art of traditional Chinese culture and art, and the art plays an active role in the promotion of modern graphic design. So, the where is to worth learning and carrying forward in traditional Chinese folk paper-cut art?

First of all, with wide range of topics and rich contents, the traditional Chinese folk paper-cut can provide ample visual art resources for modern graphic design.

The authors of folk Paper-cut, mostly are rural laborers coming from remote areas in mountains, so most of their works' theme is taken from their own real life, such as the bumper grain harvest, festival seasons, feed chickens and ducks, cattle and sheep, engage in family sideline production, and participate in field work and so on. And some expression used by plants is often seen in life, such as: plum, orchid, bamboo, chrysanthemum, peony, lotus, narcissus, and various fruits, vegetables and so on. Because these themes are derived from life, so the contents of paper-cut works' expression are very diverse. Paper-cut means that people play their hopes on for a better life and reflect the desire of the Chinese nation to look forward to and pursue the true, the good and the beautiful.

There are also the themes of historical stories and myth legends, such as "Butterfly Lovers", "Madam White Snake ", "A Dream of Red Mansions ","The West Chamber "and other story plots. In addition, "The Eight Fairies", "Moon ", "The heavenly maids scatter blossoms "and other folk tales are the popular themes of paper-cut.

These traditional folk paper-cut images are not incidental creation, but go through after thousands of years, and are created by continuous machining of many folk artists. It is the crystallization of human wisdom. Learning from these images with distinct personalities and the characteristics of a strong folk customs in modern graphic design, they not only greatly enrich our graphic design language, but also have the fairly stable audience, because they come from life, which are known by the public and have a very wide range of social universality.

Secondly, these traditional folk paper-cut images are of profound meaning and oriental aesthetic implication, drawing lessons from modern graphic design to further the sublimation of its contents and details.

The traditional Chinese paper-cut pursue the beauty of artistic conception of "conception outside the image", stressing on pictures must have meaning, and the meaning must be lucky ", which is because of the aesthetic ideas of people to pursue of perfection and appiness. Such as some images:

Plum: Plum is divided into five limbs and plum has the metaphor of five kinds of blessing "blessing, pay, longevity, happiness, and fortune"; peony: the king of flowers, Because of its beauty, it means wealth and worldly glory. Five venoms: toads, scorpions, lizards (gecko), centipedes, snakes. In civil society, when every fifth day of lunar month comes, people will in their houses scatter lime, burn mugwort (a kind of very choking herb), drink realgar wine, and paste the paper-cut carved Five venoms. People hope that these five kinds worms can combat poison with poison and ward off diseases and evils; Magpie: it is said that people hearing the sound of magpie will have good luck; Mandarin: it considered as a loving couple in the same camp ; Pine, crane: they are liken as macrobian by the words of Longevity Crane; Deer, crane: use the homophonic of "land ", the "land" and " six" are homonyms, Crane has the homophonic of "together ", the homophonic of deer, and cranes is "Liu He", Liu He means heaven, earth, east, south, west, and north.In addition, deer also has the homophonic "Pay", drawing deer and magpies together with the "Magpie deer" homophonic of " rank of nobility and its salary ", so deer and crane symbolize the official seal and promotion and high post with matched salary and so on.

This aesthetic thought of People pursuing perfection and happiness constantly affect our art design, while if we make a further research on implication of traditional folk paper-cut art and add modern elements, it will produce a spectacular artistic effect. Modern graphic designers in this regard have many successful references, for example: Tsinghua University Professor Xiao Hong in Macau emblem design has successfully use the image of lotus, taking the auspicious meaning to wish Macao's prosperity and development of motherland. (Figure 3)

Fig. 3. The emblem of Macao

Another outstanding design example is Bank of China's corporate logo designed by Hong Kong designer Kan Tai Keung (Figure 4), and this logo makes full use of ideological expression style from the traditional ideas in paper-cut---- " Heaven is round and Earth is square " Borrowing shape of ancient coins implied " Heaven is round and Earth is square ". The money hole and red string constitutes a word "in", showing that the Bank of China is an internationally financial institution facing the global, but also marks out the national concept at the same time.It is Not only highlight the features of banking industry, but also gives the bank more traditional culture connotation, and endow the bank with the auspicious meaning " safely complete " as folk paper-cut. These characteristics of paper-cut have been fully reflected in this logo design.

Fig. 4. Bank of China's corporate logo

This reference and application make the above design works have more profound cultural background, and also generate more association of national sentiment.

Next, the visual patterns of mixed space-time of traditional Chinese folk paper-cut can bring infinite inspiration to modern graphic design.

Chinese traditional paper-cut art can freely constitute images 'organization and image contents, rather than rigidly adhere to the real world, with great flexibility and creativity, so images in different space-time can be organized into the same frame, and we can see the eyes decorated with a peony, the fish decorated with lotus, small animals decorated in the animal stomach, the roof of house decorated with peony flower and other technique of expression. Therefore it can mutually connected images in different space-time according to the need of paper autistics, such as this paper-cut work entitled "immediately make a fortune, " In the painting, it is show that a newly Number One Scholar who just passed civil examination, standing in front of their own loft, both sides of the two horses, the horses laden with coins coming slowly. But here the flowers are not really flowers, but the flowers derived by coins from horseback. With a strong modern style, this design thinking has gone far beyond the thinking of European classical realism painting. in this piece of paper-cut, there have been horses, coins, flowers, attic, bats, the Number one scholar and other elements, and the content is very rich. Number One Scholar who just passed civil examination is one of the highest frequency characters in traditional Chinese ancient dram, because passing the civil service examinations to step into bureaucracy in the ancient feudal society has been the dream of the Chinese people. These elements in everyday life are very difficult to appear in the same space-time, however, in order to express the yearning for a better life, paper-cut artists skillfully blend those elements with specifically good meaning into the same space-time, whose ability of controlling images is marvelous. Meanwhile, it really express that the ancient Chinese people pursue the concept of peace, wealth, good fortune and other conceptions.

The Chinese traditional paper-cut creation is no principle of perspective and modeling concept. But through the self-confidence and talent sensitivity, it use the frame of different bright, vivid symbols in two-dimensional plane composition to form the visual shock beyond three-dimensional or even four-dimensional space, resulting in full, harmonious and brilliant pictures. Paper-cut only has a piece of paper and a pair of scissors, so it is difficult to show three-dimensional, 3D scene, but because of this limitation, creating the richness of paper-cut artistic expression. In the paper-cut world, existed or not existed in real life, are all included as one to communicate the heaven and the earth, so that man and nature can be merged into one, which is the visual of mixed space-time of Chinese paper-cut art, it can giving the infinite inspiration to modern graphic design.

Chinese folk paper-cut art belongs to China, also belongs to the world. Recalling the development of art history, many artists are enlightened in the folk paper-cut art: The French painter Henri Matisse boldly learns the nutrition of traditional folk art, getting the unique brutalism from traditional art.

2   Conclusion

Learning and drawing lessons from the traditional folk paper-cut art, and applying them into modern graphic design, can continuously enrich our creative thinking ability, inspire the designer inspiration of designers to design better graphic design art works, which will effectively promote the development of modern graphic design.

References

1. Wang, S.: A History of Graphic Design. China Youth Press (1998)
2. Jing, C.: Chinese folk paper-cut art study. Beijing Arts and Crafts Press (2001)
3. Lu, S.: Chinese folk paper-cut. Hunan Fine Arts Publishing House (2000)
4. Lan, Z., Pu, F.: Chinese Folk Art. China Textile and Appare Press (2003)
5. Li, Y.: The history of Visual Communication Design and Aesthetics. China Renmin University Press, Beijing (2000)
6. Xu, Z.: Overview of Chinese Folk Art. People's Publishing House (2006)
7. Zhu, G.: Modern poster art history. Bookstore Publishing House, Shanghai (2000)
8. Chang, Z.: Outline of Chinese traditional culture. Law Press, Beijing (1999)
9. Wang, X.: Modern advertising design. Fudan University Press (2002, 2004)
10. Wang, X.: Modern advertising design. Fudan University Press (2002, 2004)
11. Zhong, R.: Trace and Image. People's Fine Arts Publishing House (2004)
12. Wang, G.: Graphic design. Tsinghua University Academy of Fine Arts Press, Beijing (200)

Analysis of Chinese Tea Packaging Design

Liu Hui

Institute of Art and Fashion, Tianjin Polytechnic University, Tianjin, China
llhhliuhui@126.com

Abstract. Tea has a long history in China. Tea history is not only a history of Chinese nation's humanity development, but also a colorful folklore history. This paper mainly analyzes and studies on materials selection in tea packaging design, configuration, picture, color, character and other elements. Excellent tea packaging not merely protect the tea from several problems encountered in the environment, but also can transmit product's information accurately and rapidly.

Keywords: Material structure, Graphics, Color, Character, Tea culture.

1 Introduction

Tea is the great contribution that Chinese nation make to world civilization, welcomed by the world for its magic effect and charm. In thousands of years, tea is not only a material enjoyment, but also a spiritual pleasure.

With the times progress and rapid economic development, the types of tea are various in modern times, its packaging style is also different, but there were many tea packaging defects in these packaging, such as tea backward packaging design, out of date, behind the times or " virtual standstill "; and tea manufacturer pay inadequate attention to tea packaging, resulting in single creative tea packaging. Many tea packaging designs are very similar, lacking of unique feature of their business enterprises and the unification is very serious. Tea personalized and creative packaging has become a major trend in future's brand competition. China's tea industry is lack of innovation strength, which will inevitably be involved in the whirlpool of the unification. How to design a more excellent tea packaging works are practical problems our modern designers should be resolved as soon as possible. So our designers should study the package of tea from the following points.

2 Materials Selection and Configuration of Tea Packaging

Superiority and inferiority in materials selection and configuration of tea packaging are not only having a direct impact on the tea bloom, protection and transportation, but also provide a solid surface for the presence of tea packaging design.

Tea is China's traditional export commodities. Tea's characteristics are decided by the physical-chemical composition and its quality, such as moisture absorption,

J. Luo (Ed.): Soft Computing in Information Communication Technology, AISC 161, pp. 269–274.

oxidation, adsorption, fragility, volatility and so on. With the prosperity of foreign trade, tea packaging varieties are increasing. Recently, many materials are suitable for tea packaging, mainly including three categories: paper, plastic film, metal and ceramic; as well as a small amount of bamboo, wood and other natural materials, various composite packaging materials, such as paper and plastic composite and aluminum-plastic composite are also widely used in tea products. We should pay attention to the choice of tea packaging materials as the followings:

A.   Tea Packaging Materials Must Be Sanitary, Non-toxic and Food-Grade Packaging Materials

Due to packaging's hazardous substances can be transferred to the tea, causing pollution of tea, so tea packaging materials should be sanitary, non-toxic, free from pollution of fungicides, pesticides, preservatives, fumigants and other items, besides these, auxiliary agent, such as printing inks and adhesive materials in the selected packaging are also required to be not harmful to human body.

B.   The Tea Packaging Materials Had Better to Select the Durable, Moisture Proofing, Shading and Good Oxygen Barrier Properties Materials

Concerning tea has characteristics of moisture absorption and odor absorption which can be highly oxidative deterioration, so requirements should be higher on water vapor permeability and oxygen permeability of the tea packaging materials.

C.   Tea Packaging Materials Must Be Considered to Use Green Packaging Materials

With the progress of modern society, people's demand on living quality and standard become higher and higher. Concern for the environment and environmental protection of packaging materials have become an inevitable trend of social development. Especially for the natural, pollution-free tea products, the sense of environmental protection should be strengthened, in line with people's needs to return to nature.

In fact, when we design the tea packaging, no matter what kind of materials to select, above all we must know the characteristics of tea itself, considering using appropriate materials for packaging. We can select appropriate material in reasonable use according to these characteristics, so appropriately, only after we fully understand the tea characteristics and factors causing tea deterioration.

At the same time, tea packaging design is a medium to convey information. If it has a unique shape, which can lead to visual pleasure for the consumer, in the design of shape and structure, we can draw lessons from traditional tea culture, designing and making a mood and unique feature box type. We should pay attention to the shape's beauty, uniqueness, ingenuity, the structure of the scientific and solid in the design, such as the tea box may be designed into polygon, or shape of tea, bamboo-like, or finding design inspiration from the ancient tea art and tea ceremony directly to design creative freshness packaging type.

3   Tea Packaging Graphic Design

The aim of tea packaging graphic decoration is intended to provide a more prominent image of products, so consumers have a rich psychological association and affect the feelings of consumers, stimulating consumers' desire to buy. Although these graphics have realistic, abstract, exaggerated artistic techniques. No matter what kinds of artistic technique are used, we should consider tea products' characteristics first, and then beautify them with a variety of artistic techniques to make people leave good impression on the goods.

At the present time, there is many tea packaging's design level lagging behind. Some graphic decoration look old, excessively detailed and weak commodity characteristic and design style lack of sense of times, for example, some tea packaging copy traditional patterns without creating to express the national style. I think the modern tea packaging must use new art form to modify, even with the traditional patterns, which should also be entrusted some new contents, determined by the essence of modern tea packaging.

Tea is a kind of national characteristic goods. We can use some traditional patterns full of strong national cultural spirit, such as Chinese paintings, decorative patterns, auspicious patterns, the folk paper cuttings, ethnic minorities' patterns to express tea tradition. Tea is a famous traditional commodity in our country, its packaging and decoration design should focus on the nationalization, but nationalization does not mean ancient style, because tea is not unearthed cultural relics, the misappropriation repetition of ancient patterns such as no character dragon and phoenix patterns. Tradition is not retro and not copy. National characteristic is not only represented by drawing a dragon, a phoenix, so we can not just stop at the copy and copy some traditional patterns, we should integrate aesthetic characteristics of national cultural spirit spiritual essence such as spreading feelings, implication and delicate into tea packaging resign. The traditional patterns can be deformed by using modern techniques to make it have the feeling of modern flavor, symbol, conciseness, so that it will full of both national characteristics and sense of modern at the same time to shine with attractiveness, that is, implicitly represent the traditional Chinese culture through modern design techniques in tea packaging designs.

4   The Color of Tea Packaging Design

In the tea packaging design, color that affect visual perception, is one of the most active and most sensitive elements of the vision. Color plays a very important role in the psychological influence of the customer, because color has strong visual impact and is likely to cause people's mental changes and emotional reactions. One of the important factors for a successful tea packaging design is the rational application of packaging color. It requires the designer having a wealth color theory knowledge and careful and acute observation to color, and fully understanding different appreciation habit and aesthetic psychology of customers, grasping psychological law in customers' cognition of tea packaging color and appreciation of tea packaging, through which means making color to become more attractive.

At present, the species of tea in the market can be divided into green tea, black tea, yellow tea, dark tea, white tea, green tea, etc. according to different tea colors. We should use different colors for different tea species, such as black tea whose tea infusion color is red and bright, when imported like alcohol, fresh lasting, memorable, taste in fragrance, so we should use warm tones, appropriately warm, which make people feeling mellow and interest richness; the green tea has jade-green color, spiciness, with four unique features of " beauty form, green color, sweet-scented, pure taste" which is the tea masterwork, so we had better to choose green, blue and other cool colors, giving fresh and mellow feeling; white tea is characterized by covering pekoe, its soup color being apricot with a fragrant flavor, the taste of alcohol, so we should use light, gentle color… but this only present general rule of tea packaging color, rather than absolute formula.

Therefore, in design, tea packaging coloring should have goods' individuality, unique features, giving association to people. Only if we have some knowledge of color association, we can design tea packaging with artistic charm to promote the sale of tea.

The so-called color association is referred to the psychological process of other things which provide consumers color perception to image by the goods, advertising, shopping environment or other conditions. People's associations to color are divided into concrete association and abstract association.

A.   The Concrete Association Is Referred to the Color One See Associating with That Concrete Thing

Such as when one sees the color white associating it with the snow, paper, clouds and sugar, etc; when one sees the gray associating it with rat, concrete, etc; when one sees the black associating it with carbon, night and hair, etc; when one sees the red associating it with apple, red flag, blood, lipstick, etc; when one sees the orange associating it with mandarin orange, persimmon, carrot; when one sees the brown associating it with tree trunk, chocolate, etc; when one sees the yellow associating it with banana, sunflower, lemon, etc; when one sees the green associating it with leaf, grass, etc; when one sees the blue associating it with sky, marine, etc; when one sees the purple associating it with grapes, seaweed and so on.

B.   The Abstract Association Is Referred to the Color One See Associating with Certain Abstract Concept Directly

Such as when one sees the red thinking of passion, enthusiasm, revolution, warmth, happiness and dangerousness; when one sees the orange thinking of happiness, warmth, clearness, positive and delight; when one sees the yellow thinking of lucidity, brightness, energetic, pleasure, richness, excitement and attention; when one sees the green thinking of peace, quiet, growth, equity, security, cool and fresh; when one sees the blue thinking of quiet, calm, cool, rational, irrational, speed, young, youth, honest, truthful, trustworthy and cold; when one sees the purple thinking of noble, lofty, mysterious, quaint, elegant and delicate; when one sees the black thinking of calm, heavy, energetic, serious, dark, cold, heavy, sad, despair and death; when one sees the white thinking of pure, innocent, pure, clean, pure, neatness, lucidity, peaceful, sacred and blank; when one sees the gray thinking of modest, frustrated, mean, melancholy, despair, dark, deserted, ordinary, quiet and died ash.

5   Character Design in Tea Packaging

Character is acting as linguistic signs to record human being's history, becoming a symbol of human civilization. The character is an important communication tool for people to transmit information and communication, representing a cultural symbol code. In the modern packaging design, consumers have to generate thought relying on appeal of character's message, enhancing awareness towards products. The character not only transfer service function of product information's appeal, but also contains clear forming aesthetic feeling and forming semantics.

Packaging character is also an important part of the design, a package can have no decoration, but not without words, as a person, he must have a name. Characters in tea packaging must be concise, clear, and fully embodies the attributes of goods, easy to use too complicated Lock is not easy to identify the font and the words, not using too fussy and illegible characters. Tea is a goods having strong sense of tradition and nationality, so it is not suitable to use incondite and difficult character. The art of calligraphy has a long history in our country and has high artistic decorative effect. People have a strong ability to accept and love on the traditional art of calligraphy. The visual effect of traditional calligraphy art has become an identical form in designing national style. Various letterforms of calligraphy's structure have different forms, different styles, and its historical development process is a history of letterforms' design. Such as large seal script: stagnation, clumsy, messiness, plain, patterned decorative beauty; small seal script: a tightly knit structure, a Chinese character, well-proportioned blank, gently and mellow; clerical script: good stretch, stroke having round in square, lines with balanced beauty; regular script: well-behaved, preciseness, emphasizing the temper rolling of the structure and so on. There are also some letterforms such as eaves tiles, coins, seals and so on. These letterforms design are more abundant and having various styles.

When design for the tea packaging, we can use calligraphy to reflect the rich cultural connotations of tea properly, embodying a long history of Chinese culture, but by using some letterforms easy to understand, easy to read, easy to identify, avoiding using the letterforms are too grass or less clear, we must take consumer discretion into consideration and make the letterforms clear.

6   Conclusion

In conclusion, materials selections in tea packaging design, configuration, picture, color and character and other elements are not working in isolation, but having correlation, mutual influence and combined action. We should carry out general comprehensive study and discussion to achieve the best effect of tea packaging design, better to carry forward the national drink and put forward Chinese tea culture to the world gradually.

References

1. Yin, Z.: Packaging Color Design. Chemical Industry Press, Beijing (2005)
2. Chen, D.: Design for Market Positioning. Tianjin People's Arts Publishing House, Tianjin (2001)
3. Wang, Z.: Packing Classification Design - Design Basis. China Light Industry Press, Beijing (2001)
4. Wang, A.: Packaging Upholster Design. Henan University Press, Henan (2004)
5. Culver, G. (ed.), Wu, X. (transl.): What Is the Packaging Design. China Youth Press, Beijing (2005)
6. Lin, Z.: Package Design. Guangxi Fine Arts Press, Guangxi (2003)
7. Fan, K.: Package Design. Shanghai Pictorial Publishing House, Shanghai (2005)
8. Zhu, H.: Package Design. Hunan University Press, Changsha (2006)
9. Liu, X.: Basic Knowledge of Tea and Tea Culture. China Labor and Social Security Press, Beijing (2003)
10. Wang, A.: The Visual Design of the Packaging Image. Southeast University Press, Nanjing (2006)
11. Liu, Y., Mao, D.: Packaging Design. China Academy University Fine Arts Press, Hangzhou (1997)
12. Lu, J.: Traditional Chinese Tea Art. Oriental Press, Beijing (2010)
13. Wang, L.: Chinese Tea Culture. World Publishing Company, Beijing Company, Beijing (2007)
14. Hua, B.: 150 Years of Packaging Design. Hunan Fine Arts Publishing House, Changsha (2004)

On Undergraduate Management of Research Colleges

Fan Tian

School of Materials Science and Engineering, Tianjin Polytechnic University, Tianjin, China
243815161@qq.com

Abstract. The innovative nature of the research colleges decides its personnel cultivating objective- innovative talents. The college student management plays a very important role in cultivating innovative talents. This article describes the content and feature of research colleges, analyzes current problems of undergraduate management in research colleges, and finally proposes the countermeasures to improve the undergraduate management in research colleges.

Keywords: Research, undergraduates, management.

1 Introduction

The nature of research colleges is knowledge innovation, and its talents training objective is to cultivate innovative talents, which is the essential difference between research colleges and other colleges. During the growth process of innovative talents, undergraduate education plays a foundational role. The college student management is an important part of the college talents training system, having a decisive effect on the realization of talents training objectives and the quality of talents training. The construction and development of research colleges make new requirements on the student management; therefore, it is an inevitable trend of student management reformation and development to improve the undergraduate management of research colleges, which is also the objective requirement to realize the innovative talents training aims of research colleges, even of research universities. Moreover, it has the important practical significance to achieve college talents training, implement the function and role of higher education, and then to serve the socialist modernization better.

2 The Content and Feature Analysis of Research Colleges

The research college is an emerging concept, and it is also an effective way for many major universities to achieve the goal of constructing research universities. To the non-research colleges, research colleges are, in the growth process of university, based on the existing management system at both levels of school and college, and they can become discipline colleges, which are prominent in the personnel training, science research, social services, etc., by integrating their existing superior disciplines.

J. Luo (Ed.): Soft Computing in Information Communication Technology, AISC 161, pp. 275–281.
springerlink.com

According to the actual situation in China, the research colleges are divided into two types: the basic type and mature type. The basic type of research colleges has the preliminary capacity and feature to innovate knowledge. This kind of research colleges ranks from the top 10% to 20% in the discipline evaluation, and the proportion of undergraduates and postgraduates should be less than 2:1; besides, there are degrees for international students to study. While the mature type of research colleges, they relatively develop more stably, and possess the capacity and advantage to innovate knowledge. This article focuses on the innovation and improvement of undergraduate management in basic research colleges. For the basic research colleges, especially for those universities which are not strong in the overall strength but have some colleges that possess superior disciplines, they will be able to play the demonstration and radiation effects in the construction of research colleges, thus boosting the development of the whole university. Meanwhile, the process of building research colleges is also the process of exploring and forming the differentiation and specialization of education goals.

The innovation of knowledge constitutes the essential feature of research colleges, which is also the sign different from other colleges. Research colleges have the following general features: highly qualified teachers and better teaching conditions, which is helpful for students to lay a solid foundation of knowledge; high-level disciplines and rich academic resources, which is conducive to students to broaden their knowledge; high-quality and active-thinking students; innovative knowledge as the essential feature of research colleges and students as the subject of innovative knowledge. Students cultivated by research colleges are innovative thinking and have the spirit of being brave in creation and development. Therefore, they have a solid foundation of knowledge, a strong sense of innovation and possess the outstanding practical ability.

3   Problems of Current Undergraduate Management of Research Colleges

The student management of research colleges are constantly facing new challenges. There are some factors of research college undergraduate management which constrain the cultivation of innovative talents.

3.1   Student Management Philosophy and System Construction Lagging Behind

In recent years, the actual principle of college student management in China is generally still not out of the main mode of preaching. The education reform and the guidance to lead students to self-study and self-cultivate are both not in place, which cause that students cannot fully develop their personalities and their creativity ability also can not get fully strengthened. Students are accustomed to accept being managed, while their awareness and ability of self-education, self-design and self-control are not strong; besides, scientific and technological innovation are relatively backward.

The student management system and the construction of evaluation system are lagging behind. At present, the influence of student work in the cultivation of innovative talents has not attracted adequate and substantial attention, and related policy measures are still not formulated. Besides, it is also not optimistic about the situation of college students. However, the enhancement of students' innovation awareness also needs scientific system construction, which requires to not only strengthen the system construction, establish effective measures to inspire students to participate in the work of innovative talents training, but also establish incentives to stimulate students to actively participate in innovation activities.

3.2 Releasing the Student Management Process

Under the influence of the examination-oriented education, teachers, students, parents, and even the whole social environment always take the exam results as the sole purpose of learning, which has greatly affected the education management development of university students. Some students do not pay attention to the daily accumulation of knowledge, the master of learning methods and the training of their overall quality, while some other students simply learn to pass exams, burning the midnight oil before the examination, so some students are eventually qualified in the exam without daily hard study. If the situation continues, it would have an adverse effect on the students' education quality and overall quality development. Thus, the student management of research colleges should pay more attention to the guidance and test of the process so that the developments of students, in terms of process and results, relatively achieve the unity.

3.3 The Absence of Students as the Subject in the Student Management Process

The development of the innovative ability is bound to depend on individual's initiative. In the process of college education management, the excessive dominant play of student administrators would lead students to lose their subject position; therefore, the full play of their subject will be suppressed. The lack of subject is not conducive to the free development of innovation awareness and innovative thinking of college students, thus college students subjectively lack active innovative thinking, strong innovative motivation, strong innovative will and healthy innovative emotion, which can be seen from that they are used to indoctrination education, and their sense of learning competition is not strong, independent thinking and the spirit of hard study are not enough, not to mention the spirit of innovation.

3.4 The Perseverance and Adaptive Capacity Cultivation

For students unable to meet the changing situation currently, the social competition is increasingly fierce, so it is necessary to cultivate the persistence and adaptive capacity of college students. Some students lack of perseverance, entrepreneurial spirit and adaptive capacity, when they encounter setbacks, they have no will to fight and become frustrated, and some even want to commit suicide. However, at present, the perseverance and adaptive capacity education in colleges are still very poor.

3.5 Construction of the Student Management Team Lagging Behind

Features of research colleges and the characteristics of research college students require research college student teams not only possess the management capacity to organize students, but also own the research and innovation guiding capacity to cultivate innovative talents.

First of all, there is a lack of full-time student management staff; moreover, many counselors consider their future, so they just take it as a temporary job, not regarding it as a life-long position. However, the class adviser is usually adapted as a part-time job and they transfer and change quickly, besides, their treatment can not be perfectly solved from the student management itself, from which it can be apparently seen that student management team is still in the relatively unstable situation. Second, with the full implementation of quality-oriented education, the student management gives prominence to comprehension, hierarchy and modernity, which requires student managers have more extensive management knowledge and modern managing methods. From the current situation, the opportunities are few for student personnel sent to study abroad, pursue further education and get improved. Finally, the features of research college student work require student personnel have high-quality competence, while the development of research colleges and the characteristics of college students also propose many higher requirements to them.

4   Countermeasures to Improve Undergraduate Management of Research Colleges

Research colleges provide a better platform for student personnel to carry out their work. The student work can play an important role in the construction of research colleges, guaranteeing and guiding the growth and success of students, then contributing to the construction of research colleges. It is an inevitable trend of reforming and developing undergraduate management by re-examining the status and influence of student work in the innovative talents training, and improving and enhancing undergraduate management of research colleges. The undergraduate management can get improved and enhanced from the following aspects:

4.1   Improvements of Educational Management Philosophy

The key of promoting educational reform and development is to update the philosophy, which requires to fully implement the education policy and guidance of the Communist Party, take the Scientific Outlook on Development as the guidance, implement the spirit and educational plans of the National Education Conference, adhere to the principles of innovative education and make full use of the advantages of student management, then to serve the innovative talents training. The student management of research colleges requires to update management philosophy and to change the traditional management concepts. First, the relationship between the managing and the managed should be equal and interactive; second, the administrators ought to make full use of school resources, mobilize all the staff to get involved in student management, then to realize the goal of the education through all-staff; third, the administrators should manage by taking advantage of social forces, taking the school management into the scope of social management; fourth the administrators ought to make full use of the students' self-management, improving their own conscientiousness, initiative, activeness and creativity, and the administrators can also invoke the backbone of students to manage others. The administrators should change the traditional mandatory and rigid management methods, re-establish the dominant position of students and treat students as equals, handling the relations among education, service and management.

4.2   Further Establishing "Human-Oriented" Student Management Thought

The talents training objectives of research colleges are to cultivate innovative talents who possess creative spirits and practical abilities, which require that the student management model of research colleges should fully embody the concept of student-based management, giving prominence to the influence of non-intellectual factors, and promoting students' personality development.

Attach importance to the dominant position of students. In the student management of research colleges, the administrators should respect students, from the system to ensure the students to have the autonomy to choose. Administrators and students are equal partners, and students have the right to know, participate, and select the school teaching programs, training programs, rules and regulations, curricula and other information related to their life and study. In addition, the college administrators also may select some student representatives to participate in the college student management, by holding hearings about student management systems, thinking symposia, seminars of student work, etc. helping them participate in the formulation of student management regulations, evaluation of teachers and the collection and feedback of teaching information. That will not only enhance the students' initiative to acquire and use knowledge, but also develop their practical abilities, effectively expanding college students' rights of participating in the process of student management and teaching management.

4.3   Making Full Use of the Ideological Education Influence in Cultivating Innovative Talents, Shaping Students' Creative Personalities

The ideological and political education should be penetrated into the entire process of talents training. For the research colleges, the ideological and political education provides the orientation, motivation and guarantee for the cultivation of innovative talents, so the administrators should make full use of the ideological and political education influence. The ideological and political education can help students establish scientific world outlooks, views of life and values, then set socially required and proper goals. Meanwhile, through spiritual motivation, the ideological and political education could inspire and encourage students to innovate, improve the innovative characters of research college undergraduates in the process of education management and services, and enhance the will and perseverance which are conductive to the cultivation of innovativeness.

Through advising students' study and career planning, conducting psychological counseling, doing career guidance services and other aspects of talent training, college student administrators helps the students gradually develop strong will and perseverance which help them improve their self-control and difficulty-overcoming abilities,   so that students' sound personalities become an important guarantee for the cultivation of creative spirits; they also need to carry out a variety of science and technology student clubs activities and formulate corresponding incentive measures to reward innovation, guiding the healthy development of students interests and hobbies, making that the internal driving force to develop the spirit of creativity; besides, they ought to create more space for the teacher-student interaction and more opportunities for the students to do things by themselves, strengthening the construction of the research team, then stimulating the students to think independently and develop innovative awareness; finally, they should focus on creating the innovative environment and enhancing the influence of the innovative spirit on the students. One of the basic functions of the ideological and political education is to create a soft environment conducive to the growth of innovative talents, and ensure the realization of training objectives and the healthy growth of talents. Extra-curricular activities, community culture, professional societies, campus atmosphere, etc. are the important part of the campus innovative environment. A variety of innovative activities and report meetings will subtly mobilize the initiative of the students and arouse their sense of innovation and form a creative atmosphere.

4.4 Further Reform and Optimization of Internal Management System of Colleges

The reform and optimization of internal management system of colleges are the important measures to improve the student management and main approaches to improve the overall quality of students. In the process of construction, when facing the individualism of students' development, autonomy of students' learning and diversity of students' needs, etc., research colleges should establish and improve the linkage mechanism of management, implement team development, strengthen the education through all-staff, then form the powerful resultant force to cultivate high-quality innovative talents.

First, according to the activities and functions of students' life groups and learning communities, the administrators establish a management system mainly based on the line of "school — class —league—student community", forming a class management system of "class instructor—counselors—head teacher", then to create the good situation---"integration of upper and lower levels and making concerted efforts". Student management of colleges is mainly responsible for students' caucus construction, professional development guidance and capacity improvement and evaluation, undergraduate tutorial system, students' psychological health, career planning, science & technology competition, employment guidance and services, and daily student affairs, working from two aspects: the students' major and their class. Second, the administrators should establish the linkage mechanism between student management and teaching management, making it the decision-making core, guiding center and coordinating center of student management. Third, they should establish a scientific evaluation system for university students. The quality assessment system, as the baton, has a strong influence on students' learning, life and work because students would adjust their own development according to the content of the evaluation system. Research colleges should have the all-around development of moral, intellectual and physical education, establishing a scientific evaluation system for students based on the overall quality of innovative talents from the objectives of cultivating innovative talents who have creative thought and abilities.

4.5 Strengthening the Team Development, Constructing the Research Student Management Team

In accordance with the requirements of innovative talents training objectives, the research colleges must establish a student work team, especially the counselor, class instructor and leader teacher team, who expert at ideological education and management and have the appropriate academic background, high overall quality and strong leadership. The majority of the student work team should be full-time, combined with part-time student workers, forming a substantial number of student work team. They should continue to enhance the team's research and innovation abilities, explicate the function orientation of innovative talents training which the student management team carries out, formulate guarantee policies and make full use of the advantages of student work, to serve the cultivation of innovative talents.

The student management team of research colleges asks for appropriate policy supports and the positive training platform construction, further clarifying the responsibilities of full-time counselors, head teachers, class instructors, coordinating their relationship, giving full play to their respective roles, and making sure that: first, further making clear the function orientation of the innovative talents training that the student administrators carry out, to ensure students administrators are highly efficient in promoting students to do research, providing the research training for students and instructing the students for their scientific research; second, from the policy level, providing financial support for the students to carry out innovative activities; third, formulating incentive policies to encourage the college students' managers to carry out the cultivation of innovative talents.

References

1. Ji, X.: On the Developmental Student Management Pattern of Colleges. Journal of Higher Education (6) (2006)
2. Yan, Y., Pang, F.: On Features of Innovative Talents and Training Key. Jiangsu Higher Education (2) (2004)
3. Long, L.: The Research of Countermeasures to Improve College Counselors Based On Role-oriented Theory. Journal of Xiangtan Normal University (3) (2008)

Study on Upgrading Construction Industry
Take Jiangxi Province for Example

Junxiao Lin and Zhenning Liu

Economics and Management School, JiangXi University of Science and Technology,
Ganzhou, China
lin_junxiao@sina.com, Liuzhenning1987@163.com

Abstract. This paper analyzes the existed defects and the causes of construction industry structure in Jiangxi Province at present stage. Focusing on the various kinds of problems for the structures of construction industry, this paper puts forward concrete measures and strategies to achieve the optimization of construction industry structures in Jiangxi province.

Keywords: engineering management, industry upgrade, present structure.

1 Introduction

The construction industry as a mainstay industry, stimulating the national economic development plays an important role in Jiangxi province's social and economic development. The added value of construction industry arrives at 64.795 billion yuan in Jiangxi, 2008, accounting for 10.0% of GDP; it has become the third mainstay industry behind the industry and agriculture. The construction industry has made a significant contribution in the promotion of social and economic development, improvement of the urban landscape and living standards, and increase of employment opportunities [1].

However, the construction industry has also exposed a lot of problems in Jiangxi province. In 2008, the national construction industry output value was 6.2037 trillion yuan, the construction industry output value of Jiangxi province was 103.3 billion yuan, accounting for only 1.7% of China, 12% of Jiangsu province. In 2008, national construction labor productivity was 161,804 yuan / person, Jiangxi construction labor productivity was 151,597 yuan / person, the construction industry labor productivity in Jiangxi province is lower than the national average. Overall, the construction industry in Jiangxi province still has space for improvement in industrial structure and labor productivity, management level and innovation is not good. Therefore, the research and analysis of the construction industry structure of Jiangxi can help to grasp the structural problems of the construction industry in Jiangxi and have a good reference to structural adjustment of Jiangxi construction industry.

2 Definition of Construction Industry Structure

Before analyzing the industrial structure in Jiangxi, the author believes that it is necessary to clarify the definition of construction industrial structure, which helps focus on the study, and will not lead to the fuzzy research direction.

J. Luo (Ed.): Soft Computing in Information Communication Technology, AISC 161, pp. 283–290.
springerlink.com   © Springer-Verlag Berlin Heidelberg 2012

Industrial structure is also known as the sectoral structure of national economy. According to the theory of industrial economics, industrial structure refers to the production, technical, economic ties and number proportional relationship between industries of national economy and within the industrial enterprises in the process of social reproduction [2]. From this basis, we can define the construction industry structure as the number ratio and the competition and collaboration relationship of all sizes and types among the construction enterprises.

3   Existing Problems of Construction Industry Structure in Jiangxi

A.   Capital Structure Imbalance of Construction Industry Structure

Since the eighties of the 20th century, China began to promote the reform of ownership of construction industry. So, the corporate ownership of construction industry evolved into today's state-owned enterprises, collective enterprises, private enterprises, joint-stock enterprises, Hong Kong, Macao and Taiwan enterprises, foreign invested enterprises and other construction companies, which made up to the eight basic forms, from the original state-owned enterprises and collective enterprises.

Table 1. Different ownership construction enterprise in Jiangxi (unit: Home)

Years	Total	State-owned	Collective	Stock cooperation	Joint management	Limited liability	Stock limited	Private	Hong Kong, Macao and Taiwan	Foreign merchant
2003	1106	213	317	39	2	208	122	189	11	5
2004	1387	226	300	44	13	344	93	348	13	6
2005	1362	213	279	44	9	354	100	344	15	4
2006	1328	190	260	29	4	334	113	343	15	3
2007	1325	187	251	25	5	350	117	367	15	3
Data sources from Jiangxi Province Statistical Yearbook (2004 to 2008)										

Table 2. Construction enterprises of different ownership in the number of employees in Jiangxi (Unit: Million people)

Years	Total	State-owned	Collective	Stock cooperation	Joint management	Limited liability	Stock limited	Private	Hong Kong, Macao and Taiwan	Foreign merchant
2003	50.96	16.38	16.49	1.25	0.1	8.7	4.14	3.71	0.15	0.04
2004	53.89	15.96	17.36	1.09	0.22	12.55	2.6	6.3	0.23	0.04
2005	61.9	17.75	16.51	1.35	0.36	13.59	3.87	8.19	0.24	0.04
2006	61.98	16.29	16.41	0.57	0.12	14.59	4.19	9.12	0.47	0.03
2007	62.82	16.86	15.7	0.53	0.06	14.58	3.88	10.25	0.72	0.01
Data sources from Jiangxi Province Statistical Yearbook (2004 to 2008)										

Table 3. Construction enterprises of different ownership in Jiangxi (Unit: Billion yuan)

Years	Total	State-owned	Collective	Stock cooperation	Joint management	Limited liability	Stock limited	Private	Hong Kong, Macao and Taiwan	Foreign merchant
2003	359.9	147.1	90.9	8.2	0.5	66.1	23.8	21.7	1.3	0.3
2004	462.2	149.3	108.7	8.6	1.9	124.3	22.5	45.1	1.5	0.3
2005	566.5	192.3	117.4	11.2	2.9	147.6	28.9	64.2	1.8	0.2
2006	667.1	216.8	129.2	5	1.5	188.7	34.5	87.6	3.6	0.2
2007	784.43	247.1	151.9	5.9	0.5	227.4	42.1	104.4	5.1	0.03
Data sources from Jiangxi Province Statistical Yearbook (2004 to 2008)										

The author, according to the Jiangxi Province Statistical Yearbook (2004 to 2008), sorts out the data about number of enterprises of different ownership, corporate employees and corporate data output, respectively as shown in Table 1, Table 2, Table 3.

In order to facilitate the observation and analysis, the data in Table1, Table2, and Table 3 are drawn up scatter diagram as shown in Figure1, Figure2 and Figure3.

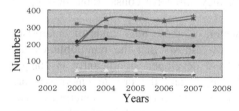

Fig. 1. Different types of ownership of enterprise quantity change

Fig. 2. Different types of ownership of enterprise employees

According to the data in Table 1, Table 2, Table 3, and combined with Figure 1, Figure 2 and Figure 3, during the 5 years period from 2003 to 2007, it is easy to find the number of state-owned enterprises has showed a downtrend, but the number of employees were keeping stable and the total output value presented an uptrend. The total output value of state-owned enterprises accounted for the proportion of total output value of all enterprise were respectively 40.87%, 32.30%, 33.95%, 32.50%, 31.50% during the 5 years period. Generally speaking, it presented a downtrend. In general, the state-owned enterprises present the contraction tendency, but it is

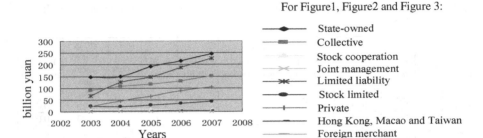

For Figure1, Figure2 and Figure 3:

———●——— State-owned
———■——— Collective
          Stock cooperation
——✕—— Joint management
——✕✕—— Limited liability
———●——— Stock limited
———+——— Private
——————— Hong Kong, Macao and Taiwan
——————— Foreign merchant

Fig. 3. Different types of ownership of enterprise value change

occupying the superiority status at present stage. During the 5 years period from 2003 to 2007, the number of collective enterprises and employees assumed a declining trend. And the total output value occupied the proportion of total output value of all enterprise respectively were 25.26%, 23.52%, 20.72%, 19.37%, 19.36% during the 5 years period. As a whole it presented the declining trend. The main cause of having this kind of change tendency is that it has carried on the various degree reforms to the property system of collective enterprises in recent years and has resulted the quantity of the type of other system of ownership enterprises to increase largely. It can foresee that the occupied proportion of collective enterprises in the construction enterprise will be less and less along with the further deepening of property system's reform in the future. [3] As for the other enterprises in addition of state-owned, collective enterprises, the development of the best are limited liability, the stock limited and the private enterprises. These three increase year by year in the aspects of enterprise quantity, the employees and the total output value in the recent 5 years. And they are gradually becoming the major ingredient in the construction enterprises. Because stock cooperation, joint management, Hong Kong, Macao and Taiwan and foreign merchant account for the proportion to be too small, they still have considerable developing spaces.

The change tendency of the three indicators of these nine kinds of enterprises during the 5 years period from 2003 to 2007 shows that the non-public-owned enterprises in the construction enterprise have developed significantly in Jiangxi province. The limited liability, the stock limited and the private enterprises are gradually becoming the major ingredients in the construction enterprise. The proportion of state-owned construction enterprises and collective enterprises is declining. The structure adjustment of the ownership capital has initial effect. However, the proportion of state-owned and collective enterprises declined. seen from the present, the proportion of state-owned enterprises is still too high, which hinders the construction enterprise to introduce foreign capital and private capital. The reform of state-owned enterprises still faces many problems. The transition of state-owned enterprises to a limited liability company or a stock limited company still has a long way to go. The ownership diversified pattern of construction enterprise also needs further improvement in Jiangxi province.

B. The Unreasonable Organization Structure of Enterprises in Construction Industry

In 2008, there were 936 general contracting of construction enterprises, 399 specialized contracting enterprises and 45 labor subcontracting enterprises. Their proportion accounted for the total number of enterprises respectively were 67.83%, 28.91%, 3.26%. There are 598155 employees in general contracting enterprises, 60584 employees in specialized contracting enterprises and 2501 employees in labor subcontracting enterprises, which respectively account for 90.46%, 9.16%, 0.38% of the total employees. And the total output value of general contracting enterprises is 92,457,910,000 yuan, the total output value of specialized contracting enterprises is 10,836,310,000 yuan and the total output value of labor subcontracting enterprises is 106.99 million yuan, which respectively account for 89.42%, 10.48%, and 0.1% of the total output value of all enterprises (shown as Table 4). It can be seen that the organization structure of construction enterprises in Jiangxi province presents the type of "inverted pyramid" distribution [2]. And the adjustment of organization structure is insufficient.

The engineering project is the complex which is made up of many specialized kinds of work. The production of architectural production usually uses multi-level specialization contracting. The imbalance of proportion of general contracting enterprises specialized contracting enterprises and labor subcontracting enterprises (that is, the type of "inverted pyramid" enterprises organization structure) can lead to the following questions. It does not favor the development of maturity specialization contracting enterprises; restricts the development of the labor subcontracting enterprises' standardization; causes no difference in market segmentation and restricts the construction industry to drive up employment [4].

Table 4. General contracting enterprise, specialized contracting enterprise and the subcontractor enterprise mainly index in Jiangxi

	Number of enterprises (number)	Proportion (%)	Employees(person)	Proportion (%)	Total output value(ten thousand yuan)	Proportion (%)
general contracting enterprises	936	67.83	598155	90.46	9245791	89.42
specialized contracting enterprises	399	28.91	60584	9.16	1083631	10.48
labor subcontracting enterprises	45	3.26	2501	0.38	10699	0.1
Data sources from China Statistical Yearbook (2009)						

Table 5. Construction enterprise average size and number in Jiangxi

Years	2002	2003	2004	2005	2006	2007
Employees(ten thousand)	42.73	50.96	53.89	61.9	61.98	62.82
Number of enterprises (number)	1047	1106	1387	1362	1328	1325
The average industry employees (person)	408	461	389	455	467	474
Data sources from Jiangxi Statistical Yearbook (2003-2008)						

Table 6. Construction enterprise average employment and scale in China

Years	2002	2003	2004	2005	2006	2007
Employees(ten thousand)	2245.19	2414.27	2500.28	2699.92	2878.16	3133.71
Number of enterprises (home)	47820	48688	59018	58750	60166	62074
The average industry employees (person)	470	496	424	460	478	505
Data sources from China Statistical Yearbook (2003-2008)						

C.   The Unreasonable Scale Structure of Construction Industry

The Table 5 indicates the average number of employees and scale of construction enterprises in Jiangxi province from 2002 to 2007. The Table 6 is the average number of employees and scale of national construction enterprises from 2002 to 2007. It can be seen through comparison that although Jiangxi Province slightly bellows the national average, due to the average size of China's construction enterprises is nearly 50 times more than that in developed countries and the average size of China's large construction enterprises draw 20 times the size of the developed countries. [5] The problem of slanting large scale is quite serious, and also the trend has been increasing year by year.

In addition to the problem of large scale of individual enterprises in construction enterprises in Jiangxi province, it also exist the problems of structural similarity and without forming layers. According to the new qualification management method (2002), construction enterprises began to implement the qualification management of the top level, first-level, second-level and third-level to ascertain the operating scope of different levels of qualification enterprises. In 2008, there is one enterprise which has the top level qualification of general contracting enterprises. And there are 142 first-level enterprises which account for 9.4% of total construction enterprises, 472 second-level enterprises which account for31.4% of total construction enterprises, 890 third-level and enterprises without levels classification which account for 59.1% of total construction enterprises. But the phenomena of bid competition at the same levels and the same kinds of business in large and small and medium-sized enterprises in Jiangxi province are very common. [4] Most of the enterprises' internal technique management and labor homework layer mix and make mixed construction contracting. The gap between enterprise's production efficiency and operation effectiveness is very small. The large enterprise is the simple accumulation of small and medium-sized enterprises, and it does not form the scale economy.

D.   The Unreasonable Industry Structure of Construction Industry

In 2008, Jiangxi completed the construction output value 103294.22 million yuan, including house and the civil engineering architecture industry 91007.56 million yuan, account for 88.11%; the construction and installation industry 7597.97 million yuan, account for 7.36%; the construction decoration industry 2430.44 million yuan, account for 2.35%; the other architecture industry 2258.26 million yuan, account for 2.18 % (shown as figure 4). The house and the civil engineering architecture industry occupy the overwhelming superiority of entire profession total quantity. Due to slow

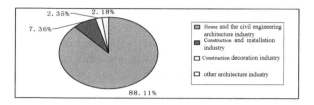

Fig. 4. According to the construction industry points in Jiangxi

development of the specialized construction enterprise, few quantities and small scales, some specialized constructions are empty. Therefore, house and the civil engineering architecture industry enterprise's proportion are excessively high. The proportion of house and the civil engineering architecture industry and the installment industry, the decoration industry and other architecture industry does not match, which will cause the malformed development of architecture industry.

4   Countermeasures of Construction Industry Structure Adjustment in Jiangxi Province

To address the above problems raised, the author believes that industrial structure can be adjusted from several countermeasures of construction industry in Jiangxi province:

- Increasing the intensity of the reform of state-owned enterprise ownership, and promoting the ownership structure adjustment and upgrade. To make full use of policy-oriented, lower entry barriers, and encourage foreign and private capital into the construction field. In addition, a modern enterprise system should be established when adjust the property rights of enterprises.
- Promoting the restructure of organizational structures, and controlling the ratio of general contracting business, professional contracting business, labor subcontracting companies in a reasonable range and changing the current "inverted pyramid" structure. To strengthen and expand a small number of general contracting businesses, and to develop mature professional contracting business, and to guide the standardization development of labor subcontracting companies. Finally, making Jiangxi general contracting construction company develop into a leader and taking specialized construction enterprises as main part and making labor subcontracting companies as the mainstay, which develop a good separating industrial structure chain between enterprise management and operation layer[1].
- Guiding the development of small and medium sized construction companies and changing the current situation of large scale of single building enterprises. To make small and medium sized construction companies take the road of specialization and cooperation, and develop to the small but excellent direction. Different levels and qualification levels of the enterprise should form a competitive situation by different level.

Regulating patterns of the construction market to change the current focus on building products in Jiangxi and civil engineering and housing situation. Promoting reasonable match development of the housing and civil engineering construction, construction and installation, construction decoration industry, and other construction industry.

References

1. Statistics of Jiangxi. On the analysis of jiangxi province construction development and thinking, http://218.87.32.231/News.shtml?p5=14174
2. Xie, Y., Liu, H.: Industrial economics. Huazhong university of science and technology publishing house, Wuhan (2008)
3. Li, Z., Fan, J., Wang, Y.: Construction of economic benefits and the ownership structure of influence. BBS Industry Development (10), 5–9 (2008)
4. Jiang, X.: Construction enterprise organization structure adjustment countermeasures in Jiangxi. Enterprise Reference (6), 79–81 (2008)
5. Min, Y.: Construction enterprise scale problems about China. East China Economic Management (4), 32–34 (2001)

The Innovative Thinking in Book Binding Design

Yanjun Wang

Institute of Art and Fashion
Tianjin Polytechnic University
Tianjin, China
yanliangdu@163.com

Abstract. Book binding design can advance the cognition of the artistic quality and design language, improve the taste of readers and promote the development of books industry and publishing in China. Excellent works of book binding have unique innovation. Having been regarded as vision carries of plane, book binding is limited into old mode of thinking for a long time. Designers should use innovative mode of thinking to conceive and transform the value for design. Making full use of the different type of format, designers innovate and produce the external styling and pages of solid of books. With the development of modern science and application of innovative materials in book binding, we will explore a new way together the material language and the green books as a whole new philosophy of creative thinking so as to display the charm of book arts.

Keywords: book design, innovation, materials, type, environmental protection.

1 Introduction

With the advancement of social civilization, book binding has gone into the thick of life and became the mental nourishment of human beings as a unique designing type.

As a carrier of materialization of cultural spirit, book binding processes certain format and expressive type. A successful innovation can move readers and convey intention to them. In all, book binding should be conceived through using innovation angle of view and technique of expressiontechnique of expression.

2 Innovative Mode of Thinking in Book Binding

As an all-around thinking pattern, the production of innovative mode of thinking is high-level type of psychological activity of human beings. Innovative mode of thinking guides people's thinking from different points of view and levels, and then breaks through strategic thinking, sparks inspiration. The innovative mode of thinking produces novel creative concepts during book binding. There are several typical patterns in mode of thinking.

J. Luo (Ed.): Soft Computing in Information Communication Technology, AISC 161, pp. 291–295.
springerlink.com © Springer-Verlag Berlin Heidelberg 2012

A.   Logical Thinking

It is significant to master logical thinking style during cultivating innovative design thinking. Logic is one of the oldest arts in the world. Logical thinking is defined a certain trends associated reasonable result is educed through using deduction, generalization and other logical methods according to the cognition on basic principles of objects. In the light of the process of reasoning in logic, the two basic tools of logical thinking can be divided into inductive logic and deductive logic. Inductive logical means summing up the general principle through a series of specific fact while deductive logical refers deducting result of individual cases by the general principle. The excellent works with logical thinking can combine subject with the process of thinking, and then take revitalizing visual effect.

B.   Contrary Opinion

Contrary opinion is a kind thinking method which opposite with the direction of thinking of the public. In the daily life, almost people are accustomed to the traditional thinking which is straightforward thinking. Suppose we have an adverse thinking to replace the traditional thinking toward ruddy, he would obtain the unexpected effects. Contrary opinion is widely applied in designing, for instance: elevator in industrial design, pencil-sharpener design. Usually, it is also used in book binding. Origin from life, adverse innovative thinking inspiration is from daily life and spiritual requirements, reflects the design of life and care, tenderness.

C.   Imagery Thinking

As a distinguished scientist, Qian xuesen said "as an important target in researching logical science, because imagery thinking is so wide and involves in significant part knowledge and spiritual values. But we have little information about it." Imagery thinking is used image as the element and mean of thinking to think. "Image" refers to all the living beings presents all forms specific types in certain space and time. As a kind design through passing inducing, refining and sublimation, the creative imagination of imagery thinking has unique creativity.

The innovative modes of thinking also include: dialectical thinking, divergent thinking, convergent thinking, system thinking and change of variable thinking.

3   Innovative Expression of Format Type in Binding Design

Format is the type and dimension of a book. Nowadays, with the appearance of all kinds' popular international formats, a traditional format has been replaced. In order to draw readers' attentions, there are many types' books, such as sidelong open and longitudinal open, rectangle, triangle, the square or circle and so on. In modern life, with intervene of other media; books are adversely affected, so reading books is not the only way to obtaining mental nourishment of human beings. Therefore, with the purpose of attract views of readers; the type of format has developed from plane to three-dimensional. As the development tendency of book design, text implication is reflected through the exaggerated changes of external modeling and solid page designing of books. In the form, books use the subjective and real space, real layout physical model or light relief solid sculpt as the new type of the morphology of books attracts attentions and imaginations of readers. If type and content matches appropriately, this three-dimensional real type books binding design will attract readers' attentions for the first time, because its real spaciousness is more direct than the type of traditional book design.

4   Innovative Materials in Book Binding Design

Modern books binding design should be set by the discovery own beauty of materials which will be interpreted on the basic of types of creative work. The positive use of material thinking is the key of design innovation in book binding. As the carrier of book, materials have restrictive function to books. The performances and characteristics of books determinate craft of work material and affect modeling of books. Therefore, designers should chose materials according to different design subjects in order to bring into play the characteristics of material texture and the best expressive force. For example: designer Liu Yuanyang uses plastic cover and injects liquid and duckweed, which has a perfect explanation on the conception of "flutter" in the book Duckweed.

In book designing, designers can break the routine and choose other materials which are used for other purpose. Perhaps, they will try to combine the regular materials without relevance between them in order to make an impact on stimulates thinking track and vision. It makes readers find everything fresh and new.

As the main using paper for newspapers and periodicals, newspapers also know as newsprint which is used to printing newspaper. In The Selection of Lv Shengzhong Line Drawing apart from cover and ring lining, inner sheet is printed newsprint. The faded yellow paper, traditional book publishing expresses the folksy line drawing of artist incisively and vividly so as to make the material and work style in harmony and unity. The reasonable creativity of corrugated paper and industrial sponger paper also take unique artistic effects.

On the use of paper for daily book designing, designers are accustomed to make a different between inner sheet and cover. Usually, cover uses a paper and seldom uses some combination of paper. If designer uses different kind and height paper respectively through sections of content in book, readers may feel rhythm sensation when they are reading. Common materials will take special effects if readers consider problems unconventionally.

5   Innovative Printing Technology in Binding Design

In book binding design, designers should follow the principle "make art according to different art ware's texture, modeling according to need" and choose different printing technology on the basis of characteristic of material in order to finish a charming art working of book. With the reformation and development of printing ink, books have presented growing visual arts. For example: in the use of the new material LCD ink books can appear change of rainbow colors with changes in temperature and perspective. There is also fragrance ink, it is using a micro capsule control smell, making the books were released to a different flavor. Pearlescent inks can reproduce natural pearls, and fish and shellfish, butterflies, and metal's own bright and color. In addition, luminous ink can absorb and store any luminous energy 10 to 20 minutes, it can keeping lighting more than 12 hours in the darkness. There is no difference between reading in the day and in night if the book is mixed into fluorescence color. It is easy to read, because the book can present a fluorescent effect or special Escher-esque effects. In addition, there is also white water-based ink which becomes transparent when it meets water. Innovative technology makes design of books step into a new field. The First common craft is used primarily in comprehensive and polish local processing which is divided into types of high sheen and semi-luster. They include polishing, laminating worn, and dumb, rubber and UV oil and dumb, printing gold or silver, fluorescent oil and so on. Second, the modeling art of paper trims includesconcavoconvex Nano-imprint lithography, gilding, perforating and open the window, sticking pv and so on. These crafts make the texture of books richer and more brightly and also make personalization of books possible.

6   Innovative Green Initiative Books in Binding Design

With the development of book binding and increasing of classification in books, people are not content with stereotyped types of books. With rich content of the way to express content breaks propositional thinking of book design, designing form and material according to the contents makes people feel fresh. enter the 21st century "green" has become a mass of the publishing industry, green books design has become a design tide both in publishing and book designing. As a books design professional, he should establish green design view. Firstly, designer should make a moderate consumption on materials and workmanship and not be extravagance and waste. Secondly, designer should advocate ecological design and try to choose material which can be the recycled to use and easy to degrade so as to protect resources renewable. Thirdly, they should choose the material with non-toxic, unstimulating, radiation-free and easy to degrade as the appearance in the printing and publications lest creating hurt to people. There are not sufficient evidence can prove Part of the books using synthetic material is harmful to the people and takes undesirable impact on children. Last, designers should learn adept in salvage, wastes and off cuts producing with the process of cutting papers is the valuable resource. The design of green books can protect the environment, promote the sustainable development of the whole society and solve the crisis of resource, lighten pressure of environment.

In conclusion, in the book binding design, we should combine subject, and organize wholly then hold every element with special viewpoint and innovative method so as to combine them into an organic whole, finally, express the content and beauty of book best.

References

1. Management Writing Group, Theory and Method of Creative Thinking. Economic Management Press (May 1999)
2. He, Z.: Walking in the New Thinking World. Central compilation & translation press (May 2005)
3. YiTang, R.F.: Binding Design. China Textile &Apparel Press (January 2004)
4. Yu, B.: Book Design. Hubei Fine Arts Publishing House, Wuhan (2001)
5. Lv, J.: JingRen Book Design. Jilin Fine Arts Publishing House, Jilin (2000)
6. Berson, H.: Material and Memory. Hua Xia Press, Beijing (1999)
7. Ling, Q.: Brief History of Book Binding Design Art. HeiLongjiang People's Publishing House, Haerbin (1984)

Analysis about Value Evaluation of Technology-Based Small and Medium-Sized Enterprises

Yan Li

Institute of Accounting
Hubei University of Economics
Wuhan, Hubei, China
452018116@qq.com

Abstract. Fund shortage is one of important restraining factors in technology-based SME development. Overall value evaluation of these enterprises can realize financing smoothly. Combined with features of technology-based SME, this paper builds value evaluation model for these enterprises based on the general evaluation of tangible assets, growth competence and core competency. Meanwhile the concrete analysis on the model is given.

Keywords: tech-based SME, value evaluation, growth competence, core competency.

1 Introduction

The development of technology-based SME based hi-tech is always a major concern for governments and theorists since 1980s. A wealth of facts indicates that tech-based SME are principal part in developing hi-tech industry and building of a national innovation system.

Chinese enterprises value evaluation always applies the cost increment law based on balance sheet for a long time. It becomes a growing problem that enterprises' book value departs from marketable value. Creating tech-based SME value evaluation system to attract more investment is an important task in current tech-based SME value evaluation.

2 Tech-Based SME Value Evaluation Index System

According to the features of tech-based SME, this paper presents that tech-based SME value evaluation model is consists of tangible assets value, growth competence and core competency, that is V=TV+GV+CV. V presents tech-based SME value. TV presents tangible assents fair value. GV presents growth competence. CV presents core competency.

For TV, it is easy to be determined by using history cost or assessment value based on future expected cash flow as its fair value.

J. Luo (Ed.): Soft Computing in Information Communication Technology, AISC 161, pp. 297–303.
springerlink.com © Springer-Verlag Berlin Heidelberg 2012

3   Analysis of Growth Competence (GV) Evaluation

GV of tech-based SME is mainly determined by its owned patented technology, human resource, and so on.

A.   Patented Technology Evaluation

Real option is an effective method. The growth opportunities brought by patented technology can be looked as call option based on real assets. Per unit's investment cost for patented technology at different points will bring further development to enterprises, so it can be look as a single growth option. Therefore, real option created by technology development is a compound option and it can be calculated with application of Geske compound option model.

B.   Human Resources Evaluation

Human resources evaluation of tech-based SME is mainly the value created by core personnel. The evaluation of this index can be described as P= S×β. P presents total predicted value created by core personnel in their tenure. S presents predicted value of enterprises future total cash flow. β is overall contribution rate for enterprises created by core personnel. S can be calculated by Black-Scholes pricing model. $\beta = v \times \hat{\beta} \times i$. It means β is the present value (i presents discount rate) of the product of core personnel comprehensive quality evaluation value v and core personnel contribution average value $\hat{\beta}$. $\hat{\beta}$ can be got according to experience data or drawing on the experience of other countries. The calculation of v is the key point in the measurement of core personnel human capital value.

In this paper, analytical hierarchy process is used to determine comprehensive quality evaluation value of core personnel.

First, according to the features of the enterprise, setting up an evaluation team formed by human capital experts and determining all competency factors which should be possessed by core personnel. Subdividing these competencies depends on the situation, sorting all influence factors with AHP and determining a very influential factor set. Human capital value of tech-based SME mainly reflects in technical research capabilities (such as research capabilities, professional experiences, innovation capabilities) and management ability (such as decision-making ability, marketability, risk-avoidance ability) of managers including core technician, and it is a comprehensive reflection of enterprise sustainable development capacity (such as gross profit margin, sales growth, rate of return). So evaluation system should be built according to technical research capabilities of core personnel, managers' entrepreneurship and enterprise's developing index.

The relationship of administrative subordination between upper and lower levels can be determined after setting up the hierarchical structure. If A is the criteria, A and its lower elements A1, A2 and A3 have dominance relationship. Our purpose is giving corresponding weight to A1, A2 and A3 according to their relative importance under criteria A. AHP method uses scale ratio of 1~9 as showed in table1.

If the criteria is enterprise actual index, then sub-criteria is enterprise total capital A1, enterprise asset-liability ratio A2 and total real asset A3.If experts think is obviously important than , then their scale ratio is 5, so scale ratio of asset-liability ratio to total real asset is 1/5.

Table 1. Scale ratio of AHP

value of scale mark	scale meaning
1	one element is as important as another
3	One element is more important than another
5	One element is obvious important compared with another
7	One element is intensely important compared with another
9	One element is extremely important compared with another
2, 4, 6, 8	if differences within things in pair between the two, we can select adjacent judgment medium value
inverse	if the importance ratios of element i to element j is aij, then the importance ratios of element j to element i is aji=1/aij

To give the reciprocal matrices of the above problem as showed in table 2.

Table 2. Reciprocal matrices

A	A_1 (research capabilities)	A_2 (professional experiences)	A_3 (innovation capabilities)
A_1(research capabilities)	1	5	3
A_2(professional experiences)	1/5	1	1/3
A_3(innovation capabilities)	1/3	3	1

For n elements, we can get judgment matrix for each other $A = (a_{ij})_{n \times n}$. A should has the following properties.

$$a_{ij} \rangle 0 \quad ; \quad a_{ji} = \frac{1}{a_{ij}} \quad ; \quad a_{ii} = 1$$

On the basis of reciprocal matrices, making calculation of single-sort for all elements to get the maximum eigenvalue λ_{max} of reciprocal matrices A. Use $AW = \lambda_{max} W$ to get eigenvector W corresponding to λ_{max}. To normalize W then it is the relative importance corresponding to one upper factor in the same level. Use the same method as above example to calculate maximum eigenvalue of above reciprocal matrices.

First normalizing judgment matrix for all columns, then we get the following:

$$\begin{bmatrix} 0.652 & 0.556 & 0.693 \\ 0.131 & 0.111 & 0.076 \\ 0.217 & 0.333 & 0.231 \end{bmatrix}$$

After normalization, making judgment matrix summation according to every row:

$$\overline{W}_1 = \sum \overline{a}_{1j} = 1.901 \quad ; \quad \overline{W}_2 = \sum \overline{a}_{2j} = 0.318 \quad ; \quad \overline{W}_3 = \sum \overline{a}_{3j} = 0.781 .$$

normalizing factor ($\overline{W_1},\overline{W_2},\overline{W_3}$) T, then getting eigenvector: W=(W_1,W_2,W_3)=(0.634,0.106,0.260)T. That is relative importance of A1, A2, A3 in A. Calculating the maximum eigenvalue AW of judgment matrix:

$$AW = \begin{bmatrix} 1 & 5 & 3 \\ 1/5 & 1 & 1/3 \\ 1/3 & 3 & 1 \end{bmatrix} \begin{bmatrix} 0.634 \\ 0.106 \\ 0.260 \end{bmatrix}, \quad (AW)1=1.944; \ (AW)2=0.319; \ (AW)3=0.789. \text{ as}$$

showed in formula (1):

$$\lambda_{max} = \sum_{i=1}^{3} \frac{(AW)_i}{4W_i} = 9.110 \tag{1}$$

To test the consistent of judgment, we should calculate its consistent index $CI = \frac{(\lambda_{max} - n)}{(n-1)}$. To judge whether the matrix has satisfying consistency, we also should judge the average random coincident indicator RI of matrix. For judgment matrix of 1~10, RI value is showed in Table 3.

Table 3. RI value

rank	1	2	3	4	5	6	7	8	9	10
RI	0	0	0.58	0.9	1.12	1.24	1.32	1.41	1.45	1.49

Table 4. Total factor weight set

level I / level II	A research-and-development capabilities of technicians W_1	B management ability of managers W_2	C enterprise sustainable development capacity W_3	total ordering
A_1 research capabilities	a_1	0	0	$W_1 a_1$
A_2 professional experiences	a_2	0	0	$W_1 a_2$
A_3 innovation capabilities	a_3	0	0	$W_1 a_3$
B_1 decision-making ability	0	b_1	0	$W_2 b_1$
B_2 marketability	0	b_2	0	$W_2 b_2$
B_3 risk-avoidance ability	0	b_3	0	$W_2 b_3$
C_1 gross profit margin	0	0	c_1	$W_3 C_1$
C_2 sales growth	0	0	c_2	$W_3 C_2$
C_3 ate of return	0	0	c_3	$W_3 C_3$

Take the above data as example: CI=3.055, CR=$\frac{CI}{RI}$=5.27>0.10. The judgment matrix has not satisfying consistency. So we should adjust the judgment matrix until it has satisfying consistency. That is letting CR<0.10.

According to above sorting results, we make more influential factors as evaluation factor set: U=(U$_1$, U$_2$, ...U$_m$). Make its corresponding weight number as X=(x$_1$, x$_2$...X$_m$) as showed in Table 4.

After determining the influential factors and its weight, we can use fuzzy math to evaluate entrepreneurial comprehensive quality. First, determining reviews set V according to experts' evaluation by survey sampling. Meanwhile, comprehensively determining reviews weight set B according to importance of every index. The review set can be divided by five levels: very good, good, fair, poop, very poor. Its corresponding reviews weight set is (B1, B2, B3, B4, B5). Then we calculate single-factor evaluation value and build single-factor evaluation matrix(R)to describe the relationship between evaluation factor U and reviews set V.

$$R=\begin{bmatrix} R_{11} & R_{12} & R_{13} & R_{14} & R_{15} \\ R_{21} & R_{22} & R_{23} & R_{24} & R_{25} \\ \vdots & \vdots & \vdots & \vdots & \vdots \\ R_{m1} & R_{m2} & R_{m3} & R_{m4} & R_{m5} \end{bmatrix}$$

R$_{ij}$ is the reasonable degree when evaluating U$_i$ using V$_j$ as review. For example, when evaluating the innovation capabilities of entrepreneur, there are ten relative experts, in which five thick it is very good, two fair, two good and one very poor. Then the evaluation of entrepreneurial innovation capabilities is (0.5,0.2,0.2, 0, 0.1).On the basis of calculation of single-factor evaluation value, making fuzzy comprehensive evaluation for entrepreneurial human capital and getting fuzzy comprehensive evaluation matrix E=X·R=(E1, E2, E3, E4, E5). E$_1$ is the very good weight, E$_2$ is the good weight, E$_3$ is fair weight, E$_4$ is the poor weight and E$_5$ is the very poor weight. In the end, calculating the value adjustment coefficient of entrepreneurial human capital, that is comprehensive quality evaluation value $v = \dfrac{\sum\limits_{j=1}^{5} E_j B_j}{\sum\limits_{j=1}^{5} E_j}$.

4   The Evaluation of Core Competence (CV)

In this paper, grey evaluation method can be used to evaluating CV. The method can consider all factors to the highest degree and has thorough calculative process. This can help to improve the Science-based and precision of evaluation.

For concrete enterprise, an index system must be built to evaluate the enterprise strength. CV is a comprehensive embodiment of multi-dimension degree. We can make a comprehensive designing according to CV or make respectively designing according to one factor. For instance, when evaluating brand strength, it can be divided with brand relation dimension and marketplace dimension. They measure brand's inner strength and outward manifestation respectively. We take brand as example to analyze the evaluation process in the following.

In order to make a comprehensive study on the constituent elements of brand value and reduce subjective evaluation factors, we should use quantifiable and easy-obtained evaluation index. Brand strength is a dimensionless value, and it can be presented as utilization coefficient in $0 \sim 1$. Its value can be determined by evaluation index value. The greater the brand strength value, the greater relative brand value in the same industry, and vice-versa. In the concrete utilization it can be selected according to situation of the evaluation industry, or can be added correspondingly, but it should maintain unity within industries which is good for standardization, comparability and conscientiousness.

The first step is standardizing the index value. For the evaluation of brand capital value, at first the index value must be true and reliable. In enterprise there must be special person or special agency to collect and process all index value according to a certain period. In order to eliminate the effects cause by units, order of magnitude differences among index on comprehensive evaluation results, we must standardize the index value. That is dimensionless.

The second step is determining evaluation grey and whiteness weighting function. Because of the dimensionless for index sample value and all grey criteria is determined by utilization coefficients, so all index has untied standard belong to one kind and has united whiteness weighting function.

When we use grey comprehensive evaluation method to evaluate the brand strength, first we should determine weighting vector of element level corresponding to criteria level in brand value evaluation index system. Weighting vector can be calculated by using analytic hierarchy process, but it is a subjective evaluation method, so it must destroy objectivity and standardization of evaluation. Therefore, analytic hierarchy process makes calculation according to actual data of the enterprise. It chooses actual data in several periods of the enterprise and determines weight value according to relative analysis of all evaluation index value and brand value (Brand Earnings × price/earning ratio). On the basis of this, calculate element level weight corresponding to criteria level.

$$\mathbf{B}_H = W_L \times R_L \text{ (H=1, 2)} \qquad ; \qquad \mathbf{B} = W_L \times R_L$$

In the formula, W_L is weighting vector of element level corresponding to criteria level L. $R_L = \left(\gamma_k^{(L)} \right)$ is judgment matrix which is Decision-making power coefficient of criteria level L belongs to grey kind k. $\mathbf{B}_H$ is Comprehensive evaluation vector of factor H (relational dimension and Marketplace Dimension). W_H is the weighting vector of target layer corresponding to brand evaluation factor level H. its calculation is as the same as that of W_L and can be got according to relative analysis of historical data of the enterprise. B= $(b_1, b_2, ..., b_k)$ is comprehensive evaluation vector for brand strength's status. It reflects the degree that the brand value belong to grey kind k. the maximum coefficient $b_k^* = \max_{1 \leq k \leq 5}\{b_k\}$ corresponding to grey kind owned by brand strength. The grey kind presents general status of the brand in the industry. Convert the comprehensive evaluation vector into comprehensive evaluation coefficient b, then can get brand strength coefficient as showed in the formula: $b = \sum_{k=1}^{5} b_k \times \lambda_k^{(2)}$.

Above we use Grey Evaluation Method to make a basic analysis of evaluation for brand strength which is one element of tech-based SME core competency. We can also use the method to evaluate other elements. In some cases, it can be make comprehensive evaluation directly according to representative indicator of core competency.

5   Conclusion

The value of tech-based SME is a comprehensive embodiment of real assets, growth competence and core competency. The latter two play an important role in Enterprise Value and further development. According to features of all assets of tech-based SME, combining the three to evaluate Enterprise Value is a more comprehensive and foresighted method. The method has important meaning for helping tech-based SME to realize financing.

References

1. Copland, T., Koller, T.: Valuation. Publishing House of Electronic Industry (2002); Hao, S. (transl.)
2. Chen, K.C.W., Chen, Z., Wei, K.C.J.: Disclosure, Corporate Governance, and the Cost of Equity Capital: Evidence from Asia's Emerging Markets. Working paper, p. 44 (2003)
3. Xiang, X.: Researches on the Methods of Hi-tech Enterprise Value Evaluation. The Journal of Quantitative & Technical Economics (2), 102–105 (2003)
4. Zhang, H., Chen, Q.: A Study on Valuation of Venture Capital Company based on Neusoft Venture Capital. Economy and Management 24(3), 67–71 (2010)

The Impact of the Traditional Female Aesthetic Ideas on Visual Communication Design

Hongjiao Duan and Li Zhang

Art design, Tianjin University of Technology, Tianjin University of Technology,
TJPU, Tianjin, China
duanhongjiao816@163.com, zhangli6458@126.com

Abstract. The traditional Chinese culture stresses "harmony between man and heaven"and "combination of softness and hardness", and a variety of visual expression techniques emerge one after another. The traditional female aesthetic ideas significantly influence the innovation of visual communication design, and provide more valuable design space for visual communication design.

Keywords: the traditional female aesthetic ideas, visual communication design, aesthetic ideas.

1 Introduction

Today's society is known as "an era of reading pictures". With the constantly updated media tools, the visual expression techniques become rich and varied, and people's awareness on "beauty" is also rising. Therefore, the means that the designers simply rely on creating the screen impact to attract people's attention is gradually eliminated. China has several thousand years of history, and the traditional female aesthetic ideas that go with it have already been seen in the creative process of visual design patterns. This design approach of "revitalizing archaic Chinese rhymes" is a dialogue between past and present, as well as a conflict and inheritance. Just as Jin Daiqiang has said, who is a leading Graphic Designer, there are three principles of virtue: first, establishing an idea——the idea goes first to take the spirit by the form; second, innovation——inherit the past and usher in the future, and destroy the old and establish the new; third, applying with imagination and ingenuity——fit for one's own use, flexible and lively.

The graphic language is the most expressive in visual communication design, which can not only fully reflect the unique charm of the visual language, but also achieve the purpose of information communication. A good visual works can allow people of different borders and nationalities to break through the language barriers and cultural differences among various ethnic groups to smoothly achieve the purpose of the effective information communication and to share its artistic charm and meanings. However, the creation of a visual graphic language is inseparable from our profound traditional aesthetic concepts, especially the traditional female aesthetic ideas, which create a beauty of ideas and also play an important role in the formation

J. Luo (Ed.): Soft Computing in Information Communication Technology, AISC 161, pp. 305–310.
springerlink.com   © Springer-Verlag Berlin Heidelberg 2012

of other graphic languages. Aesthetic appreciation in modern visual communication design is a continuation and development of our ancient traditional aesthetic concepts, so however rich and varied visual images should be on the basis of the traditional aesthetic ideas. The Chinese culture has a long history, and this paper will give a systematic elaboration on the traditional female aesthetic ideas to reflect its impact on visual communication design.

2   The Concept and Characteristics of Visual Communication Design

What does visual communication design mean? In short, it is a modeling activity that uses visual elements to communicate with others and convince them. In addition to point, line, surface, sphere, and their combination with brightness, saturation and hue, there is also a need to make a discussion on the psychological level and the level of cultural knowledge. On the psychological level, it means the ability to capture beauty. On the cultural level, it is to use the public cultural practices to complete the beauty capturing, through the mobilization of the human psychological factors, so as to generate a process with the mutual communication and interaction of information between people and works. For a successful aesthetic idea, addition is usually applied to its design concept, while subtraction is widely used in the modeling language. So the successful aesthetic idea has simple design elements, and allows people to produce a deep impression and read the information in a short period of time.

In visual communication design, we also need to "search our roots", and to search for the advantages of thinking and unique elegance that other nations lack in our traditional culture. From beginning to end, art focuses on its internal continuation. The production of an art form and its acceptance requires a specific historical and cultural background, including a nation's way of life, customs, ethics, aesthetic habits, etc. These constitute a potential deep cultural structure, and are locked in the minds of various ethnic groups, which not only regulate and constrain the development of national culture, but also promote the innovation of visual languages and forms. Just like modern design, the art of visual communication, as a separate discipline, is gradually involved in all areas of our lives. Therefore, the impact of traditional concepts on people's minds and their combination with the visual arts seem very important today. The works of the Chinese designers often show the subtle influence of the traditional aesthetic concepts. Whether it is the concept of "One is the child of the divine law; After one comes two, after two comes three, and after three comes all things" in Tao Te Ching, or the beautiful verse of "a slender fine lady is every gentleman's heart desire" in the traditional female aesthetic culture, they are both widely adopted by designers in their own works, from which it can be seen that the traditional aesthetic concepts are very important to modern visual design.

3   The Traditional Female Aesthetic Culture

Since ancient times, women have always been an important aesthetic object in Chinese aesthetic culture, so the aesthetic ideas about women have gradually formed a certain pattern. For example, men's pursuit of a beautiful gesture of "a slender fine lady" is a reflection of the building of a woman's "submissive" image by the culture of sexual distinction in the ancient patriarchal society. These "gentle and submissive" images of women have become an ideal female aesthetic paradigm in the gender concepts of the later generations.

Records of female aesthetics present a splendid sight in Chinese classical literature. There are depicts of "her smiles are bright and her eyes are limpid" in "Shuo Ren" of "The Book of Odes", and a literary rhetorical device that uses "a beauty of perfumed grass" to substitute the virtue as metonymy in "The Odes of Chu", which open the literary tradition in China that take a beauty to compare virtue. What's more, the later men of letters continuously chant the praises of a beauty. Zhang Chao, a poet in the early Qing Dynasty, had a statement:"The so-called beauty is fair as a flower and beautiful as the moon, as pure as jade and as clean as ice, and also educated and reasonable, to which I have no objection". This means that bringing up a beauty requires enriching her mentally and physically, which is not easy. Therefore, it can be shown that women are an important aesthetic object in Chinese aesthetic culture.

The beauties these men of letters praise mostly have delicate and soft bodies, and subtle and quiet temperaments. In China's traditional cultural concepts, these female aesthetic ideas have no separate existence, which are a reflection of the gender system of the patriarchal society in the ideological aspect. This system reflects the male's dominant and superior position, and the female's submissive and marginal status. In addition to refuse women to intervene in the public domain in terms of the political system, culturally patriarchy also spread the concept of strong men and weak women, and what is reflected in aesthetic ideas is a strong praise for the masculine beauty of men and the feminine beauty of women. This feminine feature does not mean the weakness biologically, but refers to the "weak" female image and ethos. In short, a weak, quiet and submissive female image is the mainstream aesthetic idea in the gender system of the patriarchal society. These female aesthetic ideas have already been gradually formed in ancient China, and have influenced thousands of years of the later female aesthetic ideas in China.

Although it is impossible for us to make a scientific and quantitative definition of "Yao Tiao", such as a certain proportion, size or range, it is clear that "Yao Tiao" is used to describe a beauty. The use of "Yao Tiao" and people's understandings on "Yao Tiao" do not have data the same as scientific formulas as a reference, but the aesthetic communication and exchange are obtained through the unique Chinese "body language". The expressive party is obscure and vague, leaving much room for imagination, and the receiving party also takes "to understand tacitly" as the target without seeking the one to one correspondence of words and objective things. The sympathy and tacit understanding aroused from both parties' aesthetic communications are based on common cultural traditions, in which a number of conventional views are formed. Especially, many aesthetic ideas and ethical concepts are common views reached in common cultural traditions. The author who enjoys his singing of "a slender fine lady" in folk songs will not worry about other people's incomprehension of its meaning.

From the view of aesthetic features of the Chinese culture, it is also very difficult to give a strict quantitative interpretation to some terms and concepts that describe the beauty. It is consistent with the idea of "words don't exhaust ideas" in Chinese culture. In the relationships among languages, things and ideas, things and ideas are richer and more fundamental. A language is only a medium that allows you to understand the objective world and subjective minds, so you cannot regard a language as a thing or idea. To give an explicit explanation with the example of Buddhism, the thing and idea are like the moon in heaven, and the language is like fingers. When someone points to the moon and says "the moon is over there", he does not mean that you should watch his fingers, but you should see the moon through his fingers. Also, when someone wants to express his thought in a language, he does not mean that you should focus on the language, but you should comprehend and obtain its meaning through his language. For westerners, if the development of the process of cognition is from the perceptual to rational knowledge, the rational knowledge at the top level is a clear definition of visual language symbols. But, for the Chinese people, the process is from the perceptual knowledge to language symbols, and then to comprehension. Language is not at the highest level, for languages cannot correspond exactly to things, and cannot reflect the most exquisite and the most subtle aspects of a thing. In order to grasp the most exquisite and the most subtle aspects of a thing, one has to grasp the extra-linguistic understandings. On this basis, change them into graphics, and then present them to viewers in visual forms, so as to convey the meanings they want to express. Therefore, visual communication design also focuses on achieving "understanding tacitly" that the Chinese people pursue through tangible and intangible languages, which is a realm that "Having no wings, I can't fly to you as I please; our hearts at one, your ears can heart my inner call".

4   The Traditional Female Aesthetic Idea in Visual Communication Design

In the traditional female aesthetic ideas, many points of view give much imaginary space to modern designers, and allow them to find out a language that belongs to their hearts. Then pass it to the audience through the appropriate visual expression means, which may strike a sympathetic chord in their hearts. Since ancient times, in many visual design works, the expression of female subjects still has not been separated from our traditional female aesthetic ideas, no matter how unusual the visual elements, or how fresh the design forms. The designers just make innovations in visual languages on the basis of carrying on our traditional culture, and under the visual principle of beauty in form they disarrange, reorganize and recreate visual elements, such as line, shape, brightness, color, quality feeling and space, so as to form the current rich and colorful visual images. Therefore, in many modern design works, the modeling language used in those expressing female subjects still serves for the typical female characteristics such as softness and delicacy, focusing on curves and soft color tones.

Figure 1 shows, this is a perfume posters, the main modeling element of this picture is curve-based, the refinement of gyrosigma is contrast to the unreal background. From the overall composition, the picture shows the beautiful curve of the female body through that cut an "S" shap in this picture only by use the women. The picture hue is purple and it rendering a unique mystery of women. Besides, the use of cool colors contrast with the color of women skin. The methods of contrast and unity enhance the image of the rhythm. In the view of the characteristics of perfume bottle, the female characteristics of "a slender fine lady" have been implied in which the shape of spiral-shaped curve around the bottle, shows a woman's graceful everywhere. The use of black and purple color are reflecting the feminine personality.The detail of whole bottle looks luxurious, unique character, without losing the women's delicate and exquisite. So that this perfume of audience clearly know that its attributable items are women. Do the intuitive visual communication, not need any writing.

Fig. 1. Example of perfume posters

5   Summary

Above is the impact of the traditional female aesthetic ideas on visual communication design, which should influence the aesthetic ideas in our modern design. The emphasis of traditional Chinese culture is harmony between man and heaven heaven and combination of softness and hardness, but we are pursuit the highly combination of the natural world, traditional aesthetic culture and our creative design. This traditional female aesthetic ideas is not just stand for a culture, it is also an importance element of the visual communication language. It often implies a kind of shape, a meaning and a beauty mark, the emphasis is an idea of harmony and unity that both the design and traditional Chinese. The gender system in the ancient, these "delicate, gentle, submissive" image of women are builded, and gender gradually has become the concept of future generations of female aesthetic ideal, and developed in many visual communication design. Only when we constantly excavate and use the traditional female aesthetic ideas so that the design is truly innovative, and the overall level and value of visual communication design are improving.

References

1. Li, Y.Z., Lu, Y.: The history and aesthetics of visual communication design. Renmin university of China publishing house, Beijing (2000)
2. Wang, Y.: The principles of visual communication design. Higher Education Press, Beijing (2008)
3. Hu, J.: Aesthetics. Beijing University Press (2010)
4. Arnheima, R., Yao, T. (trans.): Visual Thinking. Sichuan People's Publishing House (1998)
5. Liang, Y., Hu, X., Gong, C.P.: Aesthetic psychology of the Chinese people. Shandong People's Publishing House (2002)

Building of the Green Space Plants and Architecture

Mingzhi Feng and Baiyang Jin

Institute of Art & Fashion, Tianjin polytechnic university, Tianjin, China
{447130331,873494544}@qq.com

Abstract. Based on the building and plant as the research object, this paper expounds the plants and architectural design association. Through the analysis of the ecological effect of plants, the combining forms of building and plants and adornment meaning, this paper emphatically expounds plants and architectural profundity contact. Finally from the perspective of the ecological design, this paper discusses the bionic building plant application. This paper fully lists the outstanding case at home and abroad, summarizes the sustainable development of plants and architecture combination, making design much closer to the nature.

Keywords: Plants, Architecture, Green space.

1 Introduction

As China's urbanization is speeding up and people's living standards are increasingly improving, instead, the environment is deteriorating. These changes are extremely affecting the quality of our life. We should increase investment in environment, improve them gradually. As for how to let us have a good ecological space conforming to our country's low carbon policy implementation, we should have a profound analysis and research on the renewed concept of green.

2 Green Space

Green space can be understood as an ecological, resource efficient way to design, make building space. The coverage is more extensive, big to an area, space, small to a bedroom environment. This paper mainly discusses about the plant in the architectural design of green applications. Green space is an organic whole, rising to living environment level, which is to create a harmonious and peaceful coexistence of space.

When mentioning "green space", people will think very easily of green plants shaping space, such as such flowers, plants and trees. But, are plants really green? Of course not!

Therefore, we must first make sure the "green" is not those "green". Green space is to create a colorful world through the organic combination of the plant and building, and making buildings well suit to natural space, leaving people a as a colorful architecture scene.

J. Luo (Ed.): Soft Computing in Information Communication Technology, AISC 161, pp. 311–316.
springerlink.com

3   Plants Environmental Benefits and Energy-Saving Benefit

Plant photosynthesis, under the function of chlorophyll, makes water and carbon dioxide into organic material which exhales oxygen to maintain the atmospheric balance of carbon monoxide and oxygen (Fig-1). In addition, plants can absorb some toxic atmospheric substances so as to relieve urban air pollution.

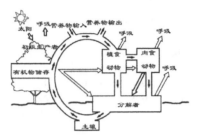

Fig. 1.

Therefore, it is obvious that plants are playing an important role in the coordinated development between building and natural environment. It is an effective means to improve the building environment quality. The following are key analysis:

A.   Adjust Building Microclimate

Greening plants is one of the effective measures to realize the summer cooling, eliminate heat island effect. It is a natural moisture regulator. Plants' roots in the rained soil are natural reservoirs. While the air is dry, through transpiration of the leaves, it can increase air humidity. This is very favorable for the dry climate area. Other plants also can adjustable the speed according to the plant disposition. It can intentionally lead the summer dominant, blocking the cold winter winds so as to achieve the purpose of regulating microclimate.

Thus, reasonable configuration in building with wind plants environment has a pivotal role.

B.   Absorb Dust and Harmful Air

Plants are natural air conditioners. Floral blade has very good adsorption effect on dust. The rougher the surface of the blade is, the stronger ability of dirt adsorption is.

For example, cordate telosma has very good role in driving away mosquitoes, tree privet can absorb sulphur trioxide etc. In the interior, reasonable plant configuration can purify the air; such plants can make these buildings more livable.

C.   Reduces the Noise

The use of green plant to reduce noise around the buildings is effective. Dense planted green belts can not only be an isolated urban noise barriers, floral blade also has a certain sound-absorbing function.

4   Plants and Architectural Combining Organically

Green plants symbolize vitality, which can make people feel a kind of hope in life. Because of this, it can adjust the person's nervous system, make the stress and fatigue mood relieved.

Especially in high-rise buildings, being full of green will become the dialogue breakthrough point between people and the environment. People open doors and windows to breathe fresh air and can enjoy nature quiet life on the balcony. As for the relationship between plants and building, this paper discusses it as following,

A.   Roof Afforestation

The history of earliest roof greenings is listed as "one of the seven wonders of the ancient world," ancient Babylonian "air garden". In developed countries, roof greening technology and building graft technology has been mature. Oakland, California, 1959 had a six-story building built on the roof of the world. It is the biggest sky garden covering an area of 1.2 hectares.

Roof greening environmental benefits:

The roof garden not only beautifies the environment, but also provides people with pleasure, fitness place. For a city, it is a natural green roof of air conditioning; it can maintain an ecological balance of residential environment and a good artistic life conception.

a). The Insulation:
Green roofs can warm winter and cool summer. It can always keep more than 20 degrees, making a comfortable environment of great benefit to live in.

b). Protection Structure:
The roof garden is a building structural protector. Usually through the roof greening, its' soil can protect the building itself with basic building blocks, prevent buildings cracks, to a certain extent, prolong its service life.

c).Capturing Rainwater:
The roof garden and stored rainfall function can be good soothing effects on reducing urban drainage system pressure, reducing the sewage disposal expenses, returning to natural ecological area, planning perfect ecosystem circulation. The roof garden can not only serve the bees and butterflies with new living space, endangered plants, reduce man-made interference in providing their homes with freedom.

Fig. 2.

Fig- 2 is an exampled contrast of an office's building roof greening projects in Guangdong. Roof greening projects not only beautify urban landscape, but also has very good ecological benefit and cater to today's "energy saving, low carbon" theme.

B.   Metope Greening

Metope greening refers to three-dimensional greening form of plant creepers building adornment facades and various fence. For exterior of vertical virescence, to beautify the facade, increase the greenbelt area and forming a good ecological environment has great significance.

a). Metope Greening Environmental Benefits

The vertical virescence application in metope is to prevent "morning sun" and "flower", which is an effective method. It can be used to effectively use their position of shading and transpiration to moderate sunlight for architecture point-blank, reduce wall to the periphery environment of radiation.

Metope greening can, according to people's intentions, ensure building facade are occluded and beautified. It can also reduce the noise of wall facing the reflector, absorb dust, and reduce dust into the interior. It will be a very good building "skin" combined with plants, and has good ecological effect.

b). MetopeGreening and Beautifying Function

Metope greening can not only strengthen artistic effect, but also make the environment more neat and beautiful, make inflexible gray metope full of vitality, enlarge the urban greening of planar greening.

Using wall to afforest and decorate can increase the beautifulness of city , promote architectural noumenon greening. Metope greening can make flat-panel single building and various rigid structures become natural, pure, fresh and pleasing to eyes. Delicate plants modeling can increase people's appealing, promote the urban landscaping art level.

Fig. 3.

Fig. 4.   **Fig. 5.**

Metope greening still can bring building surface season on change, make buildings change their beauty in according to different time and different climate condition. (Fig-3~Fig-5) is for metopes' greening different effects.

C.   Window and Balcony Greening

The balcony and window's area is small, but People's daily life really plays an important role in them as its use frequency is also high. In designing, decorating the balcony, window with plat by the balcony, window's narrow space can help to create "mini garden". People do not need to leave home to appreciate the green plants, gorgeous flowers because this is like the garden into the home, and like in the balcony, windows installed with the fresh air coolers and silencing precipitator. To ease the pressures of work and study, have a stable mood, reduce diseases, such greening is very beneficial to people's physical and mental health.

5   Plant Adornment Effect

By referring to the east and the west famous books, I noticed that every one of them has integrated harmonious relations between the inside and outside of the building. Even a tree planted in front of building can have great changes.

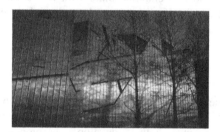

Fig. 6.

Fig-6 is the Jewish holocaust memorial in Berlin. Winter branches reflected in the memorial wall, adding to the peace. Seeing afar, you will find the branches and distorted building wall form an organic whole to awaken people to the Jews deep holocaust remembrance. Below we will take the trees as our focus to discuss its presence in adornment effect of buildings.

A.   More Vivid Facade

Trees can make buildings inflexible appearance become vivid and interesting. Trees' tiny branches look like shoes lace, making the whole building more dynamic. As it is the same effect for a man to wear.

Imagining trees are constantly changing leaves all the year round. Buds will emerge green and became strong green. In autumn, color began to become the dazzling red. Soon leaves rustle falls, twigs swaying in the cold. This can give a person the change throughout the obvious seasonal feeling. In the house, looking through the window to the outside of, the feeling is not the same, such as building the dead branches around waterscape with indoor greenery. In contrast, this mixed style fusion, the red and blue water branches of photograph reflect a chic building space.

B.   Shadowing Effect

Trees will cast shadows on on metope and ground. They will be different because radial lights giving all sorts of lively form. Imagining in the sunset on white metope, the branches in the next left and a bird fall quietly silhouette in the branches habitats. Such situation today in can be seen in Suzhou garden.

C.   For People to Realized Time Goes By

The trees planted can make the architectural space produce a realized time feeling. In many Chinese ancient building, there were many towering trees and architectural combination of cases. Many ancient village entrances have more historical flavor with ancient trees. Meanwhile, towering trees would be features of the landscape.

Thus, in architectural design, in the right place, plant on the appropriate place does bring environment vitality. Man is a natural product but also a social product. Therefore in architectural design, the addition of natural plant can help to create green space, which has a positive influence o humanities landscape and cultural tradition. Discussion above all is how to combine plant with the architecture. The following message is for study plant itself, discussing plant physiology elements in design of direct embodiment.

6   Conclusion

Through such main line, the relationship between plants and the architectural design is generally elaborated. Plants and human beings are all belong to nature and building is artificial nature. In essence, how to develop a harmonious combination is the problem the architect must face. Summary of this paper makes me more deeply appreciate important role of plants in the future design practice. I will keep the environment as the first concern and make efforts to combine buildings with nature an truly building "green space" as this is also our endowed task and responsibility.

References

1. Li, B.: Green building introduction. Chemical industry press, Beijing (2007)
2. Liu, Z., Liming: The green house greening environment technology. Chemical industry press, Beijing (2008)
3. [Jan] Jian, G.: Architecture and greening. China architecture & building press, Beijing (1992)
4. Li, M.: Urban green space system and living environment. China architecture & building press, Beijing (1999)
5. Simens, J.O.: Landscape planning— field plan and design manual, 3rd edn., p. 294. China architecture & building press, Beijing (2000)

Integrating Object Detection and Tracking in Outdoor Environment

Ziwei Ni, Meixiang Zhang, Jin Li, Quan Zou, and Qicong Wang[*]

Departments of Computer Science
Xiamen University, XMU
Xiamen, China
{zwni,qcwang}@xmu.edu.cn

Abstract. Object tracking is one of the key technologies in smart visual surveillance. However, due to the existing of varying sunlight, shadows, plants swaying, the difficulty of object tracking is greatly increased in outdoor environment. To tackle this problem, a new algorithm based on the augmented particle filter is proposed in this paper. It is able to combine object detection and tracking. Here, we employ Gaussian mixture model to model the background of the monitored environment. Furthermore, we extract the binary image of object by background subtraction as the corresponding observation. In the subsequent tracking phase, we use Kalman filter to introduce the most recent observations into particle filter and produce the suboptimal Gaussian proposal distribution. Experimental results demonstrate the proposed algorithm can handle a certain degree of complexity in outdoor environment.

Keywords: object tracking, particle filter, complex environment, observation likelihood, Kalman filter.

1 Introduction

Object tracking problem can be described as a process, which solves the maximum posteriori probability (MAP) of target state constantly with full awareness of the priori probability of object state, after obtaining new observations. In other words, the visual tracking problem is considered as optimal estimation problem under the framework of Bayesian theory. Since the unknown noise distribution in the image is not necessarily a Gaussian distribution and the projection process is non-linear, particle filter based on Bayesian theory is applied to deal with such a non-linear, non-Gaussian visual tracking problem [1]. However, since it is very difficult to accurately model movement of object from the image sequence, the conventional particle filter, using the system transition as the proposed distribution, without taking into account the latest observation of the object, may produce poor priori probability. An optimal Gaussian proposal distribution is shown in [2]. It considers the latest observation, meanwhile minimizes weight variance. However, it is still difficult to get the optimal proposal distribution. For the purpose of avoiding the defects resulting from the conventional particle filter's adopting state transition as proposed distribution, contour

---

[*] Corresponding author.

J. Luo (Ed.): Soft Computing in Information Communication Technology, AISC 161, pp. 317–324.
springerlink.com   © Springer-Verlag Berlin Heidelberg 2012

tracking method based on improving Kalman filtering and particle filter is proposed in
[3, 4]. The improving particle filter updates each state of particle set with a Kalman
filter, thus produce and propagate a suboptimal Gaussian proposal distribution. In
visual tracking based on particle filter, acquisition of reliable object observation is an
important guarantee of robust visual tracking. However, under clutter scene, for the
impact of the external environment, such as changes in illumination, shadows and
scenery disturbance, it may cause false detection of foreground object. Therefore, it is
necessary to eliminate the interfering factors when tracking object in outdoor
environment. Background modeling is an effective way to deal with these interfering
factors. Gaussian mixture model as background model is a more reasonable method of
background modeling at present [5], which describes the mixed-signal consisted
of background, small amplitude background movement and shadows as the form of
Gaussian distribution. It adopts maximum likelihood probability for realizing
background modeling and uses learning factor to update background Gaussian model
in real time. At present, most of tracking algorithms aim at tackling indoor and less
scene interference cases [9-11]. In this paper, we propose a visual tracking algorithm
based on the augmented particle filter for outdoor environment. Aiming to eliminate
possible interference in the background and obtain the current reliable target
observation, we employ Gaussian mixture model for modeling the background in
outdoor scene. Furthermore, we use on-line updating, shadow eliminating and
morphological operations to achieve adaptive detection of the foreground target. In
the subsequent tracking process, the Kalman equation is introduced into particle filter
to update the particle state with the latest observation. Unlike the idea in [3], here
Kalman filter is applied to the mean and covariance of the particle set to reduce
computation. Benefited from the acquiring of more reliable object observation and the
introducing of Kalman equation, the proposed algorithm can approach Bayesian
optimal estimate without using a large number of particles.

2   Particle Filter

When we employ particle filter to carry out visual tracking, tracking process is seen as
an optimal estimation problem which is always computed by state-space method. We
define the state vector of object as x. x_k is the output state vector of the tracking
system at time k. So visual tracking can be modeled by a state equation as the
following

$$p(x_k|x_{k-1}), \quad p(x_0) \quad for \quad k \geq 1 \tag{1}$$

Denote the target observations as z_k which indicate various target features extracted
from the images. The observation equation is expressed by

$$p(z_k|x_k) \quad for \quad k \geq 1 \tag{2}$$

The state transition prior $p(x_k|x_{k-1})$ shows that the evolution of each state is a
Markovian process. The observation likelihood $p(z_k|x_k)$ indicates that the

measurement procedure is mutually independent, and it only depends on the current state. Besides, there are always noise in the above two formulas, which are commonly assumed to be independent and identically distributed. We can predict the state at time k based on the measurements before time $k-1$, and use the measurements which are obtained at time k to update the prediction. Suppose the state prior probability at time k is $p(x_{k-1}|z_{1:k-1})$, where $z_{1:k-1} = \{z_1, z_2, \cdots z_{k-1}\}$ is a set of all of observations down to time $k-1$. Then visual tracking is to compute posterior density $p(x_k|z_{1:k})$ and mathematical expectation of some integrable functions $f_k(x_k)$. For example, if $f_k(x_k) = x_k$, mathematical expectation is the mean value of system states. Then tracking process can be divided into two steps as follows.

Predict:

$$p(x_k|z_{1:k-1}) = \int p(x_k|x_{k-1})p(x_{k-1}|z_{1:k-1})dx_{k-1} \tag{3}$$

Update:

$$p(x_k|z_{1:k}) = \frac{p(z_k|x_k)p(x_k|z_{1:k-1})}{p(z_k|z_{1:k-1})} \tag{4}$$

Then mathematical expectation of $f_k(x_k)$ is

$$E(f_k(x_k)) = \int f_k(x_k)p(x_k|z_{1:k})dx_k \tag{5}$$

In this way, we can estimate the posterior probability at time k by means of the discretization method [2], where the state transition probability models the system dynamics. For a given particle set $\{x_k^{(i)}, w_k^{(i)}\}$, $i = 1, 2, \cdots N$, where x_k is particle state, w_k is particle weight, and N is particle number, we can use the formula as follows to approximate the posterior probability

$$p(x_k|y_{1:k}) \approx \sum_{i=1}^{N} w_k^{(i)} \delta(x_k - x_k^{(i)}) \tag{6}$$

where $\delta(g)$ is dirac function, $w_k^{(i)}$ is the weight of discrete points $x_k^{(i)}$, $w_k^{(i)} \geq 0$, $\sum_{i=1}^{N} w_k^{(i)} = 1$.

3 Obtaining the Reliable Observation

To obtain the reliable image observation of the tracked object at current time, we adopt Gaussian mixture model to handle a variety of the background changing, such as varying illumination, image noise, and plant swaying, and use background subtraction to extract the foreground object. Before background modeling, the video images are converted from RGB space to HSI space. Denote the feature vector of a pixel in the image at time t as

$$X(t) = \left[s(t), r(t), g(t) \right]^T \tag{7}$$

where s is intensity lab, r and g are chroma labs. Here $\eta(X, \mu, \Sigma)$ is used to represent the Gaussian distribution of which the mean is μ and covariance matrix is Σ, then the history value of the pixel before time t can be represented by a K Gaussian mixture model(GMM)

$$P(X(t)) = \sum_{i=1}^{K} w_i(t) \eta(X(t), \mu_i(t), \Sigma_i(t)) \tag{8}$$

Where K is the Gaussian kernel number, $w_i(t)$ is the weight of the i-th Gaussian, and $\eta(X(t), \mu_i(t), \Sigma_i(t))$ is Gaussian probability density

$$
\begin{aligned}
&\eta(X(t), \mu_i(t), \Sigma_i(t)) \\
&= \frac{1}{(2\pi)^{\frac{n}{2}} |\Sigma_i(t)|^{\frac{1}{2}}} \exp\left(-\frac{1}{2}(X(t) - \mu_i(t))^T \Sigma_i(t)^{-1} (X(t) - \mu_i(t)) \right)
\end{aligned}
\tag{9}
$$

where n is the dimension of a feature vector. At time t, for an input pixel feature $X(t)$ and the i-th Gaussian, there is

$$\beta_i(t) = \begin{cases} 1 & |X(t) - \mu_i(t)| < D\Sigma_i(t) \\ 0 & \textit{otherwise} \end{cases} \tag{10}$$

where D is belief parameter whose value is from 2 to 3. If the input pixel feature $X(t)$ satisfies $|X(t) - \mu_i(t)| < D\Sigma_i(t)$, then $\beta_i(t) = 1$, this means that the input pixel is matched with GMM of background and should update the corresponding model parameters. If $X(t)$ does not satisfy the above condition, the pixel is considered as a foreground pixel, and the corresponding parameters are not updated.

In outdoor environment, another important problem is shadow detection for moving object extraction. The shadow is different from the background. It moves with the tracked object, so it is easily considered as a foreground object. We use the formula as follows to judge whether a pixel belongs to the shadow

$$\alpha_s \le \frac{I_s(x, y)}{B_s(x, y)} \le \beta_s \tag{11}$$

$$\left| I_r(x, y) - B_r(x, y) \right| \le \gamma_r \tag{12}$$

$$\left| I_g(x, y) - B_g(x, y) \right| \le \gamma_g \tag{13}$$

where $I_s(x,y), I_r(x,y)$ and $I_g(x,y)$ are the intensity and chroma of a foreground pixel feature $I(x,y)$ at coordinate point (x,y) respectively, and $B_s(x,y)$, $B_r(x,y)$ and $B_g(x,y)$ represent the intensity and chroma corresponding to the component whose weight is the maximum of GMM at coordinate point (x,y) respectively. Parameter α_s and β_s satisfy $0 < \alpha_s < \beta_s < 1$, and we should take the shadow intensity into account when choose α_s. The more the shadow intensity is, the less α_s is, and β_s is used to increase robustness to noise. Furthermore, the shadow intensity should not be too similar to the background, otherwise the background pixel can be mistakenly considered as a shadow pixel. Parameter γ_r and γ_g are selected empirically.

After eliminating the shadow, there are still some noise points in the image. To segment the moving object more effectively, we employ Morphological operation to the primary object region segmented from the differential image. Then we can obtain the reliable observation which is a binary image of moving object via erosion, dilation, filling operation.

4   Integrating Object Detection into Tracking

The conventional particle filter based on importance re-sampling commonly takes system transition as proposal distribution, and does not take the latest observation into account. Moreover, the difficulty of modeling an exact system transition further results in a poor prior distribution. In other words, the proposal distribution is not optimal. In this paper, we introduce the most recent observations into importance sampling process via Kaman equation. After updating the states of particles, we construct a Gaussian distribution based on particle set of object, and further obtain a suboptimal Gaussian proposal distribution through Kaman filter transmit. In the tracking phase, assume that the object keeps the same motion direction and velocity in consecutive two frames. Denote the state of object as $x_k = (a, b, h_x, h_y, l_x, l_y, o_x, o_y)^T$, where: (a, b) is the object center, (h_x, h_y) is velocity at x-axis and y-axis respectively, (l_x, l_y) is the width and height of object area, and (o_x, o_y) is the change rate of width and height. So we can model the system transition equation as follows

$$x_k = Fx_{k-1} + \varepsilon_k \tag{14}$$

Where F is the state-transition model, $Q_k = E\left[\varepsilon_k \varepsilon_k^T\right]$ is the covariance of the Gaussian process noise ε.

And the observation equation can be expressed by

$$z_k = Lx_k + \gamma_k \tag{15}$$

Where: L is observation matrices, $R_k = E[\gamma_k \gamma_k^T]$ is the covariance of the Gaussian observation noise γ. Here we only employ the sample mean $\hat{x}_k$ and associated covariance $\hat{P}_k$ to perform the Kalman update

$$x_k = \hat{x}_k + K_k(\hat{z}_k - L\hat{x}_k) \tag{16}$$

$$P_k = (I - K_k L)\hat{P}_k \tag{17}$$

$$D_k = L\hat{P}_k L^T + R_k \tag{18}$$

$$K_k = \hat{P}_k L^T D_k^{-1} \tag{19}$$

Where x_k, P_k are Kalman predicted estimate mean and covariance respectively, $\hat{z}_k$ is the latest observation, K_k is Kaman gain. To obtain reliable observation, we adopt the object detection algorithm in Section 3 to eliminate influence of complicated environment, and then segment the foreground object from background. After that, we can get the similarity of the current observation $\hat{z}_k$ and the observation $z_k^{(i)}$ estimated by particle filter via Mahalanobis distance Ma, where observation vector $z_k = (a, b, l_x, l_y)^T$ is the geometrical feature of foreground object, and Ma is

$$Ma_k^{(i)} = \left[\hat{z}_k - z_k^{(i)}\right] D_k^{-1} \left[\hat{z}_k - z_k^{(i)}\right]^T \tag{20}$$

Then we can formulate the likelihood of observation as

$$p\left(z_k \middle| x_k^{(i)}\right) = \frac{1}{\sqrt{2\pi}\sigma} \exp\left(\frac{-Ma_k^{(i)}}{2\sigma^2}\right) \tag{21}$$

where σ is variance. Besides, weight of each particle is

$$w_k^{(i)} \propto p\left(z_k \middle| x_k^{(i)}\right) \tag{22}$$

then the system output at each time writes

$$x_{out} = \sum_{i=1}^{N} w_k^{(i)} x_k^{(i)} \tag{23}$$

Sampling from the Gaussian proposal distribution $N(x_k; x_k, P_k)$, we obtain new particle set, of which the sample mean and covariance are x_k' and P_k' respectively. Then the Kalman prediction can be represented as follows

$$\hat{x}_{k+1} = Fx_k' + \varepsilon_k \tag{24}$$

$$\hat{P}_{k+1} = FP_k' F^T + Q_k \tag{25}$$

5   Experimental Results and Discussion

We report the results of tracking a moving object of two video sequences in outdoor environment. In our experiment, 320×240 images collected by stationary cameras are tested with 30 particles per tracker. In Fig. 1, we show the tracking results (e) in the case of less background interference scenario. Here are some of the tracking results of the test image sequence, for the frame 33 and 149 respectively. The white box in the image represents the moving object. From this test sequence, we can see that the tracker works correctly. (b), (c), (d) are images after background subtraction, shadow eliminating and Morphological operation respectively. It is observed that the background noise influence has been eliminated, and the reliable object observation has obtained. Fig. 2 shows the tracking results in the case of much more background interference scenario, where the trees in background shake violently, and shadow of the moving object is obvious. From left to right, top to bottom, they are the original image, background subtraction, shadow eliminating, Morphological operation and

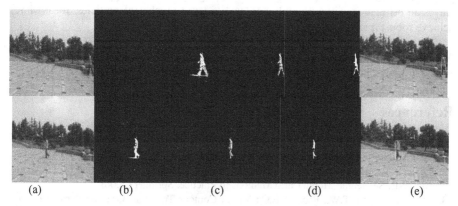

Fig. 1. Case of less background interference for frames 33 and 149: (a) original image; (b) background subtraction; (c) shadow eliminating; (d)Morphological operation; (e)tracking result

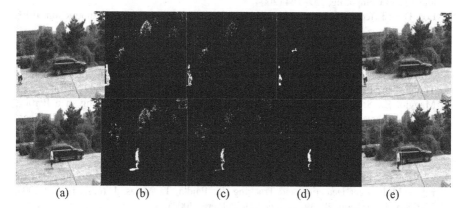

Fig. 2. Case of much more background interference scenario for frame 21 and 118: (a) original image; (b) background subtraction; (c) shadow eliminating; (d) Morphological operation; (e) tracking result

tracking results of the frame 21 and 118 respectively. As can be seen from the figure, the influence of trees swaying and shadow is greatly reduced and the tracker effectively conquers the interference of complicated scene.

6   Conclusion

Aiming to achieve robust tracking of moving objects in complex outdoor environment, we have presented a particle filter tracking method based on integrating detection and tracking. Meanwhile, in the experiment, we have tested the algorithm in two cases of varying levels of background interference. Results have demonstrated that the system can conquer disturbance (such as: trees sway and shadow of the moving object) to some extent in outdoor scene by means of object detection process. Thereby, we could obtain reliable observation in current time. To take full advantage of this observation, we fuse it into particle filter via Kalman filter. And then we generate suboptimal Gaussian proposal distribution. In this way, the tracking system can approach the optimal Bayesian estimation without employing such a large number of particles, meanwhile, achieve robust tracking in outdoor environment to some extent.

Acknowledgment. This work was supported by the National Natural Science Foundation of China (No. 61001143 and No. 61001013).

References

1. Isard, M., Blake, A.: Condensation-conditional density propagation for visual tracking. International Journal of Computer Vision, 5–28 (1998)
2. Doucet, A., de Freitas, J.F.G., Gordon, N.: Introduction to sequential Monte Carlo Methods, Sequential Monte Carlo Methods in Practice. Springer (2000)
3. Li, P.H., Zhang, T.W., Pece, A.E.C.: Visual Contour Tracking Based on Particle Filters. Image and Vision Computing, 111–123 (2003)
4. Li, P.H., Zhang, T.W., Ma, B.: Unscented Kalman Filter for Visual Curve Tracking. Image and Vision Computing, 157–164 (2004)
5. Lee, D.S.: Effective Gaussian Mixture Learning for Video Background Subtraction. IEEE Transaction on Pattern Analysis and Machine Intelligence, 827–832 (2005)
6. Huang, C.M., Liu, D., Fu, L.C.: Visual tracking in cluttered environments using the visual probabilistic data association filter. IEEE Transactions on Robotics, 1292–1297 (2006)
7. Xinyu, X., Baoxin, L.: Adaptive Rao–Blackwellized Particle Filter and Its Evaluation for Tracking in Surveillance. IEEE Transactions on Image Processing, 838–849 (2007)
8. Maggio, E., Smerladi, F., Cavallaro, A.: Adaptive Multifeature Tracking in a Particle Filtering Framework. IEEE Transactions on Circuits and Systems for Video Technology, 1348–1359 (2007)
9. Lanz, O.: Approximate Bayesian Multibody Tracking. IEEE Transaction on Pattern Analysis and Machine Intelligence, 1436–1449 (2006)
10. Okuma, K., Taleghani, A., de Freitas, N., Little, J.J., Lowe, D.G.: A Boosted Particle Filter: Multitarget Detection and Tracking. In: Pajdla, T., Matas, J. (eds.) ECCV 2004. LNCS, vol. 3021, pp. 28–39. Springer, Heidelberg (2004)
11. Zhao, T., Nevatia, R.: Tracking multiple humans in complex situation. IEEE Transaction on Pattern Analysis and Machine Intelligence, 1208–1221 (2004)

Discussion on the Courtyard under Two Levels of Management System—The Relationship between Tests Central and Faculty Working Office

Chen Sun

Tests Management Central
The Chinese People's Armed Police Forces Academy
Lang Fang, China
echo_96_2000@126.com

Abstract. After the management system of our tests central was reformed, many situations and problems first came onto being. One of the most prominent problems is how to deal with the relationship between faculty working office and tests central. Therefore, the paper discusses the problems in order to promote management systematic reform of tests central and coordinate the relationship between the development of teaching and experiment from four perspectives.

Keywords: tests central, experiment, coordinative relation.

1   Introduction

Many higher institutes, especially the ones which have passed the assessment of undergraduate teaching from Ministry of Education, all implement the two levels (college and department) of management system in tests management. Practices show that after merger and establishment from relatively small laboratory attached to faculty office with relatively small scale to large test central with a certain scale, the two levels of management system will absolutely play a significant role in improving the management efficiency, making use of resources such as laboratory assistants, equipments and labs, and taking advantages of large scale merit in long-term construction and plan of the tests central.

2   Necessity and Urgency to Carry Out the Two Levels of Management System

A.   Requirements of External Circumstances

"Recommendations of further improving undergraduate teaching in higher institutes" issued by the Ministry of Education requires that higher institutes should implement the two levels of management system according to their own needs to build comprehensive tests central.

J. Luo (Ed.): Soft Computing in Information Communication Technology, AISC 161, pp. 325–328.
springerlink.com   © Springer-Verlag Berlin Heidelberg 2012

B.   Requirements of Objective Reality

At the very beginning of establishment, the laboratory was attached to faculty working office without overall planning. Many phenomena like exclusive use of equipments, incomplete optimization of resources, relatively single model of experimental teaching, and inconsistent between faculty working office and laboratory etc, occur from time to time.

C.   Demands for Long-Term Construction and Development of Laboratory

The implementation of two levels of management system can realize the objects of planning with unification, building with emphasis, and developing with coordination. Meanwhile, it also can share resources and avoid duplication with de benefit of improving the experimental teaching level.

Under such circumstances, we set out to adjust the laboratory working from the year 2005. A college-level laboratory management agency--- tests management central for training; five department-level tests central--- tests centrals of basic courses, frontier defense, fire engineering, fire evidence identification, and fire command; and experimental teaching system of twenty laboratories have come into being at the present time. After adjustment, the advantages of laboratory management, resources sharing are extremely obvious, which can create appropriate conditions for tests teaching and reform.

3   The Relationship between Tests Central and Faculty Working Office under the New System

After the laboratory was changed from the management by faculty working office to college (department), how to handle well the coordinative relationship between laboratory and faculty working office, and then how to deal with the relationship between theory and practice, lecturers and laboratory technicians became critical questions related to the teaching quality. They need to be paid sufficient attention.

As we all know, experiment is considered to be a very important part in college teaching. In order to improve the overall teaching quality , mutual coordination among every part becomes much more critical besides ensuring quality in each part of teaching process (such as theoretical teaching, homework, experiment, curriculum design, practices and training). Whether the mutual coordination is good or bad often exerts greater impact on the overall teaching quality. The interaction and mutual promotion of each part of teaching can be collectively embodied in the organic combination of theory teaching and experiment teaching. The target and task of experiment teaching can be concluded into two points. Firstly, theoretical knowledge can be consolidated and intensified through experiment teaching. After experimental verification, it will be helpful to understand basic theories and establish basic concepts. Secondly, the theories learned from experiment teaching can guide practices. And innovation and comprehensive ability of students will be nurtured etc. Thus, experiment teaching and theory teaching are two inseparable parts in teaching process. Only the parts closely combined could receive better results. In another words, close combination between theory and experiment teaching functions as the basis of ensuring overall teaching quality. Under the system of faculty working office managing laboratory, although some kinds of defects and shortages exist, an obvious advantage shows that both theory teaching and experiment teaching can be taken into unified consideration by faculty working office. Thus it is relatively easy to realize the cooperation, coordination and arrangement of the two parts. However, after the laboratory was combined into tests central, the central and related faculty working office became separate units, and the central was no longer attached to the faculty working office. That would definitely bring inconvenience toward mutual coordination. Without special attention, former relatively closed relation will be weakened, leading to influencing the combination between theory and experiment teaching, and the overall teaching quality moreover. Our college should take it seriously after founding department-level tests central.

4   Four Perspective of Handling the Relationship between Tests Central and Faculty Working Office

A.   Correct Thinking and Coordinate Understanding

Take enough cognizance of the necessity and urgency of cooperation between tests central and related faculty working office. The coordination between tests central and faculty working office directly impacts the closed combination of theory teaching and experiment teaching, and even influences the overall teaching quality in direct way, which affects all the situations as a big problem. Whether in the past, at present or in the future, both should only strengthen the collaboration, but not weaken nor even lose connection with each other just because of changed management system that tests central is no longer attached to faculty working office. Concerning the cooperation of tests central and related faculty working office, leaders in college and departments, heads of tests central and faculty working office, and even general faculty and laboratory staff should have clear understanding towards it. Only in this way can all the colleagues forge consensus.

B.   Strengthen Consultation and Enhance Coordination

Many situations like setting up correlated experiments, adopting certain kind of system and form, and even determining, adjusting, and updating the project, content, and method of the experiment plus arranging schedules etc can be decided eventually only after tests central and related faculty working office jointly consult and achieve consensus together. Tests central has no right to decide or change the above conditions on its own because of nonexistent affiliated relationship in management system. That is important because what kind of system an experimental course adopts, the type of proof-based or design-based; how the faculty working office coordinates with tests central; whether the experimental project and content cover the main theories and technical contents are the key to the overall quality of teaching. Adopting the way of coordination between tests central and faculty working office to determine the method of putting up an experiment will provide institutional guarantee to closely combine theory teaching and experiment teaching.

C.   Exert the Strong Points to Offset Its Weakness and Assume Responsibility Jointly

The momentous reform of experiment teaching related to the overall situation has to be jointly taken charge of by the staff from both tests central and correlated faculty working office. They should take joint efforts to develop and exploit the form. For instance, the reform of experimental courses, innovation of experimental content and method, and development of design-based experiment etc need a joint seminar consisted of laboratory personnel and related staff in faculty working office. The advantages are depicted as follows:

- The combination between experimental contents and theoretical teaching functions well;
- The staff from faculty working office and tests central can coordinate with each other and exert advantages to offset weakness in order to complement each other and reform effectively with less tortuous path;
- After joint exploitation, the coordinative relationship between faculty and experimental personnel becomes more tight with purpose of making up for shortage because of reform of management system.

D.   Be Convenient to Each Other for Common Progress

Faculty staff can participate to instruct experiments and carry out teaching and research work. Meanwhile, they also can enjoy much more convenience provided by tests central. The practice needs to be fixed as a system. It is beneficial for faculty staff to closely combine theory teaching with experiment teaching, and even promote their operational capacity at the same time. Therefore, changes in management system do not influence this activity. Tests central and faculty working office are required to pay due attention to the work.

In short, it turns to be right to transform the management system of laboratory. Some newly emerging issues and problems during the procedure can be resolved with certain methods through our continuous research and exploitation. Only in this way can the reform become perfect, and the new system can operate healthily and orderly.

References

1. Xi, W., Hao, Y.: Handling the relationship between laboratory and faculty working office under two-level management system. Experimental Technology and Management 16(3), 48–49 (1999)
2. Yu, Z., Chen, K.: Enhancing functions, straightening out relations and promoting reform of laboratory system and standardization. Experimental Technology and Management 17(3), 43–46 (2000)
3. Yu, Z.: Research on standardization of laboratory in higher institutes. Experimental Technology and Management 17(5), 19–23 (1998)

Effects of Aerobic Training and Strength Training on Body Composition and Bone Mineral Density of the Middle and Old Aged People

Bo Liang

Physical Education School
East China Institute of Technology
Nanchang, China
liangbonc@ecit.cn

Abstract. Objective: In this study, a analysis was taken to compare the physical fitness average values among the 2group objects which were selected through a questionnaire survey: aerobatic training group (AT), AT combined with ST(AS). Study the effection of two kinds of exercises on elder's physical fitness. Method: One hundred and twenty elders aged 50-60 years was divide into aerobatic training group (AT) (60), AT combined with ST (AS) group (60). In addition, a training programmed lasting for 12 weeks was arranged on the objects not undertaking systematic exercises. Before and after the experimental training the fitness tests were undertaken and the indexes of the fitness tests were compared. Conclusion: Suitable and regular resistance training could enhance the elders work endurance, produce beneficial effects to their aerobic work capacity and produce optimal effects to their cardiac function.

Keywords: elders, acrobatic exercise, strength training, body composition, bone density.

1   Introduction

Participation of the middle and old aged people in physical exercise is significant for postponing of body hypofunction, prevention and control of the occurrence and development of certain vital chronic diseases (angiocardiopathy, diabetes and obesity, etc.), health promotion and improvement of life quality life. Therefore, the scientific fitness for the middle and old aged people is a problem universally concerned in the field of sports medicine at the current stage. The research of scholars in China mainly focuses on aerobic exercise with the proper strength training universally neglected. There are few researches on the effects of the exercise that integrates aerobic exercise and strength training on the body composition and bone mineral density of the middle and old aged people. There shall be further research on what differences in training effects generated from several exercising methods (aerobic exercise, strength training and the both integrated exercise) and how to arrange the fitness of the middle and old aged people more scientifically. There have been several decades of research on the effects of exercise on the physical fitness of the old aged people abroad. There are a lot of researches on the effects of aerobic exercise and strength exercise on the bone mineral density and body

J. Luo (Ed.): Soft Computing in Information Communication Technology, AISC 161, pp. 329–336.
springerlink.com   © Springer-Verlag Berlin Heidelberg 2012

composition as well as researches related to material selection for athletes and prediction of teenagers' growth and development. But there are few researches on the effects of exercise on the bone mineral density and body composition of the middle and old aged people mainly from the perspective of the mechanism that leads to changes and there is little involvement into how different the effects generated from different methods of exercise and fitness and how to rationally arrange exercise methods according to the physical conditions of the middle and old aged people. This subject plans to use the exercise method that integrates the aerobic exercise and strength training to carry out research on its effects on the body composition and bone mineral density of the middle and old aged people The research result has certain theoretical significance and actual guiding value in defining the fitness value of the exercise method in combination of aerobic exercise and strength training, increasing the recognition of the middle and old aged people for the exercise method, enriching sports exercise measures, thus the effects scientific fitness of the middle and old aged people are effectively promoted.

2   Research Object and Methods

A. Research Object
As shown by table 1, there are aerobic exercise groups and groups that integrate aerobic training and strength training. There are remarkable differences in various indexes of the physical fitness level, body composition level and bone mass status in two groups upon analysis before experiment. Age group of research objects: aged between 50 and 60, the average age is about 55. Groups: aerobic training group, aerobic training and strength training combination group, as shown in table 1.

Table 1. The Proportion Scale of Relative Importance

Group	Number
Aerobic training group	30
Aerobic training and strength training combination group	30.
Total	60.

B. Research Methods
Research method, know about the respondents' factors such as: the physical condition, living habits and disease history through questionnaire, which have effects on the experiment result. Test evaluation, contain the body composition and bone mineral density of the research objects before the experiment and the body composition and bone mineral density of the research objects after experiment, carry out evaluation and analysis again according the testing results before and after.

Experimental method, formulate different exercise methods according to different groups. Test the body composition and bone mineral density of the research objects before experiment and test that again after the experiment is finished. Aerobic training group: training forms: fast walk (40 to 90 minutes) walk fast until the target heart rate is the load criterion, the target heart rate is determined according to the corresponding test heart rate according to 40% to 60% of the maximal oxygen uptake. The exerciser tests the heart rate during exercise, but when the target heart rate can be reached, large amount of exercise is needed. Then increase the amount of exercise until the target heart rate is reached. Aerobic training and strength training combination group: the first aerobic training (30 to 60 minutes) form of the aerobic training: fast walk (walk fast until the target heart rate is the load criterion), the target heart rate is determined according to the corresponding test heart rates according to 40% to 60% of the maximal oxygen uptake. The exerciser tests the heart rate during exercise, but when the target heart rate can be reached, a large amount of exercise is needed. Then increase the amount of exercise until the target heart rate is reached. The next is the strength training (20-30 minutes), form of the strength training: dumbbell (7kg is the weight load of dumbbell training, group: 2, 8 to 12 times for each group. Test the initial 7kg of weight before training starts, test 7kg of weight after a period of time, continue training based on new load). Aerobic training group: consecutive 12 weeks, 3-4 times in each week and 40-90 minutes for each time. Aerobic training and strength training combination group: consecutive 12 weeks, 3 to 4 times in each week and 40- 90 minutes for each time. Measurement of body composition: measurement is performed by using the Inbody3.0 body composition analyzer manufactured by Korean BIOSPACE; main indexes to be measured contain percentage of body fat (PBF), fat-free mass (FFM) and muscle content, etc. Measurement of bone mineral density: ultrasound bone mineral density measuring instrument SUNLIGHT OMNISENSE700, Israel. Training monitoring: combination of professional personnel's regular centralized guide with self-monitoring of the exercise during exercise.

3   Result and Analysis

A.   Investigation into Exercise Amount

It can be seen from the table 2 and table 3 that the time of aerobic exercise per day: 180-320 minuets; step number of the walk: about 10 thousand steps/1 times; time of aerobic exercise in combination group: 90-160 minuets, step number of walk: about 5 thousand steps/1 times, strength exercise: about 30 minuets.

Table 2. Fast Walk Data in Aerobic Exercise

Time for fast walk per times	Number of exercise frequency per week	Walk distance (per times)	Step number (per times)
60-80 minuets	3-4 times	6000-8000m	8000-10000 times

Table 3. Aerobic Training and Strength Training Combination Group

Time for fast walk per times	Number of exercise frequency per week	Walk distance (per times)	Step number (per times)	Time for strength exercise
30-40 minuets	3-4 times	3000-4000m	4000-5000 times	40-60 minuets

B. Characteristics of Anthropometry Index of Aerobic Group and Aerobic Training Group and Strength Training Combination Group

It can be seen from the table 4 that there is no remarkable difference in indexes of height, weight and BMI in aerobic group and aerobic training and strength training combination group, but there are remarkable differences in PBF (percentage of body fat) and FFM (fat-free mass). Comparatively speaking, the muscle weight in aerobic training and strength training combination group is higher than that of the middle and old aged people and fat content is much higher than that of the old aged people in aerobic group. That is, strength exercise has the muscle of the strength exerciser built, and the fat free mass increased, thus the fat consumption in aerobic group is more obvious.

Table 4. Effects on Baby Composition Index of the Meddle and Old Aged People upon Exercise in Two Groups ($\overline{x}$ ±sd)

	Aerobic Group	Combination group	P
Height(cm)	164.6±5.42	165.1±4.68	>0.05
Weight(kg)	64.5±8.99	67.5±7.15	>0.05
BF(%)	26.4±5.46	32.5±4.89*	<0.05
BMI(kg/m2)	23.8±3.02	24.9±2.52	>0.05
FFM(kg)	41.8±3.87	47.3±6.08*	<0.05

Notes: *comparison of aerobic group and combination group upon training, P<0.05

Table 5. Characteristics of Changes in Body Composition Indexes before and after Exercise in Two Groups

	Aerobic group			Combination group		
	Before exercise	After exercise	P value	Before exercise	After exercise	P value
BF(%)	28.4±7.26	26.4±5.46*	< 0.05	30.6±5.09	32.5±4.89	> 0.05
BMI(kg/m2)	24.2±5.31	23.8±3.02	> 0.05	24.8±3.52	24.9±2.52	> 0.05
FFM(kg)	40.3±4.25	41.8±3.87	> 0.05	43.9±7.01	47.3±6.08*	< 0.05

Notes: * comparison before and after exercise in two groups, P<0.05

It can be seen from the table 5 that it refers to characteristics of body composition index changes before and after the exercise in two groups. There are remarkable changes in fat content before and after the exercise in the aerobic group while there is no remarkable change in other indexes; there are remarkable changes in FFM before and after the exercise in the combination group. TO sum up, aerobic exercise mainly changes the fat content of the exerciser, integrated exercise of strength and aerobic group mainly changes the content of muscle with little effect on fat change.

Characteristics of body quality index. It can be seen from the table 6 that there are remarkable differences in grip strength and reaction time and also there are extremely remarkable differences in crook before the seat body and the single foot stands with closed eyes.

As shown in the table 7, there are remarkable differences in aerobic group and combination group before and after exercise, the changes are remarkable in balanced capacity before and after exercise especially in combination group. The increase in strength in combination group is more obvious than that in the aerobic group.

Bone mineral measuring instrument uses high frequency sound wave (ultrasound) to measure the bone mass status of the heel (heel bone). The measuring result of the intensity index of bone mass is expressed with T value. Estimate the risk of bone fracture through reference crowd. T value indicates the comparison with the intensity index of reference mean value of "Yong Adult". The green regional base indicates a standard deviation （-1SD） lower than young adult value; yellow region represents -1—2.5SD scope; Red indicates scope lower than -2.5SD. The index number of bone fracture increase as index number of intensity decreases.

As table 8 figure 1 shows that it is comparison of bone mass conditions between the two groups and within the group. It can be seen from the table 8 that the overall level of the bone mass in the combination group. That is, strength practice is carried out in the combination group, saying that strength exercise has certain function of helping increase the bone mass. The further research on the specific mechanism shall be carried out. In general, the changes in the bone mass in the combination group are more obvious. Therefore, it can be seen that the effects of the strength exercise are better than the pure aerobic exercise.

Table 6. Effect on Body Quality of Old Aged Men （$\bar{x}$ ±SD)

	Aerobic group	Combination group	P
Crook before the seat body(cm)	3.75±8.41	3.2±7.95**	< 0.01
Single foot stand with closed eyes (s)	15.3±8.56	14.2±5.33**	< 0.01
Reaction time(s)	0.56±0.35	0.72±0.51*	< 0.05
Grip strength(kg)	38.9±5.62	39.9±6.78*	< 0.05

Notes: *comparison after exercise in aerobic group and combination group, P<0.05, ** comparison after exercise in aerobic group and combination group, P<0.01

Table 7. Changes of Body Quality of the Old Aged Men before and after Exercise ($\bar{x}$ ±sd)

	Aerobic group			Combination group		
	Before exercise	*After exercise*	*P value*	*Before exercise*	*After exercise*	*P value*
Crook before the seat body (cm)	3.55±9.11	3.75±8.41*	< 0.05	3.3±7.95	3.2±7.95*	< 0.05
Single foot stand with closed eyes (s)	14.8±7.43	15.3±8.56*	< 0.05	13.2±6.32	14.2±5.33**	< 0.01
Reaction time (s)	0.52±0.33	0.50±0.35*	< 0.05	0.62±0.55	0.72±0.51*	< 0.05
Strip strength (kg)	38.3±4.52	38.9±5.62*	< 0.05	36.9±7.71	39.9±6.78*	< 0.05

Notes: * comparison before and after exercise in two groups, P<0.05; comparison before and after exercise in two groups, P<0.01

Table 8. Comparison of Bone Mineral Density in Two Groups

Bone mass level	Aerobic group		Combination group	
	Before exercise	After exercise	Before exercise	After exercise
Normal	50% (15人)	60% (18人)	46.7% (14人)	63.3% (19人)
Deficient	40% (12人)	40% (12人)	50% (15人)	36.7% (11人)
Loose	10% (3人)	0	3.3% (1人)	0

4   Discussion

A.   The Influence of Aerobic Exercise on the Body Shape of the Old Aged People

With the improvement of living standards and nutrients, the old aged people lack of sufficient exercise. The accumulated fat can be easily caused across the whole body, especially the abdomen is much obvious so that the body form is deformed. Each part of the body can be fully activated trough aerobic exercise and the surplus fat in the body can be reduced, convenient for losing weight and reaching the purpose of shaping good body form. The exercise has certain effects on the body pattern of the old aged people. Aerobic exercise body fat content is obviously lower than that of the combination group, FFM (Fat-free mass) is obviously lower than that of the combination group. This manifests that aerobic exercise has certain functions of reducing the fat content of the old aged people and strength exercise has certain functions on fat-free mass. Some research shows that exercise uses fatty acid to influence the lipid metabolism through changing body composition and increasing oxidative of the organism while hyperlipemia patients come down with obesity or overweight. There are adequate evidence shows that the exercise reduces the percentage of body fat, further the circulating blood lipid decreases [1], identical with the result of this research. Thereby, the aerobic exercise has remarkable functions on changing the body pattern of the old aged people.

B. Effects of Strength Exercise on Bone Mineral Density

The metabolic activity of bone tissue is ceaseless in a life. Its main symbol is the metabolic coupling that always starts bone resorption and bone formation in different parts of the bone issue cross the whole body, namely bon turnover. The result of bone metabolic activity will lead to changes in the bone mass and bone mineral density. Bone mineral density (B) refers to the content in certain specific bone volume. There are a lot of factors that influence the bone mineral density, like heredity, meal, living habit and exercise, etc. Among many factors, the exercise has great effects on the bone. Its effects on the bone strength far surpass the effects of hormone, calcium and vitamin D related to bone metabolism on the bone strength, belonging to the active factors that have effects on the bone mass [2].

Characteristics of the effects of the strength project on the bone mineral intensity. Wolff law supposes that stress and mechanic load directly influences the bone formation and reconstruction by having functions on the bone through muscle and heel end. The strength project's stimulation on the bone gives rise to changes in absorption and formation, thus the internal structure and external shape of the bone is changed. Sun Ru, etc. [3] compare 15 professional male speed skating athletes to 15 university boys with computer major and carry out analysis. The difference in university students with strength training and common university students is much obvious and the increase in total BMD of hip joint at the left side is much obvious. The similar research is carried out abroad. Creighton [4] also assumes that the BMD of female athlete with high impact exercise is higher than that of non impact exercise such as: swimming sportswoman. What is contradicted with these research results is that the lumbar spine bone mineral density of women aged from 34 to 42 decreases after 9 months' bearing exercise training, according to discovery of Rockwell, etc. [5] (1990), which is whether related to the exercise intensity needs to be further discussed.

There is big difference in the effects of training related to endurance on the skeleton. Proved by many research results, the endurance training above moderate strength can increase the bone mineral density. It is assumed that the endurance exercise can have active effects on increase in bone mass or bone mineral density through repeated thrust load. The Continuous load at the same level has certain functions for increasing the mineral content of the trabecular bone, but the changing load only can have little effects. It is suggested that properly increase strength training while doing aerobic exercise, namely the cardio-pulmonary function of exerciser can be improved and the bone mineral density can be enhanced so as to avoid bone fracture.

References

1. Pollock, M., Foster, C., Knapp, D., et al.: Effect of age and training on aerobic capacity and body competition of master athletes. Journal of Applied Physiology 62, 725–731 (1987)
2. Wu, M.F.: The extraosseous and intraosseous arterial anatomy of adult elbow. Modern Rehabilitation 6, 170–191 (2002)
3. Sun, R., Zhao, L.J., Pollock, Foster, C., Knapp, D., et al.: Effect of recombinant human osteogenic protein-1 on healing of segmental defects in nonhuman primates. Chinese Journal of Sports Medicine 11, 703–704 (2005)
4. Creighion, D., Morgan, A.L., Debra, B., et al.: Weight bearing exercise and markers of bone turnover in female athletes. Journal of Applied Physiology 90, 565–570 (2001)
5. Rockwell, J.C., Sorensen, A.M., Baker, S.D., et al.: Weight training decreases verebral bone density inpremenopausal women: a prospective study. J. Clin. Endocrin. Metab. 71, 988–993 (1990)

Studies on the Relationship between Land Use/Cover Types and Urban Heat Island Effect in Changchun

Ming Yan[1] and Yuanyuan Song[2]

[1] Institute of Agriculture Resources and Regional Planning of Jilin Province
Changchun, China
[2] College of Urban and Environmental Sciences, Northeast Nomal University
Changchun, China
jlyanming@sina.com, songyy296@nenu.edu.cn

Abstract. In recent years, enhanced urban heat island effect is a prominent urban environmental issue that many scholars are studying. In this paper, the capital of Jilin Province - Changchun had been selected as the study area. Thermal infrared bands of Landsat 7 ETM+ daytime image(July 15,1999) and Landsat 5 TM daytime image (July 29,2007) were used to retrieve brightness temperature instead of land surface temperature. This study provides retrieved images on the UHI phenomena in Changchun and the findings can be used to guide further study integrating satellite high-resolution thermal data with land use/cover .In conclusion, the spatial structure of thermal field was obtained in different time. The NDVI, NDWI,NDBI and NDBaI indices were used to characterize the land use/cover types in the study region and to study the relationships between land use/cover types and UHI qualitatively and quantitatively. The results show that it had a significant urban heat island phenomenon in Changchun in two different years. Furthermore, according to quantitative statistics, there are close relations between the spatial structure of urban thermal patterns and land use types. The results in this study may also be used in programming of city environment and forecasting of heat island effect.

Keywords: UHI, remote sensing, NDVI, NDWI, NDBI, NDBaI.

1 Introduction

The urban heat island (UHI) refers to the phenomenon of higher atmospheric and surface temperatures occurring in urban areas than in the surrounding rural areas due to energy consumption and reduction of urban green space(Voogt et al.,2003). Urban heat island is one of the most significant features of urban climate change, it has changed the urban heat environment, affected the regional climate, urban hydrology, air quality, urban soil physical and chemical properties , the distribution and behavior of urban life, as well as a number of urban ecological processes such as metabolism, energy cycle, induced a series of ecological and environmental problems(Xiao et al.,2005) Traditionally, we monitored the UHI through temperature information and data which come from meteorological stations to study the temperature difference between Urbanand rural areas. Such conventional monitoring method not only costs a lot of manpower and material resources, but also does not have high precision.

J. Luo (Ed.): Soft Computing in Information Communication Technology, AISC 161, pp. 337–344.
springerlink.com © Springer-Verlag Berlin Heidelberg 2012

The satellite thermal infrared remote sensing just makes up for this defect. Satellite thermal infrared remote sensing can effectively and comprehensively detect the temperature characteristics of underlying surface,periodically and dynamically monitor the trends in urban thermal environment. It is an effective means to study UHI(Xie et al,2007). Most of the existing satellite-based research on UHIs can be grouped into three main themes: (1) to examine the spatial structure of urban thermal patterns and their relation to urban surface characteristics; (2) to study urban surface energy balances through coupling with urban climate models; (3) to study the relation between atmospheric heat islands and surface Uzis through combining coincident remote and ground based observations(Voogt et al.,2003).

Two remote sensing images obtained at different times were used to retrieve brightness temperature instead of land surface temperature in this paper. The purpose is to study the spatial structure of urban thermal patterns and their relation to urban surface characteristics.

2   Study Area and Data

A.   Study Area

Changchun is the capital of Jilin Province, the center of political, economic, cultural and transportation. It is located in Songliao Plain , Northeast China , Eurasia East Coast ,mid-latitude zone of northern hemisphere, ranging from 43 °05'-45 °15' north and 124 °18'-127 °02 east(Fig.1), high east and low west, is the transitional zone from mountain and hill to platform and Plain, humid climate to drought climate, mixed wood to meadow steppe, dark brown soil to chernozem. This area has a monsoon climate, with an average annual temperature about 48°C, maximum temperature 39.5°C and minimum temperature -39.8°C, sunshine time 2688 hours. The average annual precipitation is approximately 522mm–615mm, about 60% of rainfall is usually received during summer. With a total area of 20604 km^2, its population reaches 7.459 million (2007).

B.   Data

To quantitatively measure land surface brightness temperature and study the relationships between land use/cover types and UHI in the study area,Landsat 7 ETM+ image(July 15,1999) , Landsat 5 TM image(July 29,2007)(Fig.2) and basic geographic information data were selected.

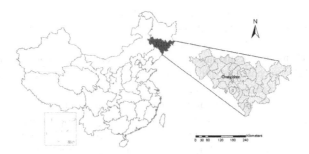

Fig. 1. Location of the study area

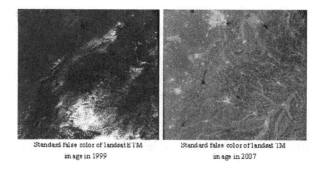

Standard false color of landsat ETM image in 1999   Standard false color of landsat TM image in 2007

Fig. 2. Remote sensing images of study area

3 Methods

A. Retrieval of Brightness Temperature

The heat infrared exploration of satellite sensor detects radiation temperature of city underlying surface. The ground objects look as a blackbody when calculation the radiation temperature. The obtained records on the detectors are at-satellite radiation values, so the radiometric calibration is needed. After that the digital numbers (DNs) are converted to radiation luminance. The formula for radiation calibration is as follows:

$$L = Gain \cdot DN + Bia\tilde{s} \tag{1}$$

Where, L is at-satellite radiation luminance; DN is the obtained radiation value of sensors; Gain is the gain coefficient and Bias is the offset coefficient. The Gain and Bias can't be obtained from the header file of the remote sensing images which were used in the study, so the formula can be expressed as:

$$L = \left(\frac{L_{max} - L_{min}}{QCAL_{max} - QCAL_{min}} \right) \times (DN - QCAL_{min}) + L_{min} \tag{2}$$

Where, L_{min} and L_{max} are the maximum and minimum spectral radiances; $QCAL_{max}$ and $QCAL_{min}$ are the obtainable maximum and minimum DN.

For the Landsat 7, (2) can be simplified as follows:

$$L = \frac{L_{max} - L_{min}}{254}(DN - 1) + L_{min} \tag{3}$$

For the Landsat 5, (2) can be simplified as follows:

$$L = \frac{L_{max} - L_{min}}{255} \times DN + L_{min} \tag{4}$$

After radiometric calibration, the next step is to convert the spectral radiance to at-satellite brightness temperature under the assumption of uniform emissivity, the conversion formula is(Tan et al.,2009; WENG Q H et al,2004):

$$T = \frac{K_2}{\ln(\frac{K_1}{L_\lambda}+1)} \tag{5}$$

Where T is effective at-satellite temperature in K,;L_λ is spectral radiance in W/(m^2 ·ster· µm); And K_2 and K_1 are pre-launch calibration constants, they list in Table 1:

Table 1. Value of K1 and K$_2$

constants	Landsat 5	Landsat 7
K_1	607.766	1260.56
K_2	666.093	1282.71

B.   Derivation of Various Indicators for Land Use Classification

In most academic documents, the traditional classification was used in the study of relationships between land use/cover types and UHI. This paper used Normalized Difference Vegetation Index (NDVI), Normalized Difference Water Index (NDWI), Normalized Difference Built-up Index (NDBI) (Zha, Y. et al,2003) and Normalized Difference Bareness Index (NDBaI) (Zhao, H. M.,2005) to classify. The conversion formula as follows:

$$NDVI = \frac{\rho_4 - \rho_3}{\rho_4 + \rho_3} \tag{6}$$

$$NDWI = \frac{\rho_4 - \rho_5}{\rho_4 + \rho_5} \tag{7}$$

$$NDBI = \frac{D_5 - D_4}{D_5 + D_4} \tag{8}$$

$$NDBaI = \frac{D_5 - D_6}{D_5 + D_6} \tag{9}$$

Where, ρ represents the radiance in reflectance units and D represents the digital numbers (DNs) of the relevant Lanasat TM or ETM+ bands.

4 Results and Discussion

A. The Spatial Structure of Urban Thermal Patterns

According to the research methods, the land surface brightness temperature images of July 15, 1999 and July 29, 2007 were retrieved as a result, they are showing in Fig.3. We compared the two images because they have almost the same obtained time, both in the summer. For Satellite orbit and imaging mechanism, the whole region of Changchun is distributed in two adjacent images. The two images can't perfectly maintain the reflectivity information in image mosaicing process. Therefore, the images which contain most region of Changchun were selected as source data in order to reduce the complex conditional restrictions in study process. Fig. 3 shows the distribution of T values in Changchun, and it is a subset of Changchun. In order to give obvious expression to the spatial structure of urban thermal patterns, the three-dimensional images of Fig. 3 are showing in Fig.4.

In 1997, the urbanized areas were found to have the highest surface temperature, the maximum land surface brightness temperature was up to 35.461°C and the minimum land surface brightness temperature was 17.08°C. The water had the lowest brightness temperature in all the land use/cover types. The rural areas far from urban areas had higher brightness temperature than water with a large number of vegetation. The brightness temperature of suburbs was higher than the rural areas'. This region greatly influenced by human activities and the process of urbanization. Comparing the urbanized areas with the highest land surface temperature to the rural areas far from urbanized areas, the maximum temperature difference can reach 15°C. This is a clear urban heat island effect. Fig.3 shows the land surface brightness temperature image of July 29, 2007, and it has the same characteristics with 1997 in generally. The difference was that a higher maximum brightness temperature can reach 41°C, and there was 20°C between minimum temperature and maximum temperature.

Take 1999 and 2007 for comparison, the spatial structure of thermal field of the same season had a greater change. The temperature difference increased clearly between the urbanized areas and the rural areas. The maximum temperature increased; moreover, the area of high temperature zone was expanding.

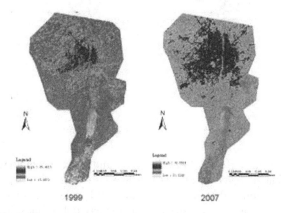

1999   2007

Fig. 3. Thermal field distribution of different time in Changchun

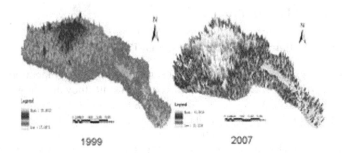

Fig. 4. Three-dimensional image of thermal field distribution

B.   Relationship between Urban Heat Island and Land Use/Cover Changes at the Regional Level

These indices mentioned in the research methods were used to classify different land use/cover types by setting the appropriate threshold values (Fig.5). Samples with 30 randomly generated points of every land use/cover type were used to investigate classification accuracy .The overall accuracy of classification is approximately 90%. Among selected 6 land use/cover types, the classification performed best for bare land, water, and vegetation, but less so for semi-bare, development and built-up area.

Combination with brightness temperature and land use/cover types, the statistical analysis list in Table 2 and Fig.6.

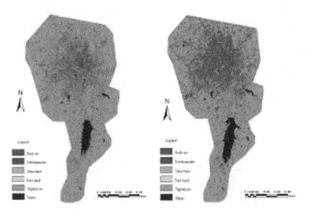

Land use and land cover map in 1999   Land use and land cover map in 2007

Fig. 5. Land use classification map

Different land use/cover types have different land surface brightness temperature, this indicates that land use/cover affects the distribution of urban heat island effect.. From Table 3, it was found that the temperature of urbanized areas increased 1.745°C , at the same time, the area of urbanized areas increased 1.6 times. Rapid process of urbanization intensified urban heat island effect; this becomes an important factor of global warming. The area of bare land which distribute in the suburb of Changchun had a significant reduction from 1999 to 2007, but the average temperature had increased and the temperature difference between urbanized areas and rural areas became smaller. For example, the temperature difference was 4°C between bare land and urbanized areas in 1999, but the difference reduced to 28.025 in 2007. The reduction of bare land transformed into urbanized areas. This added area to the urbanized areas, and enhanced the urban heat island intensify on the whole. Simi-bare and urban development areas were seen as one part. This part was a mixture of a variety of land surface types with large temperature range and irregularities. It had the most significant reduction. One Part of it transformed into urbanized areas, another part of it transformed into urban development areas. Water has the same brightness temperature distribution characteristics in two map of thermal field distribution that was water had the lowest average value of brightness temperature in all land use types, and plays a role in cooling. The value of brightness temperature in 2007 was relatively lower than in 2007, mainly due to the different precipitation of two different years. For vegetation, there was a small amount of reduction from the overall. The growing development area destroyed some vegetation. However, there was a certain amount of vegetation increased in urban area for urban greening.

Table 2. Statistics of brightness temperature between different land use/cover types in Changchun

Years	1999			2007		
	Max	*Min*	*Mean*	*Max*	*Min*	*Mean*
Water	34.066	19.638	21.726	34.999	19.384	21.067
Vegetation	26.365	18.610	22.016	37.738	19.185	22.340
Bare land	29.791	17.057	22.410	37.352	9.621	28.025
Semibare and Development	33.598	20.159	25.947	38.125	16.297	28.468
Built-up	35.451	18.078	26.507	35.394	22.902	28.252

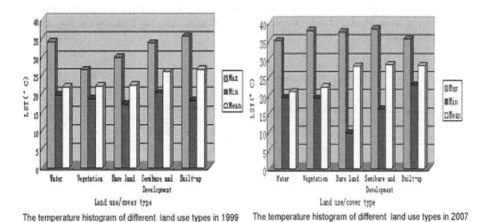

The temperature histogram of different land use types in 1999   The temperature histogram of different land use types in 2007

Fig. 6. The temperature histogram of different land use types

5   Conclusions

In the study, we try to examine the spatial structure of urban thermal patterns and their relationship to urban surface characteristics by integrating remote sensing images and basic geographic information data. Results indicated that using brightness temperature to reflect the distribution of urban heat island effect was effective. Combination of land use/cover type which retrieved base on remote sensing index and the heat island effect helped in understanding that different land use/cover types had different effects on the heat island effect. On the other hand, a disadvantage of the study was the in situ observation, a fact which limits the verification of retrieval accuracy of brightness temperature. However, despite this limitation, it is clear that the results can be effectively quantitative assessment of the urban heat island effect.

References

1. Tan, G., Cai, Z., Xu, Y.: Investigation on Heat-island Effect in Nanjing Area by Landsat TM Data. Journal of AnHui Agricultural Sciences 37, 6050–6052, 6066 (2009) (in Chinese)
2. Voogt, J.A., Oke, T.R.: Thermal remote sensing of urban areas. Remote Sensing of Environment 86, 370–384 (2003)
3. Weng, Q.H., Lu, D.S., Schubring, J.: Estimation of Land Surface Temperature–vegetation Abundance Relationship for Urban Heat Island Studie. Remote Sensing of Environment 89, 467–483 (2004)
4. Xie, X., Zhou, J., Zhang, H.: Study on City Heat Island in Xi'an Baded on Landsat TM. Journal of Hebei Normal University (Natural Science Edition) 31, 397–399 (2007) (in Chinese)
5. Xiao, R., Yang, Z., Li, W.: A review of the eco-environmental consequences of urban heat islands. Acta Ecologica Sinica 25, 2055–2060 (2005) (in Chinese)
6. Zha, Y., Gao, J., Ni, S.: Use of normalized difference built-up index in automatically mapping urban areas from TM imagery. International Journal of Remote Sensing 24, 583–594 (2003)
7. Zhao, H.M., Chen, X.L.: Use of normalized difference bareness index in quickly mapping bare areas from TM/ETM+. In: Geoscience and Remote Sensing Symposium, vol. 25-29, pp. 1666–1668 (2005)

Supply Chain Coordination Based on Buy-Back Contract under Price-Sensitive and Time-Sensitive Demand

Chenglin Shen[1] and Xinxin Zhang[2]

[1] School of Business, Tianjin Polytechnic University, Tianjin, China
[2] Tianjin Institute of Pharmaceutical Research Tianjin, China
gnuguy@163.com, tjusss@yahoo.com.cn

Abstract. In this paper, pricing and order decision problems of a two-stage decentralized supply chain was studied. Demands are both price-sensitive and time-sensitive. First, the centralized supply chain model as a benchmark, we analyzed the optimal order quantity, optimal price and lead time. Secondly, optimal decisions and coordination condition of decentralized supply chain under the buy-back contract were analyzed. Finally, a numerical example was proposed to analyze the impact of price factor, time factor and other model parameters on the profits and coordination efficiency of supply chain under buy-back contract.

Keywords: demand, centralized supply chain, decentralized supply chain, coordination.

1   Introduction

As a supply chain coordination mechanism buy back contract is widely used in practice. Through the implementation of buy-back contract, the retailer can reduce the risk of redundant inventory due to the uncertainty demand, improve supply chain coordination, and ease the supply chain 'double marginalization effect' to improve the benefits of overall supply chain[1-2].

Early studies on the buy-back contracts mainly focused on inventory decisions. Products prices are generally assumed to be determined by the market, and are predictable in advance [3]. With further research, the coordination mechanism of the supply chain based on the buy back contract with random demand is concerned widely by more and more researchers.

Pasternack [4] first studied the optimal pricing of perishable goods, and return policy, and pointed out that channel coordination can be obtained when manufacturers provide full credit to the retailers for some buyback products, suppliers. At this time buy back strategy is a function of retailer demand. Emmons et al[5] pointed out that under the uncertain demand circumstance, when demand fluctuations is consistent with uniform distribution, the supplier can increase channel revenue through the buy back contract. Bose et al [6] found that when the wholesale price is exogenous, the equilibrium buyback policy usually can not achieve the Pareto optimality, but when the wholesale price is enough high, a balanced buyback policy can achieve Pareto optimality. Yue et al [7] studied the impact of the full buy-back strategy on supply chain coordination under the asymmetric information, and find full-buy-back strategy

J. Luo (Ed.): Soft Computing in Information Communication Technology, AISC 161, pp. 345–353.
springerlink.com   © Springer-Verlag Berlin Heidelberg 2012

will benefit retailers in any cases, but only benefit the supplier in certain circumstances. Yao et al [8] analyzed the impact of the price-sensitive factors on buy-back contract operations, and pointed out that when demand volatility increases, manufacturers have to share more profits with retailers to continue the Stackelberg game of supply and demand. Even so, the supplier still gains higher benefit than the pure price contract under the buy-back contract.

However, the described buy-back contract model does not include delivery lead time, which is response time factor. The research about the impact of the response time on buy-back contract operation and the supply chain performance has not been reported. In fact, time-based competition has become the focus of competition between supply chains, and its impact on supply chain operations can not be ignored.

This paper considered the pricing and ordering decision problem of a two stage supply chain under the dual price-sensitive and time-sensitive demand. Subsequently, the centralized supply chain model as a benchmark, we analyzed the coordination conditions of the buy-back contract. Finally, through numerical examples, we investigated the efficiency of the buy-back contracts, and the impact of the price parameter and time parameter on the buy-back contract.

2   Model Description

Consider a two-stage supply chain consists of a supplier and a retailer, both parties are risk neutral. Suppliers sell products through retailers. The product demand and delivery lead time are uncertain, and the demand is affected by the product's retail price and delivery time. Before the start of the sale season, retailer decided two variables: the product order quantity Q and the unit retail price p. When the supplier receives orders from retailers they begin to arrange production. The interval between orders from retailers and receive products is defined as the delivery lead time. Suppliers determine the wholesale price p and the commitment delivery time based on the retailer's order quantity. At the end of the sales cycle, the supplier may buy back some unsold products at certain price from the retailers.

Suppose the supplier's production capacity is limited. The increase in demand will inevitably lead the actual delivery lead-time to be extended. When the actual delivery time extend, that is $t>l$, consumers will be compensated, and the compensation is provided by the supplier. b_s is defined as the cost of unit of product. When $t<l$, suppliers undertake the inventory cost of unit product h. We can write the cost of the supplier lead time $c_l(l, t)$ as

$$c_l(l,t) = h\int_0^l (l-t)f(t)dt + b_s(t-l)f(t)dt \tag{1}$$

Supplier's marginal cost of production is c. Assuming that the actual delivery time t is exponentially distributed, the probability density function and distribution function is $f(x) = \theta e^{-\theta x}, x > 0$ and $F(x) = 1 - e^{-\theta x}$, and $1/\theta$ is the average delivery time of suppliers.

Assuming that the marketing behavior of suppliers and retailers are Stackelberg Game, we get the random demand model

$$X(p,l,\varepsilon) = D(p,l) + \varepsilon \tag{2}$$

$D(p,l)$ is a decreasing function which reflects relationship of demand with unit retail price p and commitment delivery time l. ε is random variable of the distribution function $G(\bullet)$ and probability density function $g(\bullet)$ in [A,B] interval. μ and σ are the mean and the standard deviation of random variable ε. Another hypothesis is that $G(\bullet)$ and its inverse function is strictly monotonically increasing, and $g(\bullet)$ is continuously differentiable. We suppose that $D(p,l)$ is linear demand depend on price and commitment lead time.

$$D(p,l) = D_0 - \alpha p - \beta l \tag{3}$$

Where D0 is the potential demand of the market, α is the price elasticity factor, β is the time elasticity factor. The relationship of retailer's order quantity between market demands can be expressed as

$$Q = D(p,l) + z \tag{4}$$

Where z is inventory level (stocking level), which is the order quantity meet some random needs.

This paper is based on the following assumptions: the potential customer is the price/time sensitive customer. Price as well as short response time is the customer's basic requirements. Suppliers share the complete information of actual delivery time with retailers. And retailers also share the full information of customer response to supplier's commitment delivery time.

3   Models

A.   Decision Model of Centralized Supply Chain

In centralized supply chain, supplier and retailer are the same company (such as supplier is simultaneously responsible for production and sales). The decision variables of the supply chain are the quantity Q_c production/inventory, the retail prices of products p_c, and the delivery time l_c. The expected profits of the supply chain is

$$E[\Pi_M^{IC}(Q_c, p_c, l_c)] = \int_A^z (p_c[D(p,l_c) + y] - h[z - y])g(y)dy$$
$$+ \int_z^B (p_c[D(p,l_c) + z] - s[y - z])g(y)dy - [c + c_l][z + D(p,l_c)] \tag{5}$$

μ represents the expected value of random items ε, substituting (1) into (5) we obtained

$$E[\Pi_M^{IC}(z, p, l_c)] = (p_c - c - h\int_0^{l_c}(l_c - t)f(t)dt - b_s\int_{l_c}^\infty (t - l_c)f(t)dt)[D(p,l_c) + \mu]$$
$$- (c + h\int_0^{l_c}(l_c - t)f(t)dt + b_s\int_{l_c}^\infty (t - l_c)f(t)dt + h)\int_A^z (z - y)g(y)dy \tag{6}$$
$$- (p_c - c - h\int_0^{l_c}(l_c - t)f(t)dt - b_s\int_{l_c}^\infty (t - l_c)f(t)dt + \int_z^B s(y - z)g(y)dy$$

The optimal decision of the centralized supply chain is maximize the whole profit $\Pi_M^{IC}(z, p, l_c)$

$$Max_{z,p_c,l_c} E[\Pi_M^{IC}(z,p,l_c)] =$$

$$\begin{Bmatrix} (p_c - c - h\int_0^{l_c}(l_c-t)f(t)dt - b_s\int_{l_c}^{\infty}(t-l_c)f(t)dt)[D(p,l_c)+\mu] \\ -(c + h\int_0^{l_c}(l_c-t)f(t)dt + b_s\int_{l_c}^{\infty}(t-l_c)f(t)dt + h)\int_A^z(z-y)g(y)dy \\ -(p_c - c - h\int_0^{l_c}(l_c-t)f(t)dt - b_s\int_{l_c}^{\infty}(t-l_c)f(t)dt + s)\int_z^B(y-z)g(y)dy \end{Bmatrix} \text{ For given } p_c \text{ and } l_c,$$

taking first and second derivatives of $E[\Pi_M^{IC}(z,p,l_c)]$ with respect to z we get

$$\frac{\partial E[\Pi_M^{IC}(z,p_c,l_c)]}{\partial z} = (p_c - c - c_l + s) - (p_c + h + s)G(z)$$

$$\frac{\partial^2 E[\Pi_M^{IC}(z,p_c,l_c)]}{\partial z^2} = -(p_c + h + s)g(z) < 0.$$

From the foregoing, and for a given p_c and l_c supply chain revenue function is concave function of z, so in interval [A, B] z has the maximum value. We order $\frac{\partial E[\Pi_M^{IC}(z,p_c,l_c)]}{\partial z} = 0$,and substitute it into (1) we get

$$z^* = G^{-1}\left(\frac{p_c - c - h\int_0^{l_c}(l_c-t)f(t)dt - b_s\int_{l_c}^{\infty}(t-l_c)f(t)dt + s}{p_c + h + s}\right) \tag{7}$$

For a given z and l_c, taking first and second derivatives of $E[\Pi_M^{IC}(z,p,l_c)]$ with respect to p_c we yield

$$\frac{\partial E[\Pi_M^{IC}(z,p_c,l)]}{\partial p_c} = D_0 - 2\alpha p_c - \beta l_c + \mu + \alpha(c + h\int_0^{l_c}(l_c-t)f(t)dt$$

$$+ b_s\int_{l_c}^{\infty}(t-l_c)f(t)dt) - \int_z^B(y-z)g(y)dy$$

$$\frac{\partial^2 E[\Pi_M^{IC}(z,p_c,l_c)}{\partial p_c^2} = -2\alpha < 0.$$

So we conclude that for a given z and l_c, supply chain profit is concave function of p_c. In the definition interval of p, supply chain profit has the maximum value. From $\frac{\partial E[\Pi(z,p_c,l_c)}{\partial p_c} = 0$, we get

$$p_c^* = \frac{1}{2\alpha}[D_0 + \mu + \alpha(c + h\int_0^{l_c}(l_c-t)f(t)dt + b_s\int_{l_c}^{\infty}(t-l_c)f(t)dt)$$

$$- \beta l_c - \int_z^B(y-z)g(y)dy] \tag{8}$$

Similarly, for a given D and z, the centralized supply chain overall expected profit is a concave function of l_c. In the definition interval $[0, \infty)$ of l_c the supply chain overall expected profit has the maximum value. We order $\frac{\partial E[\Pi^{IC}]}{\partial l_c} = 0$ and get l_c^*

$$l_c^* = F^{-1}(\frac{b_s - \beta / \alpha}{b_s + h}) \tag{9}$$

The optimal decision of decentralized supply chain under the buyback contract

Under the buy-back contract, the supplier sold the product to the retailer at wholesale price w per unit of the product. After the sales season supplier buy back the unsold product at the transfer price b and $b<w$. When the actual demand is less than the order quantity, that is $X<Q$, the retailer will obtain buy-back benefits $b[(Q - X(p,l,\varepsilon)]$. In addition, when demand is greater than the order quantity, that is $X>Q$, the retailer bear the unit shortage cost s. Therefore, the retailer's optimal decision is

$$\underset{p,l,z}{Max}\, E[\Pi_R^{RP}(z,p,l)] = \int_A^z (p[D(p,l) + y] + b[z - y])g(y)dy$$
$$+ \int_z^B (p[D(p,l) + z] - s[y - z])g(y)dy - w[D(p,l) + z] \tag{10}$$

The first and second derivative of $E[\Pi_R^{RP}(z,p,l)]$ with respect to z can be written as

$$\frac{\partial E[\Pi_R^{RP}(z,p,l)]}{\partial z} = p + s - w - (p + s - b)G(z),$$

$$\frac{\partial^2 E[\Pi_R^{RP}(z,p,l)]}{\partial z^2} = -(p + s - b)g(z) <0.$$

For a given p and l, the retailer's expected profit is a concave function of z. We order $\dfrac{\partial E[\Pi_R^{RP}(z,p,l)]}{\partial z} = 0$ and get

$$z^* = G^{-1}(\frac{p+s-w}{p+s-b}) \tag{11}$$

When z satisfies $2r(z)^2 + dr(z)/dz > 0$ ($r(z)$ is defined as the hazard rate), and $r(z) = g(z)/[1 - F(z)]$, z^* has the sole solution. Many random distributions function, such as normal distribution and uniform distribution function meet the above constraints.

Taking the first and second derivative of (8) with respect to p yield

$$\frac{\partial E[\Pi_R^{RP}(z,p,l)]}{\partial p} = D_0 + \mu + \alpha w - \beta l - 2\alpha p - \int_z^B (y - z)g(y)dy$$

$$\frac{\partial^2 E[\Pi_R^{RP}(z,p,l)]}{\partial p^2} = -2\alpha < 0.$$

For a given z and l, the retailer's expected profit is concave function of p. We order $\dfrac{\partial E[\Pi_R^{RP}(z,p,l)]}{\partial p} = 0$, and then get the optimal of p.

$$p^* = \frac{1}{2\alpha}(D_0 + \alpha w + \mu - \beta l - \int_z^B (y - z)g(y)dy) \tag{12}$$

Where μ is the expected value of random variable ε. Joint formula (3), (4) and (10) we get

$$D(p,l) = D_0 - \frac{1}{2}[D_0 + \alpha w + \mu - \beta l - \int_z^B (y-z)g(y)dy)]$$
$$- \beta l \tag{13}$$

Under the buyback contract, after the supplier receives the order from retailers, supplier began to organize production. The unit production cost is c, the expected cost of lead time is c_l and the unit wholesale price is w. After the sale season, supplier buy back the unsold product at the transfer price b. Because the supplier can forecast that retailer will select the optimal wholesale price according to equation (10), therefore, the supplier's optimal decision in the first phase is

$$Max\{E[\Pi_s^{RP}(w,b,l)]\}$$
$$= [w - c - h\int_0^l (l-t)f(t)dt - b_s \int_l^\infty (t-l)f(t)dt][D(p,l)+z] \tag{14}$$
$$- b\int_A^z (z-y)g(y)dy$$

The above formula can be deformed to

$$Max\{E[\Pi_s^{RP}(w,b,l)]\}$$
$$= [w - c - h\int_0^l (l-t)f(t)dt - b_s \int_l^\infty (t-l)f(t)dt]$$
$$\bullet[D_0 - \frac{1}{2}[D_0 + \alpha w + \mu - \beta l - \int_z^B (y-z)g(y)dy)] - \beta l + z] \tag{15}$$
$$- b\int_A^z (z-y)g(y)dy$$

With the similar logic we can get the optimal whole price w. We have

$$\frac{\partial E[\Pi_s^{RP}(w,b,l)]}{\partial w} = 1/2[(D_0 - \mu - \beta l + \int_z^B (y-z)g(y)dy + 2z$$
$$+ c + h\int_0^l (l-t)f(t)dt + b_s \int_l^\infty (t-l)f(t)dt] - \alpha w$$
$$\frac{\partial E[\Pi_s^{RP}(w,b,l)]}{\partial w} = 1/2[(D_0 - \mu - \beta l + \int_z^B (y-z)g(y)dy + 2z$$
$$+ c + h\int_0^l (l-t)f(t)dt + b_s \int_l^\infty (t-l)f(t)dt] - \alpha w$$
$$\frac{\partial^2 E[\Pi_s^{RP}(w,b,l)]}{\partial w^2} = -\alpha < 0.$$

We can order $\dfrac{\partial E[\Pi_s^{RP}(w,b,l)]}{\partial w} = 0$, then get

$$w^* = 1/2\alpha[(D_0 - \mu - \beta l + \int_z^B (y-z)g(y)dy + z)$$
$$+ \alpha c + \alpha(h\int_0^l (l-t)f(t)dt + b_s \int_l^\infty (t-l)f(t)dt)] \tag{16}$$

From (3) and (11) we can yield

$$w = \frac{1}{\alpha}[D_0 - \mu - \beta l + \int_z^B (y-z)g(y)dy - 2D]$$ (17)

Substituting above equation into (14), when D and z are determined, we seek the first order and second order partial derivatives of Π_s^{RP} with respect to l are

$$\frac{\partial E[\Pi_s^{RP}]}{\partial l} = (-\frac{\beta}{\alpha} - hF(l) - b_s F(l) + b_s)(D+z),$$

$$\frac{\partial^2 E[\Pi_s^{RP}]}{\partial l^2} = -(h+b_s)(D+z)f(l) < 0.$$

In the defined interval $[0, \infty)$, the supplier's profit has the maximum value. We order $\partial E[\Pi_s^{RP}]/\partial l = 0$, then the optimal commitment lead time l^* can be written as

$$l^* = F^{-1}(\frac{b_s - \beta/\alpha}{b_s + h})$$ (18)

The overall profit of supply chain is

$$E(\Pi_T^{RP}) = E[\Pi_R^{RP}(z,p,l)] + E[\Pi_s^{RP}(w,b,l)]$$

4   Coordination Efficiency of Buy-Back Contracts

We use a numerical example to evaluate contract coordination efficiency. The centralized supply chain as a benchmark, we calculate the decentralized supply chain profits, optimal product decision and coordination efficiency. We assume that the stochastic market demand follow a uniform on [0,200], where $G(y)= y/200, 0 < y \leq 200$.

Consider the factors affect the efficiency of contract coordination are D_0, α, β. Studies show that when the $D_0/\alpha > 200$, this will lead to very high demand and low retail prices, while the $D_0/\alpha < 15$, the experimental will lose significance [8]. Combined with the literature, this study chose $D_0 = 200$, $\beta = 8$, $\alpha = 8,10,12,16,20$ and $D_0 = 200$, $\alpha = 10$, $\beta = 1,2,4,10,12$. We calculate the profit of centralized and decentralized supply chain and analyze the contract efficiency. We use the following parameters: $c=1$, $h=0.25$, $s=0.75$, $b_s = 0.75$. The numerical results are shown in Table 1 and 2.

From the examples, with the price elasticity factor and time elasticity factor increase, the profits of centralized and decentralized supply chain are decline. The centralized supply chain can achieve the highest profits and efficiency. In the decentralized supply chain, with the price elasticity factor increases, the supply chain's overall profits, the profits of suppliers and retailers are down. The coordination efficiency of supply chain contract also decreases. As elastic time factor increases, the overall profit of the supply chain and retail profits are down, but the profits of suppliers increase. This shows that when the price elasticity is high, the demands change largely with the price. If prices are too high, this may reduce the supply chain profits. The shorter the commitment lead time, retail price is higher and order quantity may decrease. This time supply chain gets larger profits. In all, companies may make trade-off decisions between the market share and corporate profits.

Table 1. The optimal solution of buy-back contract and contract coordination efficiency with alpha

α	Buy-back contract								Centralized system				E_f
	w^*	b^*	p^*	Q^*	l^*	Π_R^*	Π_S^*	Π_T^*	p^*	Q^*	l^*	Π_C^*	
8	13.98	6.09	20.48	89.99	1.15	428.21	862.78	1290.99	14.42	189.75	1.15	1626.30	0.794
10	12.86	5.51	15.96	86.28	1.72	246.22	686.12	932.34	11.98	175.43	1.72	1194.12	0.781
12	10.08	4.56	12.68	84.36	2.16	131.83	563.55	695.38	10.68	150.91	2.16	897.01	0.775
16	7.09	3.69	9.26	81.14	2.77	96.41	318.76	415.17	8.86	122.11	2.77	539.77	0.769
20	6.33	2.75	7.48	76.88	3.19	25.32	248.88	276.20	4.81	113.31	3.19	364.45	0.752

Table 2. The optimal solution of buy-back contract and contract coordination efficiency With Beta

β	Buy-back contract								Centralized system				E_f
	w^*	b^*	p^*	Q^*	l^*	Π_R^*	Π_S^*	Π_T^*	p^*	Q^*	l^*	Π_C^*	
1	4.18	0.89	9.18	88.86	4.65	364.02	93.86	457.88	8.89	240.59	4.65	468.26	0.978
2	4.31	0.51	9.41	87.67	3.92	375.51	110.70	486.21	9.07	236.05	3.92	528.61	0.920
4	4.67	0.36	9.70	86.73	2.77	373.97	147.21	521.18	9.39	226.77	2.77	597.38	0.872
10	5.88	0.19	10.54	78.77	0.53	263.23	179.20	442.43	10.37	200.24	0.53	589.21	0.751
12	6.39	0.8	10.81	76.89	0	228.60	184.08	412.68	10.73	184.72	0	561.04	0.736

5   Conclusions

In this paper, we study a two-stage supply chain consists of a supplier and a retailer. We use buy-back contract to coordinate the supply chain. We find that when the price elasticity factor and the time elasticity factor are small, the coordination efficiency of the contract is high; on the contrary, coordination efficiency is low. Business strategy should be based on market needs. Companies should make reasonable commitment lead time and product prices to maximize profits. At different wholesale prices, retail prices, commitment lead time, suppliers and retailers share profits under different levels. It is appropriate to shorten the lead time and raise prices to improve profits.

Future research can be extended to other contracts such as sales discount contract, price contracts to coordinate supply chain. Companies can select the appropriate contract to improve supply chain efficiency.

References

1. Lau, H.S., Lau, A.H.L.: Manufacturer's pricing strategy and return policy for a single-period commodity. European Journal of Operational Research 116(2), 293–305 (1999), doi:10.1016/S0377-2217(98)00123-4
2. Terry, T.: Supply chain coordination under channel rebates with sales effort effects. Management Science 48(8), 992–1007 (2002), doi:10.1287/mnsc.48.8.992.168

3. Pellegrini, L., Reddy, S.K.: Marketing channel: relationships and performance, pp. 59–72. Lexington Books, Lexington (1986)
4. Pasternack, A.: Optimal pricing and return policies for perishable commodities. Marketing Science 4(2), 166–176 (1985), doi:10.1287/mksc.4.2.166
5. Emmons, H., Gilbert, S.M.: The role of returns policies in pricing and inventory decisions for catalogue goods. Management Science 44(2), 276–283 (1998), doi:10.1287/mnsc.44.2.276
6. Bose, Anand, P.: On returns policies with exogenous price. European Journal of Operational Research 178(3), 782–788 (2007), doi:10.1016/j.ejor.2005.11.043
7. Yue, X., Raghunathan, S.: The impact of the full return policy on a supply chain with information asymmetr. European Journal of Operational Research 180(2), 630–647 (2007), doi:10.1016/j.ejor.2006.04.032
8. Yao, Z., Stephen, C.H., Lai, K.K.: Analysis of the impact of price-sensitivity factors on the returns policy in coordinating supply chain. European Journal of Operational Research 187(1), 275–282 (2008), doi:10.1016/j.ejor.2007.03.025

Retailer's Pricing and Channel Selection under E-Commerce Environment

Chenglin Shen[1] and Xinxin Zhang[2]

[1] School of Business Tianjin Polytechnic University Tianjin, China
[2] Tianjin Institute of Pharmaceutical Research Tianjin, China
gnuguy@163.com, tjusss@yahoo.com.cn

Abstract. In this paper, we investigated a monopoly manufacture use traditional retail channels and internet channels to sell products. Customers are heterogeneous in preference for sale channel use. Customers are divided into fashion customers and traditional customers. Using utility theory we established customer's purchase decision models. Retailer's profit maximization as the objective function, combined with the customer channel choice factors, the strategy of retail channels choice and product pricing were proposed.

Keywords: E-commerce, customer preferences, pricing, distribution channels.

1 Introduction

As market competition intensifies, more and more enterprises begin to use multiple channels to sell products and services. Mixed channel sales mode has become more popular, the rapid development of internet technology, especially the emergence of electronic commerce accelerates the application of the mixed distribution channels[1].Many world famous manufacturers such as IBM, HP, Haier and Lenovo, using of traditional brick and mortar store selling products at the same time, actively open electronic market to sell products. At the same time the world's leading retailers Wal-Mart and China, such as Gome and Suning, also opened up the electronic channels. In 2006, Online direct marketing company DELL have begun to establish stores In addition to Internet channels to meet customer 'experience shopping'[2]. Practice shows, using hybrid channels including Internet channels can help enterprises help to win different types of customers in the market to a certain extent, to better meet the needs of different channels' customers and can improve the market share [3].

Although nowadays Internet channels conjunction with the conventional brick-and mortar store channel, also known as mixed channel has become a modern marketing mode. If companies open up new marketing channels blindly, for example Internet channels, also brought new problems. Mainly for the new channel will be a competitor to the original channel, vicious competition will engulf the existing sales channels, leading to a decline in corporate profits [4].

J. Luo (Ed.): Soft Computing in Information Communication Technology, AISC 161, pp. 355–363.

Therefore, how company design channel structure? When the Internet channel is added to an existing channel system, how does it affect traditional physical channel's performance and customer welfare? How pricing decisions affect the company's profitability for different channel structure become important problems.

Current study focused on problems in multi-channel distribution system, which involves multi-channel structure arrangement and pricing, channel conflict & coordination and channel operation based on B2C E-commerce [5-6]. Most research in the past, assuming customers are homogeneous. This article assumes that customers are heterogeneous. We divided the customers into two types, fashion customers and traditional customers. The decision of channel choice for customers was analyzed based on utility theory. Finally, the product pricing strategies and mixed channel structure arrangement strategies were proposed according to these two types of customers' distribution.

2   Model Assumptions and Notations

Considers a monopoly sell the same products to a group of customers with both traditional retail sales channels and Internet channels. Assume that customer's valuation for each unit of product utility is v. Each customer either spends one unit of product or 0 per unit of product.

According to the customer preferences, in accordance with the principles of market segmentation, we divide customers into fashion (F-type) and traditional (T-type) two categories. F-type customers focus on shopping fast and convenient, and tend from network channels to buy products. T-type customers pay more attention to experiential shopping, tend to purchase products from the physical store.

Suppose F-customer ratio in the market is α, so the T-customer ratio in the market $\bar{\alpha}=1-\alpha$. Taking into account the two types of customer channel preference heterogeneity, we introduce the discount factor δ, $0\leq\delta\leq1$. T-type customer valuation of unit of product purchased from the physical channel is v. Utility of per unit product from Internet channel is estimated to be δv. With the same logic, F-type customer's utility from per unit product from Internet channel is estimated to be v, while utility of per unit product from the physical channel is δv.

We assume retailers and customers are entirely rational. First, with profit-maximizing goal, retailers determine whether to introduce Internet channels. Retailers make decision of optimal product price according to different channel structures.

When choosing to introduce the Internet channel, retailers need to decide whether to sell all of the products to these two kinds of customers or only to F-type customers. Customers decide whether to buy products and from which channel to purchase products according to consumer surplus. If consumer surplus is greater than or equal to 0 consumer choose to buy products. The channel choice is not only related with the product utility, but also related with the customer's channel preferences. To simplify the issue, the model assumes that physical channels and Internet channels are both able to meet customer needs, meanwhile without considering stock out situation. Two sales channels are equal to customer demand. Other parameters are defined as follows Table 1.

Table 1. Main parameters in the model

p_d Unit price of product of the Internet channel	S_r Sales of the physical channel
	S_d Sales of the Internet channel
p_r Unit price of product of the physical channel, $p_r > p_d$.	S_{Nd} Sale of the single physical channel
p_{Nd} Unit price of product of the single physical channel	I Market size, the number of potential customers in the market
c_d Unit sale cost of the Internet channel, $c_d < p_d$. c_r Unit sale cost of product of the physical channel, $c_r < p_r$	Π The retailer's profit of any channel

In this paper, the subscript d indicates the Internet channels (online store), r indicates physical channels (physical store), m indicates mixing channel.

3   Market Analysis of the Single Physical Channel

In the case of the single channel circumstance, all customers buy products only from the physical channel. Suppose the market size is I, the customer valuation of the products value is v, the valuation v is uniformly distributed in the interval $[0, v]$. Customers make a purchase only if the customer valuation v is greater than or equal product price p_{Nd}. Therefore, product sales/demand for single physical channel is

$$S_{Nd} = I\int_{p_{Nd}}^{\hat{v}} \frac{1}{\hat{v}} dv = I(1 - \frac{p_{Nd}}{\hat{v}}) \qquad (1)$$

Retailer's profit is

$$\Pi_{Nd} = (p_{Nd} - c_r)S_{Nd} = I(p_{Nd} - c_r)(1 - \frac{p_{Nd}}{\hat{v}}) \qquad (2)$$

Taking first and second derivatives of Π_{Nd} respectively to p_{Nd} give

$$\frac{d\Pi_{Nd}}{dp_{Nd}} = I(1 - 2\frac{p_{Nd}}{\hat{v}} + \frac{c_r}{\hat{v}}) \qquad (3)$$

$$\frac{d^2\Pi_{Nd}}{dp_{Nd}^2} = -2I / \hat{v} < 0 \qquad (4)$$

From (3) and (4), Π_{Nd} is a concave function of p_{Nd}. The first-order-condition $\frac{d\Pi_{Nd}}{dp_{Nd}} = 0$ yields the seller's optimal price in single physical channel $p_{Nd}^* = (\hat{v} + c_r)/2$.

The retailer's maximum sale quantity and profit is $S_{Nd}^* = I(\hat{v} - c_r)/2\hat{v}$ and $\Pi_{Nd}^* = I(\hat{v} - c_r)^2 / 4\hat{v}$.

4   Mixed Channel Cases

A.   Customer Utility and Choice

T-type customer's utility of one unit of product from physical store $U_r^T = v - p_r$. T-type customer's utility from the Internet store $U_d^T = \delta v - p_d$. Customer's utility is zero when they do not buy any products. T-type customers choose to purchase product from the physical channels when $U_r^T \geq U_d^T$ and $U_r^T \geq 0$. When $U_r^T < U_d^T$, there are three cases. 1) When $U_r^T < 0$ and $U_d^T \geq 0$, customer will choose the Internet channel. 2) When $U_r^T \geq 0$ and $U_d^T \geq 0$ those T-type customer who with strong purchase intensions (we define these part of customers as customers with high valuation) still choose physical channel, while the price-sensitive T-type customers turn to the Internet channel. 3) When $U_r^T < 0$ and $U_d^T < 0$ customers will not buy the product from the two channels. When $U_r^T = U_d^T \geq 0$, customer can purchase products from both channels since he/her gets the same customer utility.

For F-type customer's utility of purchasing one unit product from the physical channel is $U_r^F = \delta v - p_r$. While from the Internet channels customer gets utility $U_d^F = v - p_d$. Since for F-type customer $U_d^F > U_r^F$, when facing with the two different channels, F-type customers will choose the Internet channel as long as $U_d^F \geq 0$.

B.   Retailer's Pricing and Channel Strategy

Physical channel retailers have two options when establish the Internet channel. One is to sell products to both T-type and F-type customers, the other is to sell products only to F-type customers.

First, consider retailer sell products to T-type and F-type customers from the Internet channel. According to the customer channel choice model, there must be $U_r^T < U_d^T$. That is $v - p_r < \delta v - p_d$, we can get $p_d < \delta p_r$. Therefore, when the price of Internet channels p_d is less than δp_r, those with high valuation customer of T-type customer will choose physical channels, while price-sensitive T-type customers will shift to the Internet channels. We define the pricing strategy retailer used as the low-cost strategy.

Under low price strategy, sales quantity/demand from physical channel S_r is given by

$$S_r = S_r^T = I(1-\alpha)\int_{v^*}^{\hat{v}} \frac{1}{\hat{v}} dv = I(1-\alpha)(1 - \frac{p_r - p_d}{(1-\delta)\hat{v}}) \tag{5}$$

Where $v^* = \dfrac{p_r - p_d}{1-\delta}$ is the critical value, that is, customer get the same product utility from the physical channel and the Internet channel.

Under low price strategy, sales quantity/demand from Internet channel S_d can be written as

$$S_d = S_d^T + S_d^F = I(1-\alpha)\frac{(\delta p_r - p_d)}{\hat{v}(1-\delta)\delta} + I\alpha(1-\frac{p_d}{\hat{v}}) \tag{6}$$

$S_d^T = [I(1-\alpha)\int_{p_r}^{v^*}\frac{1}{\hat{v}}dv]/\delta = [I(1-\alpha)(\delta p_r - p_d)]/[\hat{v}(1-\delta)\delta]$ S_d^T is demand from price-sensitive T-type customers, and $S_d^F = I\alpha(1-\frac{p_d}{\hat{v}})$, demand from F-type customers. We get

$$\Pi_m^l = \Pi_r + \Pi_d = (p_r - c_r)I(1-\alpha)(1-\frac{(p_r - p_d)}{(1-\delta)\hat{v}})$$
$$+(p_d - c_d)[I(1-\alpha)\frac{(\delta p_r - p_d)}{\hat{v}(1-\delta)\delta} + I\alpha(1-\frac{p_d}{\hat{v}})] \tag{7}$$

Theorem 1. Under low price strategy, the retailer's optimal decision
$$p_r^* = \frac{\hat{v}(1-\alpha+2\alpha\delta-\alpha\delta^2)+(1-\alpha+\alpha\delta)c_r}{2(1-\alpha+\alpha\delta)} \quad , \quad p_d^* = \frac{\delta\hat{v}+(1-\alpha+\alpha\delta)c_d}{2(1-\alpha+\alpha\delta)} \quad . \text{ The}$$
maximum demand from physical channel S_r^* is given by
$$S_r^* = \frac{I(1-\alpha)((1-\delta)\hat{v}-c_r+c_d)}{2(1-\delta)\hat{v}} .\text{The maximum demand from physical channel}$$
S_d^* is given by
$$S_d^* = \frac{(\alpha(1-\delta)\delta\hat{v}+(1-\alpha)\delta c_r-(1-\alpha+\alpha\delta-\alpha\delta^2)c_d))I}{2(1-\delta)\delta\hat{v}} .$$

The retailer's maximum profit is
$$\hat{\Pi}_m^l = \frac{I(\bar{\alpha}\delta\theta c_r^2-\delta\hat{v}(\rho+\bar{\delta}\theta-\bar{\alpha}\delta)c_r-\delta(\theta-\bar{\alpha})c_r c_d)}{4\delta\bar{\delta}\theta\hat{v}}$$
$$+\frac{\alpha\delta\hat{v}(\delta-\bar{\delta}\theta)c_d+\theta(\rho-\alpha\delta)c_d^2+\delta\bar{\delta}\hat{v}^2(\rho+\alpha\delta))}{4\delta\bar{\delta}\theta\hat{v}^2} .$$

Proof: The derivative of Π_m^l with respect to p_r and p_d respectively can be written as,

$$\frac{\partial\Pi_m^l}{\partial p_r} = I(1-\alpha)(1-\frac{2p_r-2p_d-c_r+c_d}{(1-\delta)\hat{v}}),$$

$$\frac{\partial\Pi_m^l}{\partial p_d} = I(1-\alpha)(\frac{2\delta p_r-2p_d+c_d-\delta c_r}{\hat{v}(1-\delta)\delta})+I\alpha(1-\frac{2p_d-c_d}{\hat{v}}) .$$

Taking derivative again with respect to p_r and p_d respectively gives

$$\frac{\partial^2 \Pi_m^l}{\partial p_r^2} = \frac{-2I(1-\alpha)}{(1-\delta)\hat{v}} < 0, \quad \frac{\partial^2 \Pi_m^l}{\partial p_d^2} = -\frac{2I\alpha}{\hat{v}} - \frac{2(1-\alpha)I}{(1-\delta)\delta\hat{v}} < 0, \quad \frac{\partial^2 \Pi_m^l}{\partial p_r \partial p_d} = \frac{\partial^2 \Pi_m^l}{\partial p_d \partial p_r} = \frac{2(1-\alpha)I}{(1-\delta)\hat{v}}.$$

Hence we get the Hessian matrices

$$|H| = \begin{vmatrix} \dfrac{\partial^2 \Pi_m^l}{\partial p_r^2} & \dfrac{\partial^2 \Pi_m^l}{\partial p_r \partial p_d} \\[3mm] \dfrac{\partial^2 \Pi_m^l}{\partial p_d \partial p_r} & \dfrac{\partial^2 \Pi_m^l}{\partial p_d^2} \end{vmatrix} = \frac{4I^2(1-\alpha)(1-\alpha+\alpha\delta)}{(1-\delta)\delta\hat{v}^2} > 0,$$

According to Hurwitz theorem, the Hessian matrices are negative definite. Therefore, Π_m^l it is a concave function of p_r and p_d. The first-order-condition $\dfrac{\partial \Pi_m^l}{\partial p_r} = 0$ and $\dfrac{\partial \Pi_m^l}{\partial p_d} = 0$ respectively yields p_r^* and p_d^*. With p_r^* and p_d^*, solving (5) and (6) yield the desired S_r^* and S_d^*. With S_r^*, S_d^* substituting into (7) we yield $\hat{\Pi}_m^l$.

Next, consider retailer sell products in the Internet channel only to F-type customer. According to the customer channel choice model, we know that $U_r^T \geq U_d^T > 0$. Further, we know that $p_d > \delta p_r$. So if retailer pricing $p_d > \delta p_r$, all the T-type customers will choose the physical channel and all of the F-type customers will choose the Internet channel. We define the pricing strategy as high price strategy.

Under high price strategy, the demand of physical channel and Internet channels can be respectively written as

$$S_r = I(1-\alpha)(1-\frac{p_r}{\hat{v}}) \tag{8}$$

$$S_d = I\alpha(1-\frac{p_d}{\hat{v}}) \tag{9}$$

The retailer's profit is

$$\Pi_m^h = \Pi_r + \Pi_d = (p_r - c_r)I(1-\alpha)(1-\frac{p_r}{\hat{v}}) + (p_d - c_d)I\alpha(1-\frac{p_d}{\hat{v}}) \tag{10}$$

Theorem 2. With high strategy, the retailer's optimal decisions are $p_r^* = \dfrac{(\hat{v}+c_r)}{2}$ and $p_d^* = \dfrac{(\hat{v}+c_d)}{2}$. Demand of the physical and Internet are $S_r^* = \dfrac{I(1-\alpha)(\hat{v}-c_r)}{2\hat{v}}$ and $S_d^* = \dfrac{I\alpha(\hat{v}-c_d)}{2\hat{v}}$. The retailer's maximum profit is $\hat{\Pi}_m^h = \dfrac{I(1-\alpha)(\hat{v}-c_r)^2 + I\alpha(\hat{v}-c_d)^2}{4\hat{v}}$.

Proof: Taking the first and second mixed partial derivative of Π_m^h with respect to p_r and p_d , we yield $\dfrac{\partial \Pi_m^h}{\partial p_r} = I(1-\alpha)(1-\dfrac{2p_r}{\hat{v}}+\dfrac{c_r}{\hat{v}})$, $\dfrac{\partial \Pi_m^h}{\partial p_d} = I\alpha(1-\dfrac{2p_d}{\hat{v}}+\dfrac{c_d}{\hat{v}})$, $\dfrac{\partial^2 \Pi_m^h}{\partial p_r^2} = -\dfrac{2I(1-\alpha)}{\hat{v}}<0$, $\dfrac{\partial^2 \Pi_m^h}{\partial p_d^2} = \dfrac{-2I\alpha}{\hat{v}}<0$, $\dfrac{\partial \Pi_m^h}{\partial p_r \partial p_d} = \dfrac{\partial \Pi_m^h}{\partial p_d \partial p_r} = 0$. Based on these results, Hessian matrix is negative definite. Π_m^h is a concave function of p_r and p_d .

The first-order-condition $\dfrac{\partial \Pi_m^h}{\partial p_r} = 0$ and $\dfrac{\partial \Pi_m^h}{\partial p_d} = 0$ respectively yields p_r^* and p_d^* .

With p_r^* and p_d^* substituting into (8) and (9) we obtain S_r^* and S_d^*. Finally, with S_r^* and S_d^*, solving (10) yields the desired $\hat{\Pi}_m^h$.

Theorem 3. In mixed channels, there is a threshold value $\hat{\alpha}$ in the system. When $\alpha > \hat{\alpha}$, retailers can choose high price strategy. When $\alpha < \hat{\alpha}$, retailers can choose low price strategy, while

$$\hat{\alpha} = \frac{c_d^2 - 2c_r c_d \delta + c_r^2 \delta^2}{(1-\delta)(c_d^2 - 2c_r c_d \delta + c_r^2 \delta^2 + \delta \hat{v}(1-\delta)^2)}.$$

Proof: While $\hat{\Pi}_m^l - \hat{\Pi}_m^h \geq 0$, retailers will choose low price strategy, otherwise select the high price strategy. According to Theorem 1 and 2, we yield

$$\hat{\Pi}_m^l - \hat{\Pi}_m^h = \frac{-\alpha(1-\delta)((c_d^2 - 2c_r c_d \delta + c_r^2 \delta^2 + \delta \hat{v}(1-\delta)^2)}{4\delta \hat{v}(1-\alpha+\alpha\delta)(1-\delta)}$$
$$+ \frac{c_d^2 - 2c_r c_d \delta + c_r^2 \delta^2}{4\delta \hat{v}(1-\alpha+\alpha\delta)(1-\delta)} \geq 0$$

and the denominator is greater than 0. After simplification, we get the condition $\alpha \leq \dfrac{c_d^2 - 2c_r c_d \delta + c_r^2 \delta^2}{(1-\delta)(c_d^2 - 2c_r c_d \delta + c_r^2 \delta^2 + \delta \hat{v}(1-\delta)^2)} = \hat{\alpha}$, this yields $\hat{\Pi}_m^l - \hat{\Pi}_m^h \geq 0$.

From theorem 3, in mixed channel, whether retailers choose high price strategy or low price strategy are depend the number of F-type customers in the market. When the number of F-type customer in the market is small($\alpha \leq \hat{\alpha}$), if retailers sell products in Internet channel only to F-type customer, sales quantity of Internet channels will be relatively low, thus affecting the channel profits. Therefore, retailers had to take low price strategy to attract price-sensitive T-type customers in order to ensure the Internet channel sales and profits. On the other hand, the Internet channels can obtain higher sales and profits when retailers seize the numerous F-type customers in the market. Therefore, the retailer can take high price strategy to seize only the F-type customers, and leave all T-type customers to the physical channels to reduce the negative impact of channel cannibalization on overall profits.

Consider the following channel selection strategies, that is, when should retailers choose the single physical channel mode and when to select the mixed channel mode.

Theorem 4. When $\alpha > 0$, and $c_d < c_r$, which means that the cost of the Internet channel is less than the cost of physical channels, retailers select the mixed channel mode, otherwise select the physical channel. When $\alpha = 0$ and $c_d < \delta c_r$, retailers choose the mixed channel mode, whereas select single physical channel mode.

Proof: When the mixed channel profit is greater than or equal to that of the single physical channel, that is $\Pi_m^* = \max(\hat{\Pi}_m^l, \hat{\Pi}_m^h) - \Pi_{Nd}^* \geq 0$, retailers select the mixed channel mode. First consider when $\alpha > 0$,

$$\hat{\Pi}_m^h - \Pi_{Nd}^* = \frac{I(1-\alpha)(\hat{v} - c_r)^2 + I\alpha(\hat{v} - c_d)^2}{4\hat{v}} - \frac{I(\hat{v} - c_r)^2}{4\hat{v}}$$

$$= \frac{I(2\hat{v} - c_r - c_d)(c_r - c_d)}{4\hat{v}}.$$

And since $c_r \leq \hat{v}, c_d \leq \hat{v}$, when $c_d < c_r$, we get $\hat{\Pi}_m^h \geq \Pi_{Nd}^*$. So when $\alpha > 0$, and $c_d < c_r$, there must be $\Pi_m^* \geq \Pi_{Nd}^*$.

Second, consider the case of $\alpha = 0$, When $\alpha = 0$, the customers in the market are all of the T-type customers, products of Internet channels could only be sold to the price-sensitive T-type customers, the retailer can adopt low price strategy. Only if $\hat{\Pi}_m^l - \Pi_{Nd}^* \geq 0$ and $S_d^* > 0$, the retailers choose the mixed channel model. We get

$$\hat{\Pi}_m^l - \Pi_{Nd}^* = \frac{(\delta c_r - c_d)(\delta c_r + c_d)I}{4(1-\delta)\delta\hat{v}}, \quad S_d^* = \frac{(\delta c_r - c_d)I}{2(1-\delta)\delta\hat{v}}$$

according to theorem 1. So that when $c_d < \delta c_r$, we yield $\hat{\Pi}_m^l - \Pi_{Nd}^* > 0$ and $S_d^* > 0$.

Theorem 4 shows that if there exist F-type customer($\alpha > 0$), the retailer whether to choose the Internet channel are related to sale cost of the Internet channel(c_d) and the physical channel(c_r). When $c_d < c_r$, retailers select the mixed channel mode. When the market does not exist in F-type customers ($\alpha = 0$), in order to ensure the profits of the Internet channels, retailers can only take low price strategy and sell products the price-sensitive T-type customers. So the retailer need to further cut down the sell cost of the Internet channels, that is $c_d < \delta c_r$. The profit under the mixed-channel mode will be greater than the profits under a single physical channel. Retailers will choose to open up the Internet channel in addition to the physical channels.

5   Conclusion and Further Research

By utility-theoretic model we have investigated how the introduction of an Internet channel affects the channel profits and pricing decisions. Use of utility theory on customer purchasing behaviours we find that different types of customers have different channel selection preferences, resulting in differences of customer groups distribution in the market. Whether to choose the Internet channel are related to at least two factors: one is the sale cost of the Internet channel and the physical channel, the other is the type and proportion of the relevant customer. Enterprises should take reasonable marketing channel structure and pricing strategies according to customer purchase behaviours differences and the characteristics of different products.

This study assumed that customer are heterogeneous moreover, did not consider the traditional distribution channels to compete with the Internet channel (called channel cannibalization) affect sales and profits. In future research can be extended to the case of channel cannibalization and consider sale volume and profit changes.

References

1. Webb, K.L.: Managing channels of distribution in the age of electronic commerce. Industrial Marketing Management 31(2), 95–102 (2002), doi:10.1016/s0019-8501(01)00181-x
2. Zehr, D.: Dell to close all U.S. kiosks as it moves PCs into stores. Austin American Statesman (January 31, 2008)
3. Agatz, N.A.H., Fleischmann, M., van Nunen, J.A.E.E.: E-fulfillment and multi-channel distribution- a review. European Journal of Operational Research 187(2), 339–356 (2008), doi:10.1016/j.ejor.2007.04.024
4. Zhang, X.B.: Retailer's multichannel and pricing advertising strategies. Marketing Science 28(6), 1080–1094 (2009), doi:10.1287/mksc.1090.0499
5. Zhao, L.Q., Guo, Y.J.: A review on multi-channel distribution system based on B2C E-commerce. Management Review 22(2), 69–78 (2010) (in Chinese)
6. Yoo, W.S., Lee, E.: Internet channel entry: a strategic analysis of mixed channel structures. Marketing Science, Article in Advance (2010), doi:10.1287/mksc.1100.0586

Research on the Principle Parts of Ecological Technology Innovation for Recycled Economy

Yan-bing Yin[1], Shao-wei Liu[2], and Zi-yi Han[3]

[1] School of Economics Tianjin Polytechnic University Tianjin 300384, China
yyblsw@gmail.com
[2] Information Institute, Academy of Agricultural Sciences Tianjin 300000, China
collapsarliu@x263.net
[3] Qingdao University Qingdao 266071, China
Hanziyi_413@163.com

Abstract. The characters of ecological technology innovation limit the structure of innovation principle parts during the development of circular economy, which is a deterministic process including economic benefit, social benefit and ecological benefit. According to this, the principal parts of innovation cannot be single one but multielement, in which enterprise is the core with the support of government, college, institute and the public.

Keywords: Circular Economy, ecological technology innovation, innovation principle parts.

1 Introduction

Ecological technology innovation is a new type of innovation system which comprehensively implements ecology theory in the course of innovation. On the premise of economic growth, it aims at the balance of environment and society. It can lead traditional technology innovation to good interaction and synergy of natural ecology, economy and society.

ETI is the important guarantee for circular economy and sustainable development. Besides the common characters, ETI has distinct characters with traditional technology innovation(table 1):

Table 1. Difference between ETI and TTI

Difference	TTI	ETI
Innovation aim	Maximum of economy benefit	Win-win of the benefit of economy ,ecology, and society
Innovation content	Product and distribution	Whole lifecycle of product
Innovation mode	Unilateral mode	Bidirectional mode
Innovation measure index	Input-output	Sustainable of economy, ecology and society

J. Luo (Ed.): Soft Computing in Information Communication Technology, AISC 161, pp. 365–372.

The construction of ETIS should comprehensively consider the characters of ETI on the background of recycled economy. The construction should be a complex systems engineering which involves increasing the function of microeconomic body as well as the available interaction and symbiosis; exerting market mechanism, macro-control of govt. as well as promotion of public ; enhancing economic benefit, social harmony as well as environment promotion.e method.

2   Limition of the Single Principle Part of Econological Tecnnology Innovation

Innovation implies individual creation; nevertheless it generally emphasizes collective creativity. During the course of TTI, we usually consider enterprise as the main body of innovation which is "logos one" with the common goal of maximum merit.

It is a rightabout-face that changes TTI pattern into ETI one. We find that the principal part of innovation can not be single one but multielement, in which enterprise is the core with the support of government, college, institute and the public

The "ecological" character of ETI can decrease the negative domino effect of traditional technology innovation, from economic angle, it means negative externality.

Table 2. Principles of Indices Importance and Parameter λ

Demand of measurement	Importance	λ
Place importance on single one or more outstanding indices	Place more importance	Close to -
Place importance on any or several outstanding indices	Equal importance	1
Place important on equilibrium	No limitation	>0
Place importance on both equilibrium and single or more outstanding indices	Place more importance	< 0 and close to 0
Place importance on both equilibrium and any or several outstanding indices	Equal importance	

According to the principles, decision-maker can set parameter λ and importance to meet the requirements in different phrase. Moreover, based on the order change of different parameter λ and importance, it can also find the weakness and put forward the improved measures, which could promote the development of the object.

The concrete steps of improved fuzzy-integral method are as follow:

Step 1. Make quantitative indices to be dimensionless
Each index has different character, whose dimension is also different. In order to ensure the reliability and veracity, it needs to make the dimension normalization.

Step 2. Make qualitative indices quantize
It can use fuzzy semantic quantifier to make qualitative indices quantize. According to the theory of trapezoidal fuzzy number, we divide the measurement into nine grades, as follow Table 3.

Table 3. Value of Fuzzy Semantic Quantifier

Semantic variable	Fuzzy-value	Semantic variable	Fuzzy-value
worst	(0,0,0,0)	Comparative good	(0.5,0.6,0.7,0.8)
worse	(0,0,0.1,0.2)	good	(0.7,0.8,0.8,0.9)
poor	(0.1,0.2,0.2,0.3)	better	(0.8,0.9,1,1)
Comparative poor	(0.2,0.3,0.4,0.5)	best	(1,1,1,1)
common	(0.4,0.5,0.5,0.6)		

Step 3. Calculate fuzzy-value of each lay with fuzzy integral
According to the principles of λ to calculate g_λ and measure value of each lay.

$$g_\lambda(\{x_1,x_2,...,x_n\})=\sum_{i=1}^{n} g_i + \lambda\sum_{i1=1}^{n-1}\sum_{i2=1}^{n} g_{i1}g_{i2} +...+ \lambda^{n-1}g_1 g_2.....g_n =\frac{1}{\lambda}\left|\prod_{i=1}^{n}(1+\lambda g_i)-1\right|, \lambda\in [-1, \infty]$$

(1)

in which g_i means index weight.

Step 4. Calculate the integrated fuzzy-value
Based on the value of each lay, we can get the integrated fuzzy-value with the method above.

The steps above can be described with the figure 1:

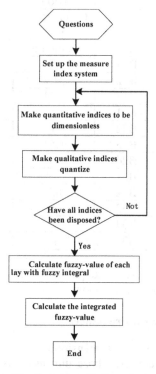

Fig. 1. Arithmetic flow chart

3　Applications Analysis

It takes MS group for example, and uses the improved fuzzy-integral method to evaluate the development of ecological technology innovation. The index system is as table 4, it includes qualitative and quantitative indices.

Table 4. Index of Ecology Technology Innovation

	Criterion	Particular index	2009	Benchmark value
A	A1	A11	Comparative good	—
		A12	good	—
		A13	5.4	2
		A14	1.24	1
		A15	12	8
		A16	Comparative good	—
		A17	Comparative good	—
	A2	A21	good	—
		A22	Comparative good	—
		A23	Comparative good	—
		A24	common	—
B	B1	B11	187	80
		B12	21.2	12
		B13	25.1	15
	B2	B21	5.55	3
		B22	4	10
		B23	1.51	1
		B24	11	3
	B3	B31	7.2	7
		B32	8.7	10
		B33	6	5
		B34	2	1
		B35	2(3年)	1

According to the indices above, it adopts the improved fuzzy-integral method to measure, the concrete approach are as follows:

1)　Confirm the values of measure indices

(1) Make quantitative indices to be dimensionless
Each index has different character, whose dimension is also different. In order to ensure the reliability and veracity, it needs to make the dimension normalization. This

article uses the method of efficiency coefficient (formula 2) to deal with the data; the values are as table 5.

$$X' = \frac{X - \overline{X}}{\overline{X}} \times 0.5 + 0.5 \tag{2}$$

Thereinto, parameter X is dimensionless value, $\overline{X}$ represents the standard value. We need to ascertain the standard value when we use the method above. In order to assure the result comparable, it should obey uniform principles.

In this article, we will obey the principles as follows:

①It will use the standard values as far as possible if the indices have national or international values;

②Take the values of national (regional) or the same type enterprises for reference;

③According to the theories of environment, society and coordinated economic development to make qualitative indices quantize;

Table 5. Dimensionless Value of ETI

Index	Dimensionless	Index	Dimensionless
A11	0.621	B12	0.883
A12	0.658	B13	0.836
A13	1.35	B21	0.925
A14	0.62	B22	0.2
A15	0.75	B23	0.755
A16	0.723	B24	1.833
A17	0.468	B31	0.514
A21	0.612	B32	0.435
A22	0.435	B33	0.6
A23	0.568	B34	1.0
A24	0.348	B35	1.0
B11	1.168		

(2) Make qualitative indices quantize

We engage experts to evaluate the indices through questionnaires. It takes the index "A_{16}"for example to show how to quantize the qualitative indices.

① Each expert measure the semantic variable, the result is as table 6.

② According to table 3, we can achieve the qualitative value through fuzzy operation.

$$S(\tilde{X}_{16}) = \frac{0.7178 + 0.725 + 0.725}{3} = 0.723$$

$$g(\tilde{X}_{16}) = \frac{0.7447 + 0.755 + 0.751}{3} = 0.750$$

Similarly, we can calculate the other qualitative values.

Table 6. Dimensionless Value of ETI

Index	Expert	Evaluate value	Dimensionless	Importance	Semantic value
A16	1	good	(0.7,0.8,0.8,0.9)	More important	(0.8,0.9,1,1)
	2	good	(0.7,0.8,0.8,0.9)	Comparative important	(0.5,0.6,0.7,0.8)
	3	good	(0.7,0.8,0.8,0.9)	More important	(0.8,0.9,1,1)
	4	good	(0.7,0.8,0.8,0.9)	Most important	(1,1,1,1)
	5	good	(0.7,0.8,0.8,0.9)	Comparative important	(0.5,0.6,0.7,0.8)
	6	Comparative good	(0.5,0.6,0.7,0.8)	important	(0.7,0.8,0.8,0.9)
	7	good	(0.7,0.8,0.8,0.9)	More important	(0.8,0.9,1,1)
	8	Comparative good	(0.5,0.6,0.7,0.8)	Comparative important	(0.5,0.6,0.7,0.8)
	9	common	(0.4,0.5,0.5,0.6)	common	(0.4,0.5,0.5,0.6)
	10	Comparative good	(0.5,0.6,0.7,0.8)	common	(0.4,0.5,0.5,0.6)

(3) Repeat step 2, calculate the attention degree of each index.

2) All qualitative data come from experts' estimation according to some relative criteria

Then according to different λ, we can get the evaluation values of MS group and integrated value with fuzzy integral (Table 7-8)

Table 7. Fuzzy Intergral Index Value of ETI

	λ=-0.99	λ=-0.1	λ=0	λ=1
$f1$	1.158	0.757	0.725	0.558
$f2$	0.589	0.503	0.394	0.435
$f3$	1.072	0.979	0.972	0.923
$f4$	1.323	0.881	0.843	0.590
$f5$	0.912	0.705	0.684	0.555

Table 8. Integral Value of Each Criterion

	λ=-0.99	λ=-0.1	λ=0	λ=1
fA	1.057	0.577	0.575	0.485
fB	1.215	0.854	0.826	0.631

3) Calculate the integrated value

According to the method, we can calculate the fuzzy-integral values of each lays, and then achieve the integrated value.

Table 9. Fuzzy Integral Integrated Value of ETI

	$\lambda=-0.99$	$\lambda=-0.1$	$\lambda=0$	$\lambda=1$
F	1.240	0.775	0.720	0.528

4   Results Analysis

From TABLE 9, we can conclude that the integrated value of MS group is still high than 0.5 under any λ, namely that the ETI level of MS is comparative good and each index reach to the lowest criteria.

But from the Table we also know that the integrated value is related to parameter λ when the weights of all indices are definite. The value becomes lower as parameter λ increases. According to the principle of parameter λ, we can get the conclusion that the group performs noticeably well in some aspects but equilibrium is worse.

In conclusion, the measurement of ETI is relative to parameter λ, and decision maker can adjust λ for a certain aim based on practical realities. If the enterprise wants to keep the state of ETI, and make one or several indices outstanding, we can let λ close to-1; if the enterprise place importance on both equilibrium and single or more outstanding indices, we can let $\lambda<0$ and close to 0; if the enterprise only place importance on equilibrium, then we can let λ close to 1.

Government and enterprise can initialize parameter λ based on the situation and development phrase of ETI. With changing parameter λ, we can reach the aim of punishing laggard and encourage each index balance. Moreover, we can also attain the goal of encouraging advanced and buoying outstanding ones.

5   Conclusion

In a word, fuzzy-integral method can be used to measure interdependent indices without particular hypothesis; it is fit for solving the evaluation with subjective factors. With this method, decision-maker can find the disadvantages of the indices according to the sort of each index with different parameter λ, based on which, they can establish and implement corresponding strategies.

Acknowledgement. National Social Science Fund 098JY020.

References

1. Chen, L.-H.: A Fuzzy Credit-rating Approach for Commercial Loans: a Taiwan Case. International Journal of Management Science (10), 408–419 (1998)
2. Sugeno, M.: An interpretation of fuzzy measure and the Choquet Integral as an integral with Respect to a fuzzy measure. Fuzzy Set and Systems 29, 201–227 (1989)

3. Lee, K., Hyung, L.: Identification of λ-fuzzy Measure by Genetic Algorithms. Fuzzy Sets and Systems, 301–309 (1995)
4. Sapp, J.: Concepts of Symbiogenesis, pp. 100–105. Yale University Press (1992)
5. Kacprzyk, J., Yager, R.R.: Advances in the Dempster-Shafer Theory of Evidence. Wiley (1994)
6. Burgelman, R., Maidique, M.A., Wheelwright, S.C.: Strategic Management of Technology and Innovation, pp. 322–324. McGrow-Hill Inc., New York (1996)
7. Burkill, J.C.: The lebesgue integral. Cambridge Unibersity Press (2004)
8. Wu, Y.-Z.: Communication Theory of Technological Innovation, pp. 44–45. Northeastern University Press, Shen yang (2002)
9. Mowery, D.C., Rosenberg, N.: Technology and the pursuit of economic growth. Cambridge University Press (1989)

Optimizing Books Purchasing Indicators Evaluation System with AHP Method

Ding Ying[1], Liu Jin[2], and Ma Yushan[3]

[1] Library Tianjin University of Science & Technology Tianjin, China
[2] Logistics Infrastructure Division Tianjin College, University of Science and Technology Beijing Tianjin, China
[3] Education Institution Tianjin Normal University Tianjin, China
{teacherding,teacher.lj}@126.com, mayushan801213@163.com

Abstract. This paper analyses the business process of books purchasing and bidding, studying the criteria for the structure, content and weight, advancing the AHP method to optimize indicators evaluation system, improving the evaluation mechanism of bidding process, which make the bidding process more scientific, fair, and form a highly efficient pattern of books purchasing.

Keywords: AHP, Book Purchase, Indicators Evaluation System, Weight.

1 Introduction

With the development of the socialist market economy system, as a method to promote economic development by competing, bidding is more and more universal in the field of books purchasing. Bidding activities should follow the principle such as equality, voluntariness, trueness, lawfulness, fairness, openness, pick out the best and so on. An ordered bidding behavior can promote the reduction of cost, increase the level of resources utilization and optimize the allocation of resources. To make up for this deficiency, several researchers integrated fuzzy theory with AHP determine the criteria weights from subjective judgment of decision makers [1].

This paper analyses the business process of books purchasing and bidding, advancing a hierarchical structure pattern of purchasing indicators evaluation system, obtaining a calculated value of each element's weight with AHP(Analytic Hierarchy Process) method, inosculating subjective judgment and objective calculation efficiency, optimizing contend, avoiding the institution that decide the highest bidder all by subjective assume.

2 Analyse of Books Bidding Business Process

Books bidding behavior is a contract behavior which tenderee sends out bid invitation to particular person or society about the books purchasing business, every tenderer makes a tender for it, then a integrated survey is carried out by tenderee with price, quality, production capacity, time of delivery, financial position, prestige and so on. Based upon balanceable, deciding the highest bidder, coordinate, bargain and a contract legal nexus is established ultimately. The supplier selection decision in books render is usually made by multiple decision-makers after considering multiple attributes of the bidders [2].

Through the analysis of bidding behavior, the whole include invite public bidding, submit a bid, bid opening, bidding evaluation, successful bid, verify and so on, as shown in Fig 1.

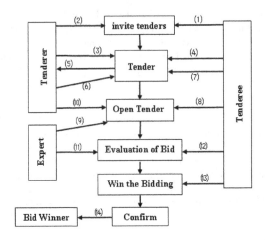

Fig. 1. Bidding Business Process

The meaning of every business step in whole process is: (1) tenderee sends out bid invitation, (2) tenderers see bid invitation, (3) tenderers submit qualification examination, (4) tenderee examines and countersigns qualification, (5) effective tenderers obtained bidding documents, (6) tenderers submit bidding documents, (7) tenderee looks over bidding institution, (8) tenderee asks questions in bid opening, (9) bid evaluation experts ask questions, (10) tenderers explain, (11) experts evaluate bidding documents (technical bid), (12) tenderee evaluates bidding documents (economic bid), (13) tenderee publishes successful bid's information, (14) highest bidder confirms notice on bid winning.

3   Study of Indicators Evaluation System

Since it is more difficult to evaluate experience goods than search products before purchasing [3], consumers become more reliant on the content element of indicators evaluation. Scientific indicators evaluation system is the cornerstone whether whole bidding evaluation will success or not. Scientific and rational decision-making can be an effective use of limited funds, make full use of valuable space, purchase the books readers need in deed, improve efficiency in the use of books information [4].

Indicators system is an indicators group which formed by several relative indicators to finish some study purpose, it should definite indicators, still more relationship called indicators structure. Indicators system can be looked as an information system, which includes the assignment of system element and arrangement of system structure. System element is indicator, includes concept, computer capacity, measuring unit and so on, the relationship between every indicator is the system structure.

The process of optimizing indicators evaluation system includes:

A. Construct Content Element of Indicators Evaluation

The first step, divide the whole indicators group element generally, a clustering is formed with indicators group which up to the special standard. The indicators group of books purchasing include: aptitude of bidding enterprise (W1), financial position (W2), interview data (W3), catalogue data (W4) and book processing (W5). These indicators group build a whole impression of bidding enterprise. The second step, complete every indicator group element, form a detail indicators element which can obtain a test of support data, assure the efficiency.

B. Construct a Hierarchical Structure of Indicators with AHP Method

AHP was developed by Saaty in 1980 [5,6]. For over 30 years, this approach has been used and studied extensively and has been applied especially to multi-criteria decision making [7]. AHP is a multi-level analysis institution pattern, which makes the problem layering, dividing it into different component element according to the characteristic and general objective, putting every element into different level clustering with the influence of relationship and membership. A three levels include overall evaluation indicators (the first level), classify evaluation indicators (the second level), and individual evaluation indicators (the third level) with every indicator group and detail indicators of books purchasing indicators system sorted out through analysis. The detailed rules of indicators (with indicators number) as shown in Table 1.

Table 1. Indicators system of books purchasing

Layer-1	Layer-2	Layer-3
Comprehensive Technical Standards	W1 Qualification conditions	W11 Industry experience
		W12 Human resources
		W13 Customer cooperation
		......
	W2 Financial position	W21 Registered capital
		W22 Financial position
		W23 Cooperation Press
		......

Table 1. (*continued*)

		W31 Book database format requirements
	W3 Interview data	W32 Average annual number of new books published
		W33 Press the number of core
		......
	W4 Inventory data	W41 Certificate
		W42 Data authority
		W43 Data accuracy
		......
	W5 Book Processing	W51 Book mark
		W52 Barcode
		W53 Magnetic stripe
		......

C. Obtain the Weight of Each Indicator in AHP Indicators System

AHP method always used to determine the relative importance of each indicator, through a normative approach of weight number, computing comprehensive weight of every indicator compared with general objective. To the accurate weight of each indicator, needs the analysis that hangs the subjective judgment and objective calculation together. Generally, comparison in one level indicators obtained by experts' subjective judgment, while, weighed value which means using numerical value to show the importance of each indicator obtained by matrix calculation. The process is:

1) Procedure one

Judge the relative importance a_{ij} of each indicator in one level subjectively. The a_{ij} means an importance that the ith. relate to the jth., it approximate the ratio of weight i to weight j, $a_{ij} \approx W_i \Big/ W_j$.

The computing method is: $a_{ij} = 1$ (indicator i and indicator j are both important); $=3$ (slight important); $=5$ (relative important); $=7$ (obvious important); $=9$ (absolute important); $=2,4,6,8$ is a compromise of consecutive values, if W_i less important than W_j, use $1 \Big/ a_{ij}$.

The relative importance a_{ij} of each indicator in "Classify Evaluation Indicators (Layer-2)" judged by experts according to their experience in books purchasing process is:

$$A = \begin{cases} a_{11} & a_{12} & a_{13} & a_{14} & a_{15} \\ a_{21} & a_{22} & a_{23} & a_{24} & a_{25} \\ a_{31} & a_{32} & a_{33} & a_{34} & a_{35} \\ a_{41} & a_{42} & a_{43} & a_{44} & a_{45} \\ a_{51} & a_{52} & a_{53} & a_{54} & a_{55} \end{cases} \approx \begin{cases} W_1/W_1 & W_1/W_2 & W_1/W_3 & W_1/W_4 & W_1/W_5 \\ W_2/W_1 & W_2/W_2 & W_2/W_3 & W_2/W_4 & W_2/W_5 \\ W_3/W_1 & W_3/W_2 & W_3/W_3 & W_3/W_4 & W_3/W_5 \\ W_4/W_1 & W_4/W_2 & W_4/W_3 & W_4/W_4 & W_4/W_5 \\ W_5/W_1 & W_5/W_2 & W_5/W_3 & W_5/W_4 & W_5/W_5 \end{cases} =$$

$$\begin{cases} 1 & 1 & \frac{1}{3} & 1 & 3 \\ 1 & 1 & \frac{1}{3} & 1 & 3 \\ 3 & 3 & 1 & 3 & 5 \\ 1 & 1 & \frac{1}{3} & 1 & 3 \\ \frac{1}{3} & \frac{1}{3} & \frac{1}{5} & \frac{1}{3} & 1 \end{cases}$$

2) Procedure two

Compute relative weight of "Classify Evaluation Indicators (the second Layer-3)".

The sum of weight in one level is 1, so $\sum_{i=1}^{n} W_{ij} = 1$ (j=1,2,......,n) . Combine the

obtained method of a_{ij} : $\sum_{i=1}^{n} a_{ij} = \dfrac{\sum_{i=1}^{n} W_i}{W_j}$ (j=1,2,3,......,n) , so $W_j = \dfrac{1}{\sum_{i=1}^{n} a_{ij}}$.

(Formula 1)

To "Classify Evaluation Indicators (Layer-2)", according to matrix A and formula 1, we have

$$W_1 = \dfrac{1}{\sum_{i=1}^{5} a_{1j}} (j=1) = \dfrac{1}{a_{11}} + \dfrac{1}{a_{21}} + \dfrac{1}{a_{31}} + \dfrac{1}{a_{41}} + \dfrac{1}{a_{51}} = 0.16$$. In a similar way, so

$W_2 = 0.16$, $W_3 = 0.45$, $W_4 = 0.16$, $W_5 = 0.07$.

3) Procedure three

Compute relative weight of "Classify Evaluation Indicators (Layer-3)". For example, each weight in "W3-Interview data" indicator system is W_{31}, W_{32}, and so on (detail content vide Table II), using the method of procedure two, so $W_{31} = 0.035$, $W_{32} = 0.149$, $W_{33} = 0.394$. Obtain others' weight in a similar way.

4) Procedure four

Form a whole indicators evaluation system that had been optimized. Compute the weight in procedure two and three, obtaining the calculated value of "weight" line in Table II, which reflect the importance of each indicator in the whole evaluation system.

However, numerical value grade is discommodious and is not intuitional. This paper arranges every parameter value, turns "weight" into fraction, which more intuitional and operable, as shown in Table 2.

Through the procedure above, optimizing books purchasing indicators evaluation system from experience judge to combine with subjective and objective and rely on AHP scientific theory, which make the purchasing more fairly.

4 Summary

This paper designs and analyses the content and weight of books purchasing indicators evaluation system, obtaining an effective parameter and standard value, forming a whole evaluation system by optimizing. Through the using on the spot in bidding process, responsive effect good, increasing bidding evaluation efficiency effectively and the optimized project is scientific. We may analyses bidding data deeply with data mining in the future, using Web technique to realize Internet bidding system, which makes the purchasing work a new step.

Table 2. Optimized books purchasing indicators evaluation system

Layer-1	Layer-2	Layer-3	Weight	Score
Comprehensive technical standards	W1 Weight 0.16 Score 20	W11	0.312	4
		W12	0.312	3
		W13	0.110	3
		......		
	W2 Weight 0.16 Score 20	W21	0.159	4
		W22	0.455	5
		W23	0.159	4
		......		
	W3 Weight 0.45 Score 25	W31	0.035	2
		W32	0.149	4
		W33	0.394	5
		......		
	W4 Weight 0.16 Score 20	W41	0.200	4
		W42	0.200	4
		W43	0.200	4
		......		
	W5 Weight 0.07 Score 15	W51	0.200	3
		W52	0.200	3
		W53	0.200	3
		......		
Total	1/100			100

References

1. Fu, H.P., Ho, Y.C., Chen, R.C., et al.: Factors affecting the adoption of electronic marketplace: A fuzzy AHP analysis. International Journal of Operations and Production Management 26(12), 1301–1324
2. Li, W., Wang, L., Rao, C.: A bid assessment decision-making model and its application. In: Proceedings of the 27th Chinese Control Conference, CCC, pp. 813–816 (2008)
3. Chen, Y.-F.: Herd behavior in purchasing books online. Computers in Human Behavior 24(5), 1977–1992 (2008)
4. Yang, M., Sun, H.: Research of prediction arithmetic in purchasing books application. In: International Conference on Wireless Communications, Networking and Mobile Computing, WiCOM 2008 (2008)
5. Saaty, T.L.: The Analytic Hierarchy Process. RWS Publications, Pittsburgh (1980)
6. Saaty, T.L.: The Analytic Hierarchy Process: Planning, Priority Setting and Resource Allocation. McGraw-Hill, New York (1980)
7. Lina, M.-I., Leea, Y.-D., Hob, T.-N.: Applying integrated DEA/AHP to evaluate the economic performance of local governments in China. European Journal of Operational Research 209(2), 129–140 (2011)

Research on Relations between Illumination Media and Dry Goods

Na Zhao and Baiyang Jin

Institute of Art & Fashion Tianjin Polytechnic University Tianjin, China
{353898232,873494544}@qq.com

Abstract. With the development of science, Light is used actively in the creation of art as materials; people do not just use it for lighting. This article is talking about how the light is skillfully used in the exhibition of dry goods, by comparing the difference of various elements, to explore a way to explore a way to widen the use of fiber art.

Keywords: light, fiber, material.

1   The Relations between Illumination Media and Dry Goods

With the development of science, people gradually found that light is a electromagnetic wave, but the wavelength of light is much shorter than radio wave .The color of the light will be different if they are not the same wavelength. Using light in the decoration existed many years, one hundred years ago, Edison invented the electric lamp, to some extent this could be treated as the start of using light in the decoration. With the rapid devolvement of society and science technology, various artificial light sources have been made .But how can we put light into the use of decoration? Decoration aims to beautify the environment and satisfy the aesthetic pleasure of people's sights.Nowdays people do not use the light just for lighting require, they use the light for creating art works. All in all, if a art work can use the light skillfully, it can be easily picked up, attracting people's eyes, giving people enough imaginary space.

2   The Art Language of Illumination Media

Light has its special aorment art language and its profound meaning, it can always gives us hopes when faced with darkness, it stands for hope strength courage and future . Adornment art language of the Light effect also has emotional changes. The light has various magic effects.

Different from traditional decorative art, the space is full of flow lighting. By this way, the effect of decoration is no longer the statically effect, we can see the dynamic effect. Where there is light, there is decoration.

J. Luo (Ed.): Soft Computing in Information Communication Technology, AISC 161, pp. 381–384.
springerlink.com   © Springer-Verlag Berlin Heidelberg 2012

3   The Difference between the Internal Structure of Fibre Can Affect the Refractive Index and Transparent Index

A. The Connection between Light and the Internal Structure of Fibre

Fiber	n//	n+	Δn=n//-n+
Ramie	1.595~1.599	1.527~1.540	0.057~0.068
Wool	1.553~1.556	1.542~1.547	0.009~0.012
Fiber flax	1.594	1.532	0.062
Vinyl on	1.547	1.522	0.025
Silkworm refined silk	1.585	1.537	0.047
viscose acetal fiber	1.539~1.550	1.514~1.523	0.062
Cotton	1.573~1.581	1.5524~1.534	0.041~0.051
Acrylic fibers	1.500~1.510	1.500~1.510	0.000~0.005
chinlon66	1.570~1.580	1.520~1.530	0.053
chinlon6	1.568	1.515	0.053
dilacerate fiber	1.476~1.478	1.470~1.473	0.005~0.006

The internal structure of fiber leads to the difference that exists in the absorption, reflection ability to the light. So we should take fiber into account.

B. The Length and Width Can Affect Its Ability to Absorb Light

When we study the fiber, we find that the Length and Width can affect its ability to absorb light. If the fiber is short, then the dry good that is composed of this fiber is bad in color and transparency If the width increase ,then the rate of fiber's surface will increase, in this case ,when color the dry good ,it will not get a satisfying result.

4   The Relationships between Fiber Art Show and Light

The fiber material's charm does not only exist in its pattern, but for itself it is a special art language .Various fiber material will give people different feelings, causing the profoundest imagination in the deep heart. The different shapes colors materials will give people various feelings

In the exhibition of fiber art works, the require for the light varies from different works. Using the light to the full extent, improving the lighting quality, we should take peoples visual requirements, avoiding light refraction and exposure.Meanwhile; we should pay attention to these points in order to avoid visitors` visual fatigue and unhealthy visual psychology effect.

A. Different Fiber Materials Need Different Extent Illumination

As the whole environmental light resource it should take luminance expressivism of light decoration and the number of lights into consideration. When people do different things, they need different extent illuminance.The luminance of the Exhibition Hall should be rational arranged, if it is not proper, it may cause damage to eyes. The rational luminance can give pleasure to visitors. Different fibers need different luminance. Comparing linen with silk, we find that, linen need high intensity light while silk do not .So only we give the rational luminance, can we get the right result.

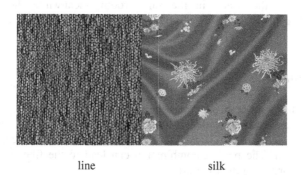

line   silk

B. Different Light Sources Give Fiber Exhibits different Effect

When determine the illuminance, the next to do is to choose the suitable Light source. When choose the light source, first we should take colored light effect into account. Lights have cold light and warm light, they give people different feelings. Warm light can give people comfortable soft feelings while cold light give people open vividness feelings. Before choosing the light source, the color is also very important, because different color give people different inner feelings. In addition, the object's color will change when colored light lights on it .Using the different colored light, we can efficiently create different atmospheres, at the same time we should think about the design of the fiber exhibits.

C. The Form of Illumination`s Expression

According to the different forms of Fiber works, the expressions of lamplights can be various. We can choose the suitable form in accordance with demands, the forms are as follows.

a) The expression of Front lighting   Front lighting refers to the light-emitting area that consist of the surfaces of ceiling wall floor. When under the light, the feature of the ceiling are homogeneous abundant multitudinous in forms. We use it for the show of co planate fiber works.

b) Sometimes we change the light sources into the shape of banding. The appearance can be various such as tetragonal ring form triangle and so on. Elongated light band has a certain guide, often designed as the light source of public places or as the decoration.

c) Point light refers to the light source has a widen range but intensive. For its strong Light luminance, it is often used in the dining room bedroom schoolroom and so on .Point Light behaves in many ways like Ron lighting, backlight, sidelight and so on.

d) Static light and movable light   Static light——the light source is settled, the lights shines on the fiber works. Most rooms use the static lights. This way can use the energy efficiently and create a stable gentle harmonious atmosphere. It is often used in the school factory office block and such places .Movable light has abundant Artistic expression. When it light on the fiber works, it will give people a very special feeling.

e) Light and shadow are special art, light can create shadow, and shadow can reflect light. They create the shape in the same room, meanwhile they create great environment. They perform on the ceiling wall floor and create wonderful effect at the same time.

5   Conclusion

Using the light has a great influence on the Fiber adornment art, it makes us treat the Fiber adornment art in a more widen vision. Reinterpret fiber art and return to the origin of fiber art——beautify the environment decorate things and satisfy people's Aesthetic pleasure and so on. The perfect combination can lead to the fiber art influence on people's spirit, improving the fiber art.

References

1. Li, Y.: Introduction to Art Design. Hubei Fine Arts Publishing House, Wuhan (2002)
2. Langers, S.: Feeling and Form. China Social Sciences Press, Beijing (2005)

Application of Virtual Reality in Tourism Management Professional Teaching

Shi-chen Fan and Feng-xia Wang[*]

Tourism College Hainan University Haikou, China
Fansc1230@126.com, Summer_wangfx@126.com

Abstract. With the rapid development of tourism industry, the demands and requirements for tourism management talents get higher and higher. The traditional teaching model can not meet the requirement, however, Virtual Reality technology can effectively solves this problem. Firstly, the paper introduces the Virtual Reality technology and its features, and then discusses the characteristics of tourism management profession and teaching status simply, which lead to the function of Virtual Reality in tourism management teaching. With the application and popularization of Virtual Reality in all fields, the application in tourism teaching will also be inevitable.

Keywords: Virtual Reality, tourism managemen, teaching, practical operation.

1   Introduction of Virtual Reality Technology

Virtual Reality (referred to as VR) technology, also known as Ling environmental technology, is an application driven by many disciplines involved in the new practical skills, including computer technology, sensor and measurement technology, fax technology, microelectronic technology. By creating a three-dimensional visual, auditory and tactile environment, allow users to use tools for man-machine dialogue with the virtual environment to interact with objects, to feel like being in the real environment to reach the real effect in the virtual realm [1]. Using virtual reality technology allows the user to enter impossible to achieve geographic or situation in real life, such as the deep desert, snow-capped mountains of the roof and so on.

Virtual reality is an exact reproduction of the real world, and has the four most prominent characteristics: sense of presence, interaction, and more awareness, autonomy. Sense of presence is present in the simulated environment as a Protagonist in the true extent. Interaction means the operating level that users can operate within the object level and natural level getting feedback from the environment. More awareness includes not only visual perception, but also the auditory, force, touch, movement, taste, smell perceptions. Autonomy refers to the objects in a virtual environment based on the degree of movement laws of physics.

2   Characteristics of Tourism Management Profession

Firstly, the Tourism Management is a comprehensive discipline which collects multidisciplinary basic theories and research methods, aims to develop the purpose of

---

[*] Corresponding author.

J. Luo (Ed.): Soft Computing in Information Communication Technology, AISC 161, pp. 385–389.
springerlink.com   © Springer-Verlag Berlin Heidelberg 2012

talent with theoretical and practical ability, both needs to define Economic Development and Tourism Law Tourism Management Mechanism by using qualitative analysis, and needs to analyze and process the objective phenomenon in tourism management with the help of quantitative analysis [2], and propose a higher requirement to comprehensive experimental technology and training.

3   Problems in Tourism Management Teaching

A. Single Means of Teaching, and Difficult to Realize Situation Teaching

Limitations of traditional multi-media teaching technology are difficult to meet the needs of tourism management situation teaching. Currently, multimedia technology in tourism management teaching only includes single images, pictures. especially the tourist knowledge for domestic and international attractions, in the current tourism management teaching only presented beautiful films, photographs, graphics, etc. to students. The aspect of scenery pieces is single with little amount of information, and the knowledge content is not enough. And the images and photos are displayed graphically and lack the three-dimensional and situational sense. The professional knowledge in depth understanding is difficult to be displayed all aspects and three-dimensional, such as in the ancient knowledge of the teaching building roof, representing different levels— wudian ding, xieshan ding, xuanshan ding, yingshan ding, and zanjian ding are all important and difficult elements. However, the teachers is difficult to describe the image by language and text, the pictures can only displayed plane. Usually, the teachers spend a long time in teaching students the knowledge points, but the students can not master them.

At present, some good conditions colleges use traditional sand table model to do practical teaching, but because of the teaching space, site constraints, high cost and other issues, it is difficult to mass production. In addition, the sand tables can not give students truly immersive feeling, for the model has no sense of realism and the scene.

B. The High Teaching Cost and Risk, Can't Repeatedly Practice on the Spot

The major of Tourism Management is a major which requests very high professional operational skill and is also the basis of taking up an occupation for students in the future, it's an effective way to improve students' occupational quality in the actual scene carrying out all kinds of operational and capable practice. However, it often can't allow students to practice in the actual training place because of the limit of teaching cost and condition. It is almost impossible to get to some remote tourist attractions or tourism enterprises. Especially in the latest years, with the rapid development of tourism, the scale of recruiting students by the major of Tourism Management constantly extend, and the number of students increase sharply, when the students attend the courses of practice, the college will spend a large number of funds renting many buses and a large number of time travelling to and fro between every tourist attractions and tourism enterprises, and the college also will think of the students' safe, the weather and so on adequately. In addition, because this style can't spend much time, although it spends large amounts of manpower, material resources and financial resources preparing for them meticulously, it doesn't gain good effect, and the students only acquire a little perceptual cognition, it doesn't really reach the goal of improving the level of practice [3].

C. Simplex the Main Part of Teaching and Function, Can't Really Exert Students' Subjective initiative

The modern education require that it can build an new model of teaching which can give scope to the main function of teachers and embody students' function of cognizing the main part fully. At present, the teaching model of the major of Tourism Management in our country still doesn't break away from the traditional model of teaching which gives first place to teachers, and teachers as the main men take control of the multi-media system to develop activities of teaching which pass on knowledge and skill to students. Although many teachers have taken notice of interactive teaching and paid attention to arousing the initiative of students by every method of teaching, the limit of teaching method still can't treat the students as the motif of classroom exerting their subjective initiative.

The current teaching model is a single-track, teachers only pass on knowledge and cognition to students by the multi-media system, students can't feed back the information by the teaching system. In addition, the style of many examination of the courses which have the strong operation about the major of Tourism Management are still only answering questions on the papers or the style that students answer questions facing to teachers, it's very rigid and students are short of the sense of actual combat.

4   The Effect of the Technology of Virtual Reality in Teaching of the Major of Tourism Management

A.   Improve the Environment of Teaching, Achieve the Teaching Scene More Reality

It can make students completely be addiction to the environment needed in teaching that moves airport, restaurants and tourist attractions to classroom and make students be personally on the scene to be familiar with the environment by the technology of virtual reality. It will realize an effect of real scene that can experience the operation of tourist reception actually never to go out. The most advantage of applying the technology of virtual reality is that it can vividly express abstract concept and principle in the teaching, and can create real learning circumstances for students, help students acquiring exemplary knowledge and grasping the essence of concept and principle. It also can show some things that students can't observe directly by a scene that is realistic and consists of the pictures, the voice. For example, it has mentioned some kinds of archaic buildings' roof before, students can enjoy the sight of them from every angles and bearings, not only can they build these roofs' spatial structure in their brains, but also they can see, listen, even touch them, so they can understand their structures and forms from every aspect. In addition, the strong sense of scene also can contribute to remembering the knowledge firmly. This teaching style by using virtual scene is unique , is a renovation and a challenge to the traditional book-education and education by giving lessons.

Finally, complete content and organizational editing before formatting. Please take note of the following items when proofreading spelling and grammar:

B. Reduce the Teaching Cost Greatly, Increase the Benefit of Running a School

It is very important and necessary to teach the students of the major of Tourism Management in the real places of tourist attractions and tourist enterprises. However, the college will spend large amounts of money necessarily (such as the cost of renting cars, road maintenance costs, the cost of admission tickets, the hotel expenses ,etc) organizing many students to carry out teaching in the real places. The system of virtual reality needn't organize students to explain in the tourist attractions and experience services and receptions and other jobs in the tourist enterprises, it not only save the teaching costs, but also it solves a series of problem such as the safe, the weather, etc during the course of organizing activities, at the same time, it also saves a large number of manpower, material resources and time, and when it make students learn more knowledge, it also increases the benefit of running the school, it is a result of two-to-win in a manner of speaking.

C. Increase the Teaching Efficiency of Tourist Teaching by a Wide Margin

In the course of teaching the major of Tourism Management, it involves in many concrete practical operational scenes, such as restaurants, airports, tourist attractions and so on, it can observe above-mentioned scenes using different styles by means of showing the virtual reality, for example, it can look from distance, look from vicinity, look up to them, look down at them, look from transverse direction, look from flank and so on . When it sets up the roaming routes , the routes can be many dimensions, they can be from the surface to the inside, can be from the inner to the external, can be from the front to the back, can be from the left to the right, can be from this route to that route, it contributes to students' understanding and grasping those correctly and make the students' remembering style change from the single remembering style that only remembers the letters to a new remembering style that combines the visual memory with the letters' memory . In the real scene of virtual reality, students can make the train imitating the real scene repeatedly, after they are skilled, they can train in the real tourist attractions, it increases the teaching efficiency vastly.

D. Arouse Students' Learning Interest

The interest and enthusiasm bring up the best students. By using the teaching system of virtual reality, students learn in a real virtual circumstance which are imitated by the real scene in the tourist enterprises or tourist attractions completely, it's no longer the traditional education of books and the ordinary teaching of multi-media, thus it will improve students' learning enthusiasm and interest. At the same time, in the virtual circumstance because it can set up the roaming route, angle and so on by itself, it can give scope to students' subjective initiative adequately, and make the learning initiative, efficient.

E. Create the Modern Model of Teaching as a Tour Guide

The technology of virtual reality changes the process of traditional education's characteristic, it makes teachers turn into the mentors and organizers in teaching activities from the role who pass on the knowledge, and it makes students turn into the person who discover the knowledge forwardly from recipient who receive knowledge passively. The teaching media are turn into cognitive tools for students from assistant methods, teaching tools, the process of teaching is turn into " create the scene, the two parties of teaching interact, explore, discuss and construct significantly " from " teachers explain, students listen to the teacher in class".

5   Conclusions

Since 1990s, Virtual Reality technology was regarded as one of three major areas of innovation of teaching with multi-media teaching and online education, with the most powerful human-computer interaction technology, and has been one of hot directions of research, development and application in information field. Virtual Reality technology will be the representative of 21st century information technology. Its development will have extraordinary significance and impact in the field of education. With the boom of tourism development, the demand for high quality service personnel grows rapidly, which also stimulates tourism college enrollment expansion to a considerable extent. Training high quality skilled professionals in tourism management and realizing the function of education to better social service are the needs of tourism industry great development and educational development. At present, many education professionals have concerned about the combination of Virtual Reality technology and tourism practice teaching, and done a lot of work in deep study and development of tourism education functions. Thus, with the applications and popularization in all kinds of fields, its application in tourism teaching will also be inevitable.

Acknowledgment. Financial supports from: National Education Ministry Planning Project unit funded-Study on Tourism Professionals Model under Hainan International Island Strategy(FFB108132), National Natural Science Foundation Project of China in 2009 (NO.40961005), the National Soft Science Item of China in 2010 (NO.2010GXS5D252), Hainan University Education Project Fund in 2010(HDJY1008), Hainan Province Education and Science, "Eleventh Five-Year Plan" of 2009 Project(QJ11512), and Hainan University "211" construction sub-project "economic management institutional mechanisms innovation in Hainan".

References

1. Zhu, X.-H., Lv, G.-N., Wang, J.: Application of Virtual Reality in Geography. Geography and Territorial Research 14(3), 60–63 (1998)
2. Zhang, J.-X.: The Status and Development Strategy of Higher Tourism Education Training. Journal of Jianghan University (Social Sciences) (2), 87–90 (2005)
3. Wei, K.: The assisting function of Virtual Reality in Tourism Teaching Practice. Journal of Shandong Youth Administrative Cadres College (4), 135–137 (2007)

Analysis on the Interaction of Public Art in Different Areas of City

Yong Ni and Ya Teng

Department of Art Design
Yancheng Textile Vocational Technology College
Yancheng, China
hicheng2011@yahoo.com.cn, 344051932@qq.com

Abstract. In recent years,with the rapid development of our economic and enlarge our city,the demands of culture are improving,more people take more focus on the development of our urban public art. As an improtant part in urban contruction, public art has its own positive effect, it will keep connection with works, designer, people and city space.The communication in these four parts is also the main form expression of interaction about public art. In system of urban construction, the interaction of public art has different meanings in defferent areas. This paper will show these differents from business district, tourist attractions and campus. And indicate the expressions of city public art in these different areas.

Keywords: public art, interaction, urban construction.

1 Analysis on the Interaction of Urban Public Art

The interaction of urban public art is mainly performance for the communication and influence of works, designers, people and urban space, the principal part of the interaction is balance. public opinions and perceptions affecting the designer's conception, which requires a public art design designers have the explicit public awareness, understand the trend of urban development, research the regional culture of different areas, in order to perfect the design work. The interaction of designers and people is the extension of works, is also the part of the work itself, the feedback of works come from the people is an important factor to checkout the public art successed or failured.

The interaction of urban public art is also performance for that it is not merely a form of art, not get the concept of personal modelling imposed on urban spaces, which require designers to consider the city space around the works and the surrounding buildings and the works foiled each other. in addition, it is a kind expression of the idea, through public participation, to proclaim values of comply with the development of society.

2 The Interaction of Public Art in Different Areas of City

According to the different division standard, the areas of city have different explanation. this paper based on the function of differences urban area, to elaborate the interaction of public art from business district, tourist attaction and campus in the city.

J. Luo (Ed.): Soft Computing in Information Communication Technology, AISC 161, pp. 391–396.
springerlink.com © Springer-Verlag Berlin Heidelberg 2012

A.   Public Art in Business District

Business district usually refers to the area of retail business together and trade frequently, business district located in the center of the city, both sides of the prosperous streets and large public facilities around. It contains superstore, comprehensive catering, financial institutions and so on, it is a place of so many people get together. Business district is a special area of the city, public art in order to achieve the purpose of interaction, it have to choose the subject pertinently, the center of business district is also the center of the city, there will have works of public art to reflect the city culture. the public art works are generally in the city sculpture, installation works, landscape design, as the main form of expression, the purpose is to convey centuries-old historical culture and rich geographic characteristics. The copper carriage beside QueYeChang in TianJin (Figure 1), it generalize the TianJin history: progress constantly under the foreign culture, Meanwhile, reflect with historic building around, show out the concession culture in space; More important is that people remember the TianJin culture clearly, at the same time, form the good interaction with work itself: foreign tourists take photos with carriage. Works spread the local history culture, at the same time, itself also is a beautiful scenery line. With the rapid urbanization, emerged a batch of public art have novel idea, abstract modelling, rich color in the new city downtown construction, to express the positive face. However, not all the form of public art can get positive interaction, the public art is often one-way conversation in the center of city, works on the subject, dimension, style got reiteration or opposite with the spatial environments, works lost the city culture, so we should pay attention to the following two points:

Fig. 1. The copper carriage in TianJin

First, avoid to performance the political leaders and the great wordsmiths blindly. Some character sculpture in the center of city always be shaped as the hero, the famous person, the great man, as the masses have been ruled out choices, we always seen the conceptualization of heroism and heroic, this form neglected the   reflection from masses.

Second, avoid to aspire the personal style bigotedly. Designers often willing to achieve self-worth through the works, to highlight the designer's personality, this kind of works lack culture, it is difficult to incarnate the connotation of city.

As an important part of urban business district, the flourishing main streets reflect the status of city's economic development. On both side of streets there are so many kinds of public logos, it is an important part of urban public art in business district. As the identification system of urban space, public logos have the role of directionality for signpost, Warning signs, identifier of shops, ornament surrounding environment.

Inchoate public logos, mostly are signboards, only focus on considering the function of identifier, so select the material on a single, rough craftsmanship and Inflexible modelling do not have interaction with the city. With the rapid development of urban construction, new buildings are springing up stands on both sides of the street, public logos developping with the changes of city's buildings, it is not only a signboard for identifying direction and recognizing public facilities, but also was recognized as an important fact of regional culture and city custom. Reasonable public logos enriched the urban space, at the same time, expanded entertainment and Playability of business district, benefit to have more interaction with people.

When people in feeling and understanding about his environment around, the subjective ideas will influence to receive the information of peripheral environment, when visual sense received the forceful objective things, it will influence people's subjective ideas. The modelling and color of indicative logos can have difference with peripheral environment, to arouse people's attention, revealed the recreation of public logos (Figure 2).

Fig. 2. The public logo of cinema

In addition to pursuit the recreation, it is important to take the regional culture in the design. the regional culture in public logos is not specific forms, not represent any specific entity, it is a product in abstract thinking come from people's living environment. When people distinguishing the public logos, people have different views in modelling, color and content, but people can feel the regional culture from existing abstract thinking, people have formed the perception in spirit.

B. Public Art in Tourist Attractions

Tourist attraction is a place to attract people from everywhere to go sightseeing; tourist attraction can satisfy the people's demand of entertainment, fitness, learning; tourist attraction is a management institution to provide necessary services. Because of the different nature, we can divide the tourist attraction into several categories: cultural relics, scenic spots, natural sceneries. Cultural relics are the places to have certain cultural and historical value since ancient time, people can enrich their knowledge and understand the history, at the same time, they can educate their children in the place; scenic spots are the places to have unique scenery, people can Relax their mood; natural sceneries are the places to have beautiful natural environment, it is a good choose to enrich people's experience and expand people's horizon.

Cultural relics built by physical architectures of artificially legacy as the main part, each cultural relic represents the local architectural style and regional culture, enery physical architecture can reflect each other in tourist attraction, it is a integrated style. Every building in tourist attraction is a public art work, from the architectural appearance tectonic to carve on details all reflect craftsmen's superb artistry. All the building on the vision in the space of tourist attraction have been formed saturated, except for some suggestive public logos, people do not need to increase more public works. In the interactive aspect, incarnated people's spiritual yearning for cultural relics.

The Imperial Palace is a good example about cultural relic(Figure 3), the tourists from far and near get together to understand chinese history, this understanding come from development of the Imperial Palace; people understand the culture of the Imperial Palace from brick and tile; people comprehend aesthetic ideas from plaque in the Imperial Palace. In addition to accept the information, the Imperial Palace caused the modern people to have new thinkings: how to preserve the buildings completely and how to receive and inherit the culture, these are all mental feedback to the Imperial Palace. This interaction not only promote the communication between people and public art, but also promote the communication between people and people about chinese cultural.

Scenic spots and natural sceneries constitute with native element. When people developed and implemented resources, they have recognized to keep the original appearance in tourist attraction, public art in tourist attraction must fit original geological environment, try to make harmony of style.

Because of the natural landscape, vegetation is an important part of the tourist attraction, the trimmed vegetation is a form of public art. People were surrounded by green vegetation, nervous mood would be relaxed, trouble would be put away, green vegetation purified people's emotion. That embodied broad scope and various interacter form of public art.

Fig. 3. The Imperial Palace

C. Public Art on the Campus

Campus have differences with other areas of city, it is a isolated place relatively, but also a free place relatively, the enclosed conditions avoid many limit in public art, people have more choose in modelling, color, materials. Because of the nature of each university have differences, campus have uniqueness in overall layout and architectural style, the form of public have been impacted with integrate environment on the campus. But whatever the campus, public art on that will choose many themes like that, combine campus space environment, outstand campus culture, emphasize important people who have the relationship with the university.

Interaction on the campus can reflect the connection in works, designers, masses, space, the designers in this particular environment can fully put their personality into the works, the designers can advocate positive life value and vigorous vitality through their works. Student is a community with good acception which have enrich ideas, public art forms should also try to show the enthusiasm of student's life, accord with the student's visual aesthetic requirements.

Campus culture do not be constracted without the development of public art, all kinds of public art form become an effective carrier of campus culture, enriched the color of campus culture, increased the contents of campus culture, replenished the flavour of campus culture. Various public art forms will get the students' support and acceptance, various public art forms will attract more students into the construction of harmonious campus actively.

Tianjin university as an example, public art forms are abundant on the campus, the forms consists of some historical architecture and sculpture constitute(Figure 4), the content publicized chinese traditional culture and the positive values, the modelling is lively, in order to encourage students to find and solve many problems in the study life, to build a harmonious campus environment.

Fig. 4. The public art in Tianjin university

3   Conclusion

Urban construction must have its own cultural contexts and development idea, public art must follow the development of local culture, only understand the local culture in essence, the culture will be inherited and developed through the forms of public art.

References

1. Ma, Q.: Public art basic theories. Tianjin University Press, Tinjin (2008)
2. Jin, B.: Public art design. People's Art press, Beijing (2010)
3. Weng, J.: Urban public art. Southeast University Press, Nanjing (2004)

The Existence for RDE with Small Delay

Xunwu Yin

Tianjin Polytechnic University
College of Science
Tianjin, China
yinxunwu@163.com

Abstract. In this article, we prove that for the classical reaction-diffusion equations (RDE) $u_t - \Delta u = f\big(u(t), u(t-\tau)\big)$ with a small delay, there exists a weak solution. Our method is Galerkin approximations which is one of the most important methods in proving the existence of weak solution. This method can be found in many works, for example [4]. It is to build a weak solution by first constructing solutions of certain finite-dimensional approximations, and then building energy estimate and last passing to limits. Imposing some condition on the nonlinear f, we first make use of Galerkin approximations to prove the local existence and uniqueness theorem for weak solutions of the Initial-Bounded Value Problem.

Keywords: reaction-diffusion equations with small delay, Galerkin approximations, energy estimates.

1 Introduction

It is well-known and not hard to prove [1,2,3] that for scalar reaction-diffusion equations

$$u_t - \Delta u = f(u) \quad \left(x \in \Omega \subset R^n \right) \tag{1.1}$$

Subject to homogeneous boundary conditions, all globally defined bounded solutions must exist. In this article, we introduce a time delay into the right-hand side so that (1.1) becomes

$$u_t - \Delta u = f\big(u(t), u(t-\tau)\big) \tag{1.2}$$

Giving certain conditions to f, we prove the existence and uniqueness of weak solution.

In fact, this problem has been settled in [5]. There the nonlinearity $f : R^2 \to R$ is assumed to be locally lipschitz and to satisfy the one-sided growth estimates

$$f(u,v) \le (u+1)\gamma(v) \qquad for \ u \ge 0$$
$$f(u,v) \ge -\big(|u|+1\big)\gamma(v) \quad for \ u \ge 0$$

J. Luo (Ed.): Soft Computing in Information Communication Technology, AISC 161, pp. 397–404.
springerlink.com   © Springer-Verlag Berlin Heidelberg 2012

For some continuous γ. Here our nonlinearity f may not too strong. To prove existence, they treat (1.2) stepwise as a non-autonomous un-delayed parabolic partial differential equation on the time intervals $[(j-1)\tau, j\tau]$ ($j \in N$) by regarding the delayed values as fixed. However, despite the vast literature on existence for such equations, they have been unable to locate a result directly applicable to (1.2) since rather mild regularity assumptions on the non-autonomous terms are required there. For example, results involving Holder continuity assumptions with respect to t as in [2] cannot be applied since there initial data, considered as functions from $[-\tau,0]$ into $L^2(\Omega)$, are only continuous. To work in $C^{0,\alpha}([-\tau,0];L^2(\Omega))$ instead of $C^0([-\tau,0];L^2(\Omega))$ is no way out of the dilemma since solutions of parabolic equations do not in general attain their initial values Holder continuously, thus time translates of typical solutions with Holder continuous initial data are not even locally Holder in t. Thus their strategy is to mimics the results of Henry [2] theorem 3.3.3 and corollary 3.3.5 but with his assumption of Holder continuity in t replaced by $p-$ integrability. They build mild solution, however we will build weak solution. As is known to all, mild solution must be weak solution. Weakening some conditions, we make use of Galerkin approximations to prove the local existence and uniqueness theorem for weak solutions of the Initial-Bounded Value Problem.

2 Preliminaries

In this article we assume U to be an open, bounded subset of R^n, and set $U_T = U \times (0,T]$ for some fixed time $T > 0$. And let $\tau > 0$ be a positive parameter, that is a delay. We study the scalar delayed initial-boundary-value problem

$$u_t - \Delta u = f\left(u(t), u(t-\tau)\right) \quad in \quad U \times R^+$$

$$u = u_0 \qquad\qquad in \quad U \times [-\tau, 0]$$

$$u = 0 \qquad\qquad on \quad \partial U \times R^+$$

As a matter of fact, we have studied the ordinary parabolic equations

$$u_t + Lu = f\left(x,t\right) \quad in \quad U \times R^+$$

The letter denotes for each time t a second-order partial differential operator, having the divergence form

$$Lu = -\sum_{i,j=1}^n (a^{ij}(x,t)u_{x_i})_{x_j} + \sum_{i=1}^n b^i(x,t)u_{x_i} + c(x,t)u$$

Obviously, $a^{ij} \equiv \delta_{ij}, b^i \equiv c \equiv 0$, it is reaction-diffusion equation. General second-order parabolic equations describe in physical applications the time-evolution of the density of some quantity u, say a chemical concentration, within the region U.

As noted for the equilibrium setting, the second-order term $\sum_{i,j=1}^{n}(a^{ij}(x,t)u_{x_i})_{x_j}$

describes diffusion, the first-order term $\sum_{i=1}^{n}b^i(x,t)u_{x_i}$ describes transport, and the

zeroth-order term cu describes creation of depletion.

To make plausible the following definition of weak solution, let us first temporarily suppose that $u = u(x,t)$ is in fact a smooth solution of our parabolic problem. We now switch our viewpoint, by associating with u a mapping $u:[0,T] \to H_0^1(U)$ defined by $[u(t)](x) = u(x,t)(x \in U, 0 \leq t \leq T)$. In other words, we are going to consider u not as a function of x and t together, but rather as a mapping u of t into the space $H_0^1(U)$ of functions of x. This point of view will greatly clarify the following presentation.

Let us similarly define $f:[0,T] \to L^2(U)$ by $[f(t)](x) = f(x,t)(x \in U, 0 \leq t \leq T)$. Then if we fix a function $v \in H_0^1(U)$, we can multiply the PDE $u_t - \Delta u = f$ by v and integrates by parts to find $(u',v) + \int Du \cdot Dv dx = (f,v)$ for each $0 \leq t \leq T$, the pairing (,) denoting product in $L^2(U)$.

Next, observe that $u_t = f + \sum_{i=1}^{n}(u_{x_i})_{x_i}$. So the right hand side of u_t lies in the

Sobolev space $H^{-1}(U)$ with $\|u_t\|_{H^{-1}(U)} \leq C\left(\|u\|_{H_0^1} + \|f\|_{L^2}\right)$. This estimate suggests it may be reason to look for a weak solution with $u' \in H^{-1}(U)$ for a.e. time $0 \leq t \leq T$; in which case the first term (u',v) can be reexpressed as $\langle u',v \rangle$, $\langle , \rangle$ being the pairing of $H^{-1}(U)$ and $H_0^1(U)$.

All these considerations motivate the following definition of weak solution. We say a function
$$u \in L^2(0,T;H_0^1(U)), \text{ with } u' \in L^2(0,T;H^{-1}(U))$$
is a weak solution of (2.1)-(2.3) provided
$$(1) \quad \langle u',v \rangle + \int Du \cdot Dv = (f(u(t),u(t-\tau)),v)$$
For each $v \in H_0^1(U)$ and a.e. time $0 \leq t \leq T$, and
$$(2) \quad u(t) = u_0 \text{ for } -\tau \leq t \leq 0$$
In section 3, we will build three steps to prove existence. One is Galerkin approximation. Another is Energy estimates. The last is Existence and Uniqueness.

3   Main Results

Step1. We intend to build a weak solution of the classical reaction-diffusion equations (RDE) with a small delay

$$\begin{cases} u_t - \Delta u = f(u(t), u(t-\tau)) & \text{in } U_T \\ u = 0 & \text{on } \partial U \times [0,T] \\ u = u_0 & \text{on } U \times [-\tau, 0] \end{cases} \tag{3.1}$$

by first constructing solutions of certain finite-dimensional approximations to (3.1) and then passing to limits. This is called Galerkin's method. Here we assume

$$f(u(t), u(t-\tau)) \in L^2(0,T; L^2(U)), u_0 \in L^2(U)$$

More precisely, assume the functions $\omega_k = \omega_k(x)(k=1,\cdots)$ are smooth, $\{\omega_k\}_{k=1}^{\infty}$ is an orthogonal basis of $H_0^1(U)$ and $\{\omega_k\}_{k=1}^{\infty}$ is an orthonormal basis of $L^2(U)$.

Fix now a positive integer m. We will look for a function $u_m : [0,T] \to H_0^1$ of the form

$$u_m(t) := \sum_{k=1}^{m} d_m^k(t)\omega_k \tag{3.2}$$

Where we hope to select the coefficients $d_m^k(t)(0 \le t \le T, k=1,\cdots,m)$ so that

$$d_m^k(0) = (u_0, \omega_k)(k=1,\cdots,m) \tag{3.3}$$

And $(u_m', \omega_k) + \int Du_m \cdot D\omega_k dx = (f, \omega_k)(0 \le t \le T, k=1,\cdots m) \tag{3.4}$

Here as before, $(,)$ denotes the inner product in $L^2(U)$.

Thus we seek a function u_m of the form (3.2) that satisfies the projection (3.4) of the problem (3.1) onto the finite dimensional subspace spanned by $\{\omega_k\}_{k=1}^{m}$.

Theorem 3.1 (Construction of approximate solutions). For each integer $m = 1, 2, \cdots$ there exists a unique function u_m of the form (3.2) satisfying (3.3) and (3.4).

Proof. Assume u_m has the structure (3.2) we first note that $(u_m'(t), \omega_k) = d_m^{k'}(t)$

Furthermore $$\int Du_m \cdot D\omega_k dx = \sum_{l=1}^{m} e^{kl}(t)d_m^l(t) \qquad \text{for}$$

$e^{kl}(t) = \int D\omega_l \cdot D\omega_k dx(k,l=1,\cdots,m)$.

Let us further write $f^k(t) = (f(t), \omega_k)(k = 1, \cdots, m)$. Then (3.4) becomes the linear system of ODE

$$d_m^{k'}(t) + \sum_{l=1}^{m} e^{kl}(t)d_m^l(t) = f^k(t)(k = 1, \cdots m) \tag{3.5}$$

subject to the initial conditions (3.3). According to standard existence theory for ordinary differential equations, there exists a unique absolutely continuous function $d_m(t) = (d_m^1(t), \cdots, d_m^m(t))$ satisfying (3.3) and (3.4) for a.e. $0 \le t \le T$. And then u_m defined by (3.2) solves (3.4) for a.e. $0 \le t \le T$.

Step2. We propose now send m to infinity and to show a subsequence of our solutions u_m of the approximate problems (3.3) and (3.4) converges to a weak solution of (3.1). For this we will need some uniform estimates.

Theorem 3.2 (Energy estimates). There exists a constant C, depending only on U, T such that

$$\max_{0 \le t \le T} \left\| u_m(t) \right\|_{L^2(U)} + \left\| u_m \right\|_{L^2(0,T;H_0^1(U))} + \left\| u_m' \right\|_{L^2(0,T;H^{-1}(U))}$$

$$\le C(\left\| f \right\|_{L^2(0,T;L^2(U))} + \left\| u_0 \right\|_{L^2(U)}) \tag{3.6}$$

Proof. Multiply equation (3.4) by $d_m^k(t)$, sum for $k = 1, \cdots m$, and then recall (3.2) to find

$$(u_m', u_m) + \int Du_m \cdot Du_m dx = (f, u_m) \tag{3.7}$$

Because of

$$(u_m', u_m) = \frac{d}{dt}(\frac{1}{2}\left\| u_m \right\|_{L^2(U)}^2) \quad ; \quad \int Du_m \cdot Du_m dx = \left\| u_m \right\|_{H_0^1(U)}^2 \quad ;$$

$$|(f, u_m)| \le \frac{1}{2}(\left\| f \right\|_{L^2}^2 + \left\| u_m \right\|_{L^2}^2)$$

So we get from (3.7)

$$\frac{d}{dt}(\left\| u_m \right\|_{L^2(U)}^2) + 2\left\| u_m \right\|_{H_0^1(U)}^2 \le \left\| f \right\|_{L^2(U)}^2 + \left\| u_m \right\|_{L^2(U)}^2 \tag{3.8}$$

For a.e. $0 \le t \le T$.

Now write $\eta(t) := \left\| u_m(t) \right\|_{L^2(U)}^2$ and $\xi(t) := \left\| f(t) \right\|_{L^2(U)}^2$

Then (3.8) implies

$$\eta'(t) \le \eta(t) + \xi(t) \quad \text{for a.e. } 0 \le t \le T$$

Thus the differential form of Gronwall's inequality yields the estimate

$$\eta(t) \le e^t \left(\eta(0) + \int_0^t \xi(s) ds \right) (0 \le t \le T)$$

Since $\eta(0) = \left\| u_m(0) \right\|_{L^2(U)}^2 \le \left\| u_0 \right\|_{L^2(U)}^2$ we obtain the estimate from the above

$$\max_{0 \le t \le T} \left\| u_m(t) \right\|_{L^2(U)} \le C(\left\| f \right\|_{L^2(0,T;L^2(U))} + \left\| u_0 \right\|_{L^2(U)}) \tag{3.9}$$

Returning once more to inequality (3.8), we integrate from 0 to T and employ the inequality above to find

$$\left\| u_m \right\|_{L^2(0,T;H_0^1(U))}^2 = \int_0^T \left\| u_m \right\|_{H_0^1(U)}^2 dt \le C(\left\| f \right\|_{L^2(0,T;L^2(U))} + \left\| u_0 \right\|_{L^2(U)}) \tag{3.10}$$

Fix any $v \in H_0^1(U)$, with $\left\| v \right\|_{H_0^1(U)} \le 1$ and write $v = v^1 \oplus v^2$, where $v^1 \in span\{\omega_k\}_{k=1}^m$ and $(v^2, \omega_k) = 0(k = 1, \cdots, m)$. Since the functions $\{\omega_k\}_{k=1}^{\infty}$ is an orthogonal basis of $H_0^1(U)$,

$\left\| v^1 \right\|_{H_0^1(U)} \le \left\| v \right\|_{H_0^1(U)} \le 1$. Utilizing (3.4), we deduce for a.e. $0 \le t \le T$ that

$$\left(u_m', v^1 \right) + \int Du_m \cdot Dv^1 dx = \left(f, v^1 \right)$$

Then (3.2) implies

$$\left\langle u_m', v \right\rangle = \left(u_m', v \right) = \left(u_m', v^1 \right)$$
$$= \left(f, v^1 \right) - \int Du_m \cdot Dv^1 dx$$

Consequently

$$\left| \left\langle u_m', v \right\rangle \right| \le C(\left\| f \right\|_{L^2(0,T;L^2(U))} + \left\| u_m \right\|_{H_0^1(U)})$$

Since $\left\| v \right\|_{H_0^1(U)} \le 1$. Thus

$$\left\| u_m' \right\|_{H^{-1}(U)} \le C \left(\left\| f \right\|_{L^2(U)} + \left\| u_m \right\|_{H_0^1(U)} \right)$$

And therefore

$$\int_0^T \left\| u_m' \right\|_{H^{-1}(U)}^2 dt \le C \int_0^T \left\| f \right\|_{L^2(U)}^2 + \left\| u_m \right\|_{H_0^1(U)}^2 dt$$

$$\le C \left(\left\| u_0 \right\|_{L^2(U)}^2 + \left\| f \right\|_{L^2(0,T;L^2(U))}^2 \right) \tag{3.11}$$

Integrating (3.9)-(3.11), we obtain our energy estimate

Step3. Next we pass to limits as $m \to \infty$, to build a weak solution of our initial/boundary-value problem (3.1).

Theorem 3.3 (Existence of weak solution). There exists a weak solution of (3.1).

Proof. According to the energy estimates (3.6), we see that the sequence $\{u_m\}_{m=1}^{\infty}$ is bounded $L^2(0,T;H_0^1(U))$, and $\{u_m'\}_{m=1}^{\infty}$ is bounded in $L^2(0,T;H^{-1}(U))$.

Consequently there exists a subsequence $\{u_{m_l}\}_{l=1}^{\infty} \subset \{u_m\}_{m=1}^{\infty}$ and a function $u \in L^2(0,T;H_0^1(U))$ with $u' \in L^2(0,T;H^{-1}(U))$ such that

$$\begin{cases} u_{m_l} \to u \ \text{ weakly in } L^2(0,T;H_0^1(U)) \\ u_{m_l}' \to u' \ \text{ weakly in } L^2(0,T;H^{-1}(U)) \end{cases} \tag{3.12}$$

Next fix an integer N and choose a function $v \in C^1([0,T];H_0^1(U))$ having the form

$$v(t) = \sum_{k=1}^{N} d^k(t)\omega_k \tag{3.13}$$

Where $\{d^k\}_{k=1}^{N}$ are given smooth functions. We choose $m \geq N$, multiply (3.4) by $d^k(t)$, sum $k = 1, \cdots, N$, and then integrate with respect to t to find

$$\int_0^T \langle u_m', v \rangle + Du_m \cdot Dvdt = \int_0^T (f, v)dt \tag{3.14}$$

We set $m = m_l$ and recall (3.12) to find upon passing to weak limits that

$$\int_0^T \langle u', v \rangle + Du \cdot Dvdt = \int_0^T (f, v)dt \tag{3.15}$$

This equality then holds for all functions $v \in L^2(0,T;H_0^1(U))$ as functions of the form (3.13) are dense in this space. Hence in particular

$$\langle u', v \rangle + Du \cdot Dv = (f, v) \tag{3.16}$$

For each $v \in H_0^1(U)$ and a.e. $0 \leq t \leq T$.

In order to prove $u(0) = u_0$, we first note from (3.15) that

$$\int_0^T -\langle v', u \rangle + Du \cdot Dvdt = \int_0^T (f, v)dt + (u(0), v(0)) \tag{3.17}$$

For each $v \in C'([0,T];H_0^1(U))$ with $v(T)=0$. Similarly, from (3.14) we deduce

$$\int_0^T -\langle v',u_m \rangle + Du_m \cdot Dvdt = \int_0^T (f,v)dt + (u_m(0),v(0)) \qquad (3.18)$$

We set $m = m_l$ and once again employ (3.12) to find

$$\int_0^T -\langle v',u \rangle + Du \cdot Dvdt = \int_0^T (f,v)dt + (u_0,v(0)) \qquad (3.19)$$

Since $u_{m_l}(0) \to u_0$ in $L^2(U)$. As $v(0)$ is arbitrary, comparing (3.17) and (3.19), we conclude $u(0) = u_0$

Theorem 3.4 (Uniqueness of weak solutions). A weak solution of (3.1) is unique.

Proof. It suffices to check that the only weak solution of (3.1) with $f = u_0 = 0$ is $u = 0$.

To prove this, observe that by setting $v = u$ in identity (3.16) we learn that

$$\frac{d}{dt}\left(\frac{1}{2}\|u\|_{L^2(U)}^2 \right) + \|u\|_{H_0^1(U)}^2 = \langle u',u \rangle + \|u\|_{H_0^1(U)}^2 = 0$$

Because of Gronwall's inequality, we can get $u \equiv 0$.

References

1. Hale, J.: Asymptotic Behaviour of dissipative systems. Math. Surv. Monogr. AMS 20 (1988)
2. Henry, D.: Geometric Theory of Semi-linear Parabolic Equations. Springer Lecture Notes in Mathematics, vol. 840 (1981)
3. Matano, H.: Asymptotic behaviour and stability of solutions of semi-linear diffusion equations. Publ. Res. Inst. Math. Sci. 15, 401–458 (1979)
4. Evans, L.C.: Partial Differential Equations. Graduate Studies in Mathematics, vol. 19
5. Friesecke, C.R.: Convergence to equilibrium for delay-diffusion equations with small delay. J. Dynam. Differential Equations 5, 89–103 (1993)

From Marxist Economics to See the Value of Labor Education

Tang Liang

School of Humanities and Laws, Tianjin Polytechnic University, Tianjin, China
tangliang@tjpu.edu.cn

Abstract. Marx's Theory of Labor Value, we can draw on the nature of educational labor the following conclusions: First, education is a labor of labor or service labor; Second, labor education, labor is a commutative; Third, education has an indirect nature of production. Education itself has great productivity features, mainly reflected in the enhancement of human labor and to promote scientific and technological progress and so on. And the educational value of the products of labor can be achieved through education market.

Keywords: labor theory of value, educational labor, value.

1 The Characteristics of Labor Education

To clear the characteristics of labor education must first clear theory of Marx's productive labor. The so-called productive labor, refers to the value can be used directly for production of material production labor. Marx believed that productive labor has three meanings: First, "the labor process from the perspective of a simple up study of the productive labor." The abstract, setting aside its various historical forms , as a process between man and nature to examine the production of labor. It applies to any form of social history of labor under. Second ,the development of collaborative study from the labor productive labor. He said: "With the collaborative nature of development work, production work and its stakeholders, namely, the concept of production workers are bound to expand. To engage in the production of labor, does not have to do it yourself. As long as the total labor of an organ, as a function of it belongs to is enough. "Third , under the conditions of capitalist production relations of production. This is regulated in the labor community, this work includes buyers and sellers of labor between a very determined relationship. Since capitalist production is not only the production, its essence is the production of surplus value, so the "only direct production of surplus value of labor is productive labor."

According to Marx's labor theory of production, the concept of productive labor on the scope of jurisdiction to investigate the nature of education, labor and education nature of the investment, several major conclusions can be drawn as follows:

First, education is a labor of labor or services of labor. The labor or services of labor, can be used for the community to provide special value. The so-called special use value is only active service, and this activity directly into the production process, there is not a production of a non-labor. This work should be said that the production of certain. As Marx said: "Some services are trained to maintain the ability to work,

J. Luo (Ed.): Soft Computing in Information Communication Technology, AISC 161, pp. 405–410.
springerlink.com

ability to change the form of labor, etc., in short, is the ability to work with a specialized, or just keep it the ability to work, such as school teachers of services (As long as he is' essential industries' or useful), the doctor's services (as long as he can to protect health, the source of all values that maintain the ability to work itself) - to purchase these services on their purchase offer 'to sell the goods, etc. other '. that provides the ability to work themselves instead of their own services, which should include the ability to work cost or reproduction cost of production."

Second, education is a commutative labor. Marx believed that education services can be exchanged. He said: "In schools, teachers, the school boss, can be pure wage laborers, such education are too many factories in the UK, these teachers for students, though not production workers, but the boss who hired him are production workers, but the boss with his ability to work the capital exchange of teachers, through this process to make their own fortune. " teachers to be exchanged for the" ability to work "through the educational activities of teachers to reflect, this special" service "commodity form can also be based on the" consumer. "Any time, in consumer products, in addition to the form of consumer goods, it also includes a certain amount of consumer goods to serve the form.

Third, education has an indirect nature of production. Education is to teach people the knowledge and skills, development of human wisdom. Thus education is a direct result of increased knowledge and ability of people, rather than direct access to economic gains. Only when the education process of production of material products and combined into productivity, in order to create a new more material wealth, education is to produce economic benefits. Cost of education is not a direct input into the production process of material, but put into the education process. Cost of education in the educational process not only can not completely get financial compensation and economic efficiency, it cannot simply go directly from the cost of education measured the economic benefits of education; only through material production, in the measurement of material goods can be tested and measured.

2 The Function of Education Productivity

Educational Labor is not only the nature of productive labor, and their great productivity features, mainly reflected in the work to improve people's ability and promote scientific and technological progress and so on.

2.1 Education in the Function of Reproduction of Labor

Social reproduction is the reproduction of material, but also the reproduction of labor. Reproduction of material information only when the reproduction of labor power in the entire production process, in proportion to the quantity, quality, adapt to each other, the social and material reproduction can be achieved. Therefore, the reproduction of labor power is the social and material information necessary for reproduction. The so-called reproduction of labor power, one that restore, maintain the existing workforce; Second, extension and training new workers. The cultivation and training of the workforce must be achieved through education, so education is a means of reproduction of labor power. Education on the reproduction of labor power has the following main functions:

1) Education for human reproduction, the ability to work

Marx pointed out: "Education will produce the ability to work," "reproduction of the working class, but also include the skills taught from generation to generation and accumulation." This shows that education is an important means of reproduction of labor power. Specifically, the meaning and characteristics of labor from these visits .First, from the general sense, the ability to work is a person conquer nature or the ability of the objective world is the human ability to work is the human physical and mental sum; Second, the modern sense, the development of human labor capacity, improve people's ability to work, mainly refers to the development of human intelligence, improve people's scientific knowledge and production technology. As Marx said: "The ability to work so production is because of its values and its difference between the value created." Education can give full play to people's ability to work out, it features the full show, developed. "Education can make the young man soon to be familiar with the entire production system, which enables them according to the needs of society or their own interests, turns from a production department to the other productive sectors." This is mainly because of people's working capacity of the production, development and improvement of social practice is to achieve the day after tomorrow, is obtained through education and training.

2) Education can change people's ability to work on the nature and form

Marx pointed out: "to change human nature in general, give it some labor sector skills and techniques, as advanced and specialized labor force, we must have a certain education and training." Nature of people's working capacity, also known as quality and quality, mainly refers to the manual to improve the intellectual level of a non-skilled labor force to raise a skilled labor force. The manual mainly depends on the intellectual level of education and production technology, including education level and level of education is essential. So, through education, training to the training of unskilled labor, skilled labor force to change the nature of people's ability to work to improve the quality of people's ability to work. Form of education as a service, it can "make work the ability to change shape," "the ability to work with the specialized." Education can be a simple labor, the general nature of the labor force, processing, training as a complex and specialized labor force, that scientists, engineers and technicians. Because the manual through education can be characterized by experience and work skills of the workforce, and to cultivate scientific knowledge into a form characterized by labor. With the development, the role of education in this increasingly significant.

2.2 Education in Scientific Knowledge Shape Productive Forces Reproduction Function

Marx made it clear that capitalist production is that the whole production process is not directly affiliated with the skills of workers, but the performance of the application of science in the process. "Fixed capital development that general social knowledge has become a large extent on how direct the productivity of the conditions of social life itself is in the process of how great the degree of control by the general intelligence and the intelligence received by this transformation. "Marxism these discussions, focus on explaining the large-scale industrial production in the machine under the conditions of social productive forces of natural science itself, is a general social productivity, the productivity of scientific knowledge forms can also be called as a potential productivity. Marx believed that productive manifestations are diverse, there is a " useful labor productivity in specific," there are "general social productive forces"; a material form of productivity, but also the productivity of knowledge, morphology, combined with the production of social workers productivity, but also personal productivity. Form of scientific knowledge productivity belongs to that used in modern production process, productivity in the past, neither the specific labor, nor is it for real process, and only penetrate into the productivity of various elements, the material can be converted into a direct productivity. Thus, the productivity generally permeability, potential and gifts of. Specifically, education is a form of scientific knowledge production, mainly in the following areas: (1) education has passed the accumulation, development and reproduction of the social function; (2) Education has to make science into the intermediary role of production technology; (3) Education is an effective form of reproduction of scientific knowledge.

3   The Realization of the Value of Labor Education

Labor theory of value is the foundation of Marxist economic theory, Marx in his discussion of the value of labor law and also reveals the forms of labor value.

3.1   Educational Expenses is an Integral Part of the Value of Labor

Marx said: "Labor's education expenses with the complexity of the nature of labor force are different. Thus, the education costs - for ordinary labor is minimal - including the cost of labor in the production value of the sum." Education as the form of a service, for educational expenses should also take the appropriate attitude. "You know, where as each phase exchange of goods and commodities, is equivalent for equivalent." With the development of science and technology and its wide application in production, labor in the total value of the ratio of the three components of the structure , will undergo major changes, education and training costs, which will account for a large proportion. Measure of the value of labor, education and training mainly depends on the proportion of the size of cost.

According to Marx's labor theory of value, production of goods and labor has simply divided labor and complicated labor. The so-called complicated labor, it is the need to go through some of the specialized education and training so that workers with certain skills and knowledge of the work of labor. Laborer and his complexity is closely related to education and training, is the performance of such a labor force, this labor-labor needs of higher education than the average cost of its production work to spend more time, so it has a higher value. Since this relatively high value of labor, it also is the more advanced work, also in the same amount of time materialized to a higher value. The cost of labor training, along with the complexity of the different labor different. Therefore, "one hour of labor complicated products work the same hours compared to a simple, is a two or three times higher than the value of the goods." Labor force due to the complex needs of labor higher than the simple cost of education, higher value and, therefore, must be matched by higher wages paid.

3.2 Educational Labor of the Formation of Social and Economic Value

Educational labor to create social and economic value, mainly refers to the fruits of labor education, the intellectual culture of labor, skilled labor and the reproduction of science, and physical production combine to create material wealth. Marx in his discussion of the role of labor under capitalism, said: "This product has a unique feature: it is the power to create value, is the source of value, and - in the appropriate use of time - is worth more than they have the source of value. "This fully shows that labor is the labor force and only the creativity of social material wealth. But labor can create new value, not the value of the natural form, is not a natural process, it is the level of educational labor force are closely related. Labor to create the social and material wealth, in general, is proportional to the degree of educational labor cases. Strengthen education and workforce training, years of education to extend the labor force and improve the education level of labor, then the reproduction of labor will be more new values, social material wealth will increase, the economy will continue to grow.

3.3 The Products of Educational Labor to Achieve Its Value through Market

According to Marxist theory, labor products to become commodities, must have a social use value, and to users through the exchange or market go hand in order to realize the value of products. Therefore, the value of the product of labor education, it should be through the establishment of educational products market - the labor market to achieve. Market is a means of optimizing the allocation of resources and mechanisms, the establishment of educational products market is supply and demand of educational products through the market to realize the value of the product of labor education to achieve optimal allocation of educational resources. Labour is the object of education of labor and product training, but also the field of material production industries and workers, is an important element of productivity. Under the socialist market economy system, the configuration of factors of production need to market regulation. Education and training of labor by the labor supply and industry demand for labor, only by the laws of supply and demand of labor market regulation to resolve. Education, labor training ability to work, must conform to the market (social) needs, combined with production and demand, supply and demand balance. This requires that education must be conducted according to the needs of society, according to the changing labor market and demand set the professional, to adjust learning content, improving educational content and teaching methods to make education work products meet the needs of society, the introduction of market mechanisms, the exchange by value regular paid training, paid use, paid services to make education the value of labor compensation.

Acknowledgment. This article by planning philosophy and social sciences in Tianjin project "Modernization of Higher Education under Globalization and Localization of Some Problems" (No.TJJX10-1-852) for their support in this special gratitude.

References

1. Yuan, Z.: Contemporary Education. Educational Science Press, Beijing (1999)
2. Fan, X.: Economics of education. People's University Press, Beijing (2008)
3. Cui, L.: The economic value of higher education. Shenyang Normal University (Social Science Edition), (6) (2008)
4. Shao, B., Zhao, Z.: Made on China's current economic function of higher education. Alienation Nanjing Institute of Technology, Social Sciences (June 2005)

Theoretical System of Logistics Research on Logical Starting Point

Fukui Li, Xin Li, Weiping Ma, Qiushuang Huang, and Hong Jiao

Military Logistical Department
Academy of Military Transportation
Tianjin, China
741169377@qq.com

Abstract. The full understanding and integration of logistics disciplines theoretical system of the present research results, combined with academic theory system requirements on the logical starting point, analysis of the logistics system as a whole emergence of the formation mechanism of the overall logistics system derived emergence of the theoretical system of the logistics discipline logical starting point.

Keywords: Logistics disciplines, Theoretical System, Logical Starting Point.

1 Introduction

"It is a top priority of Chinese logistical theory and practice to establish and improve theoretical system of logistics research in China". "According to the scientific and theoretical system of the discipline, it is commonly acknowledged that logical starting point should be determined in the first place, starting from logical starting point by virtue of logical approach. Compact logical system is constructed on the basis of the intrinsic laws of the discipline and through derivation and development step by step". Therefore, research on logical starting point of the theoretical system of logistics discipline is of great theoretical and practical significance to the establishment and improvement of the theoretical system of logistics discipline, stimulation of its progress and the instruction of logistical practices.

2 Prescription to the Starting Point of the Disciplinary Theoretical System

As for logical starting point, Hegel did specialized exposition in the 19th century. He points out that "It is a difficult task to find out the starting point of philosophy." He could not solve this problem intrinsically due to constrains of mentalistic world outlook and methodology. Successfully settlement of logical starting point was realized during the process of the creation of Capital by Marx. When confirming logical starting point, Marx absorbed reasonable factors of logical starting point theory by Hegel critically and abandoned Hegel's mentalism outer skin and took

J. Luo (Ed.): Soft Computing in Information Communication Technology, AISC 161, pp. 411–417.

commodities as logical starting point of Capital. But this process was not that smooth. It took shape in three phases which include labor, value, and commodity. By then, problems failed to be solved correctly by many bourgeois economists in spite of arduous exploration could be solved then scientifically. Owing to the confirmation of logical starting point, Capital was provided with rigid scientificity and rigorous logicality. Combined with the two's point of views, especially the process of concluding commodities as logical starting point of Capital, we can reach the conclusion that prescription of logical starting point includes the following several aspects.

Logical starting point should be the objective basis for the establishment and development of the integrated theory. "It can not take anything as premises and medium. Based on no foundation, it itself is the scientific basis of entire science." It contains potentially whole rich contents, forms, categories, laws and concepts as well as whole amount of information of entire theoretical system. Gradual stretch starting from the beginning of rich contents of entire system demonstrates potentially rich contents of the most abstract categories as logical starting point.

Logical starting point reflects intrinsic prescription of the study object. In other words, it is formed by leaving aside various complicated phenomena and extracting object's intrinsic attributes. It sets aside various single and occasional phenomenal forms to the maximum. It constitutes its internal relations departing from nature and reveals all kinds of phenomena from the basis of the essence of objective substances.

Starting point and finishing point is dialectically united. Marx agrees with this point in capital circulation: "Each point manifests starting point and finishing point simultaneously and only when it acts as finishing point it presents as starting point. The process varying from single concreteness of starting point to finishing point is not simple regression but a series of transitions and transformations of continuous expanding curve which appears like helix; the starting point of one process is at the same time the finishing point of another process. Starting point and finishing point is united.

Logical starting point is united with historical starting point. Logical starting point of one discipline should also be the historical starting point which reflects the discipline's objects. Things served as the beginning of scientific theory is the rudiments in history. Just as Engels had said: "Where begins history begins ideology course". "Process begins from simpleness to complexity corresponds with practical historical process". Therefore, coherence of the logical starting point and the historical point is fundamentally the exact expression of matter determines ideology and existence determines thinking.

With the above-mentioned prescription of logical starting point, searching for logical starting point of logistics theory system then has an instructive principle method. We abstract all kinds of complex phenomena of study objects which enable abstract categories correspond to the above-mentioned four prescriptions of logical starting point. And then logical starting point of the discipline can be grasped by and large.

3   Whole Emergence of Logistics System

Professor He Mingke points out in his representative work Logistics System Study "deep research should be conducted in several aspects like formation mechanism of the logistics' system as a whole and conditions that unity is larger than the combination of parts. This research is of greater importance to logistics system." This kind of more significant research is the whole emergence of logistics system.

3.1   Connotations of Whole Emergence

Acquiring knowledge of logistics whole emergence should at first determine connotations of whole emergence. Whole emergence simply means unity's function is larger than that of the combination of parts. It shows by mathematics formula as follows:

$$F > \sum_{i=1}^{n} f_i \tag{1}$$

In this formula, F represents functions of the entire system, f_i represents functions of components contained in the system. Systematic scientific theory and approach aims at the realization of formula (1) and the more the left side larger than the right side the better. "Larger" here can also be substituted by "much more", "higher" and "better".

Take logistics systematic optimization hypothesis as an example: suppose that the goal of one logistics system is to minimize logistics cost C under given condition. In this logistics system, there are many factors like $x_1, x_2, \cdots, x_n$ that may affect cost objective of logistics system. These factors affect the cost in various ways. Formula (2) can illustrate their influences to cost and cost objective itself:

Objective function: $\min c = d_1 x_1 + d_2 x_2 + \cdots + d_n x_n$

$$\begin{cases} a_{11}x_1 + a_{12}x_2 + \ldots + a_{1n}x_n \leq b_1 \\ a_{21}x_1 + a_{22}x_2 + \ldots + a_{2n}x_n \geq b_2 \\ \cdots\cdots\cdots\cdots \\ a_{m1}x_1 + a_{m2}x_2 + \ldots + a_{mn}x_n \leq b_m \\ x_1, x_2, \ldots, x_n \geq 0 \end{cases} \tag{2}$$

Constraint condition:

In this example, pursuit of minimum cost C is to make minimum cost C after logistics systematic integration superior to previous cost. Of course, the pursuit of the maximum efficiency index can be achieved by limiting other conditions.

3.2   Formation Mechanism of Whole Emergence

Systematic whole emergence is not mysterious. In the final analysis, it originates from four effects: constituent effect, scale effect, structure effect and environmental effect. Analysis on the cause of modern logistics concept is the result of these four effects' comprehensive action.

1) Constituent Effect Mechanism

The so-called constituent effect means that systematic whole emergence ultimately comes from its elements or constituents. Systematic whole emergence was restrained and specified by special attributes of the systematic elements or constituents. Random selection of constituents can not constitute system of specific whole emergence. Elements and constituents of the system are the basis for the establishment of and key to the improvement of the whole emergence. The quality, attributes, strong points and short points are the real foundation for the creation of systematic whole emergence. Once the constituents are fixed, possible range of the whole emergence in the system, what it may be and what it may not be is determined. Isomerism in chemistry is widely used by people to expound that systematic whole emergence has nothing to do with the constituents of the system. But the fact that the possibility of producing new chemical property is limited by simply changing structure is ignored. For example, protium and oxygen synthesize water but salt can not be synthesized no matter how the constituents are changed. Chlorin and sodium can synthesize salt rather than water. Take another example, if the constituent "helicopter" is not available, emergency materials badly needed could not be sent to the disaster-stricken area in the urgent logistics support during the time of Wen Chuan Earthquake. Comparisons on the tremendous changes taken place before and after the application of modern information technology to logistics system can better appreciates the importance of constituent effects. Emphasis on the role of constituent here may seem paradoxical to systematic whole emergence. In fact, it isn't paradoxical at all. It is because every time when we speak of systematic whole emergence, it already met the need of the formation of whole emergence basically. It is the materialistic basis for systematology. If constituent problem left unsettled, the establishment and improvement of logistics systematic whole emergence is impossible. Constituent effect mechanism is the fundamental mechanism.

2) Scale Effect Mechanism

The so-called scale effect mechanism means only when scale reached certain level can complexity and much more emergence be produced. In the first place, scale means the number of constituent types that constitute the entire system. In the second place, it means the number of individual constituent. Constituent effect plays a key role in deciding the formation of systematic whole emergence by taking advantage of attributes of the constituents; While scale effect affects systematic whole emergence with the aid of having constituents or not and its number. Systematic whole emergence is concerned with systematic scale. Different scale results different emergence. Influence from the size of the scale to whole emergence can not be neglected. Scale effect is related to scale economy, but is not the same concept. Scale economy is one form of emergence. Scale effect is an emergence law in wider range that includes scale economy. For example, conventional transportation enterprise can transform into modern logistics enterprise by integrating other logistics resources. Expansion of its own scale can it provides with integrative high standard service and obtains more profits than before. In addition, packaging containerization in logistics workings is another manifestation of scale effect. Therefore, focus should place on scale effect when observing, analyzing, devising, building, processing and operating system. Scale effect mechanism is important mechanism.

3) Structure Effect Mechanism

The so-called structure effect means when systematic elements and constituents are under certain occasion, different organizational structure between them and concomitant different action mode would produce different emergence. Systematic elements or constituents are the real foundation for producing emergence. But possibility can only turn into reality by letting different elements or constituents interact, stimulate, restrict and complement each other. Practical whole emergence is produced on the basis that different constituents correlate, interact, restrict and stimulate with each other. The way of mutual correlation, interaction, restriction and stimulation is systematic structure. The saying "systematic structure decides systematic function," goes in systems science is pointing at structure effect mechanism. Systematic structure effect is of great significance to the creation of whole emergence. One logistics concept "separation of business and commodity" which makes it possible for the rationalization of logistics flowage structure is one type of structure effect mechanism; another concept "third part logistics" which makes it possible for the optimization of logistics structure is one type of structure effect mechanism; when scanning Logistic Terminology, concepts like "logistics", "logistics center", "delivery" all embody reasonably structured concept. Various logistics technology used up all kinds of means and approach to make logistics system optimized. Structure effect mechanism is the core mechanism.

4) Environmental Effect Mechanism

The so-called environmental effect mechanism means the phenomenon of improving whole emergence during the process of the communication between the system and environment. Systematic whole emergence is largely depends on the environment the system locates. Systematic whole emergence depends not only on internal constituents and structure but also on external environment. On one hand, the formation, maintenance, evolution and development of the system requires resources and conditions coming from the environment. On the other hand, environment not only provides resources and conditions but also imposes constraints and restrictions on shaping the system. The above-mentioned two aspects would engender environmental effect which is indispensable to the formation of systematic whole emergence. Systems science holds that environment is the reference system of how systematic elements is related, correlated and interacted as well as whether one structure is valid rather than the determination of the system itself. The guiding principle is how to adapt to and make use of the environment as far as possible. When the enterprise transforms from "vertical integration supply chain management" to "horizontal integration supply chain management", relevant logistics presents supply chain logistics pattern; Restrictions imposed by national environment protection laws and regulations produce green logistics form. These all are effects environmental effect has on logistics system. Logistics system must adapt to or ameliorate the environment it locates to make it better for the performance of the above-mentioned three effects. Environmental effect mechanism is restrictive mechanism.

The reality is that influences from the above-mentioned four effect mechanisms to logistics systematic whole emergence are not sporadic. They propel the creation of whole emergence through mutual effects. This is the effect mechanisms that produce logistics systematic whole emergence.

4 Logical Starting Point of Logistics Disciplines Theoretical System Is Logistics Systematic Whole Emergence

During the process of recalling the history of the understanding and acceptance of Chinese academies of logistics, Professor Wang Zhitai said, "A very importance foundation for Chinese academies to understand and accept modern logistics lies in that Professor Qian Xuesen's idea about systematology had been widely accepted by Chinese academies previously and at the same time understanding of foreign logistics had been accepted through various medium and cultural channel. Logistics, the unique system has been understood in the perspective of systematology. In fact, new concepts embodied in logistics can be understood easily. What is the new concept? It is a systematic concept that places packaging, transportation, assemblage, delivery safeguarding and so forth which was regarded as independent and irrelevant activities untied under logistics system as a systematic elements. When integrated, people indeed felt that it is suitable for those activities which correlate with each other so closely to be put in a unified system. Though simple, the concept has leapfrogging meaning, it is the reason why logistics discipline being so charming. The concept on systematology which appears as simple but has undergone longtime practical test and theoretical deduction is logistics systematic whole emergence—the logical starting point of logistics disciplines theoretical system. Rigorous logical system of is the purpose of establishing and improving logistics disciplines theoretical system.

Whole emergence is the objective basis and foundation for establishment of logistics theory. Western scholars had put forward many meaningful theories and viewpoints worthy of borrowing and pondering in the more than one century of logistics theory. There emerged many schools of thoughts such as management school, engineering school, military school and entrepreneurial school. Though there are huge differences and disputes between those theories and viewpoints, they were found on the basis of longtime logistics management practice conducted by human beings and through theoretical summarization and abstract sublimation. From the definitions of each school of thought, it is not difficult for us to figure out there is a fundamental thing shared by all. That is "systematization"—whole emergence. Whole emergence potentially contains all the rich contents, forms, categories, laws and concepts of logistics theory as well as entire information of the whole logistics process and system.

Whole emergence is the essence of logistics theory. The essence of logistics theory is the internal and consequent prescription of the logistics theory itself, which is fundamental attributes. Starting from logistics definition and connotation analysis, we hold that the internal attribute of logistics is whole emergence. Whole emergence is the simplest form of logistics system. It is the thing that can not be divided another time. In the meantime, whole emergence is the most abstract thing. It has not any detailed prescription. It is abstracted from all kinds of detailed logistics activities and can be applied widely to every aspect of logistics activities.

Whole emergence reflects the dialectical unity of the starting point and finishing point of logistics theory. Logistics activities originate from the integration of all kinds of logistics elements. As a result, matter's integration corresponds with logistics goal and well-organized circulation formed in the process. As the logical starting point of the logistics disciplines theoretical system, whole emergence manifests the dialectical unity between the starting point and the finishing point. As the finishing point, whole emergence is the aim of starting point, as the starting point, whole emergence is realized in finishing point.

Whole emergence manifests the unity between the logical starting point and historical point of logistics theory. In light of preliminary logistics concept, analysis on logistics theory history also starts from whole emergence in the beginning. Therefore, it is beyond all doubt that regarding systematic emergence as the logical starting point of logistics disciplines theoretical system according to the principle that logical starting point is integrated with historical point.

5   Conclusion

Logical starting point of logistics disciplines theoretical system is logistics system's whole emergence. Any discipline has its own logical starting point. Only through the confirmation of logical starting point can research be conducted step by step, can the discipline has its own strict logical system and solid theoretical basis and can the discipline becomes integrated science system. This article only illustrates the logical starting point of logistics disciplines theoretical system. "Theoretical system" we mean here is only knowledge system, experience system or theory. To do logistics theoretical research in a long round is to accumulate "knowledge system, experience system or theory". Only when this type of accumulation is abundant can rigorous discipline is formed by starting from logical starting point and be expanded gradually through dependence on the accumulated materials.

References

[1] He, M.: Logistics System Study, p. 105. Higher Education Press (2004)
[2] Hu, J., Zhou, C.: New concept of Higher Education, p. 2. Jiangsu Education Press (1995)
[3] He, M.: p. 134. Higher Education Press (2004)
[4] Wang, Z.: New Edition of Modern Logistics, p. 38. Capital Economy and Trade University Press (2005)

The Deepening of International Labor's Division and Transnational Human Resource Management

Lin Zhang and Ming-fu Cao

Tianjin Polytechnic University, Tianjin, China
linlinzhang1@yahoo.com.cn,guangmingxingfu@126.com

Abstract. As the deepening of international labor's division, many of our enterprises start to "go out" and try to become multinational corporations, so how to improve our human resource management is very important. In view of this, this paper describes the new development of transnational human resource management which are promoted by the deepening international labor's division, then illustrates the challenges which also owe to the deepening the international labor's division, finally this paper gives some suggestions to improve our level of Transnational Human Resource Management.

Keywords: international labor's division, transnational human resource management, human resource management.

1 Introduction

With the development of productive forces, the traditional international labor's division began to change into the global international labor's division. The global international labor's division and the traditional international labor's division show different characteristics, the traditional international labor's division based on factor endowments, the global international labor's division based on process, segment, part of the foundation; the traditional international labor's division is mainly labor's division among industries, the global international labor's division is mainly labor's division based on product specialization; the traditional international labor's division is only involved in the production areas, the global international labor's division is involved in the production, service and other areas. The deepening of international labor's division will inevitably lead to economic ties and trade more closely between countries. In this context, foreign multinational companies have sprung up rapidly in every corner of the world, Chinese enterprises are also trying to set foot on the road of multinational operations, however, the case of business success in China has few, especially in the face of different culture traditions, values, social relations, political views, labor relations, education and other differences, Chinese enterprises have many outstanding problems in the international human resource management. In addition, the practice of multinational companies in developed countries shows that successful companies rely more than 60% on international operations management, and human resources management is the most important in management. Therefore, it is very urgent to strengthen the research of Chinese enterprises' transnational human resource management.

J. Luo (Ed.): Soft Computing in Information Communication Technology, AISC 161, pp. 419–424.
springerlink.com   © Springer-Verlag Berlin Heidelberg 2012

2   The Meaning of Transnational Human Resource Management

Broadly speaking, transnational human resource management and domestic human resource management are engaged in the same activities. However, domestic human resource management considered only employee issues within a country, transnational human resource management has to consider the operation in different countries and the recruitment of employees of different nationalities. In other words, there is no significant difference in the activities of human resource between transnational human resource management and domestic human resource management. From the perspective of international labor's division, the reason which multinationals carry out international operation is to break the home country resource constraints, get resources from around the world, and maximize the utilization efficiency of resources. Multinational corporations operating abroad are faced with differences of political, economic, cultural, market and customer conditions, so multinational corporations have to give their proper and positive response for various demands, which requires multinational companies in human resources management to take specific measures.

Therefore, transnational human resource management called international management of human resources, is the human resources activities, staff type which includes the host country staff, the home country employees and the third country workers ,and the type of country business which are the host country including enterprises' subsidiaries or branches, the home country containing enterprises' headquarters, the third-country offering labor or funds, among the three dimensions of interaction combination.

3   The Deepening of Intenational Labor's Division Promoting the Development of Transnational Human Resource Management

3.1   Accelerating the Strategic International Human Resource Management Transformation from Theory to Practice

Since 1990s, through summarizing and analysis of their own multinationals' problems which were encountered in the development, Foreign scholars turned the micro, qualitative level to the strategic level on the research of transnational human resource management. The deepening of international labor's division, making strategic international human resource management as a new research branch of transnational human resources management got the growing attention by the domestic and foreign scholars.

Strategic international human resource management is not a simple combination of human resources activities, staff type and type of the country business. and its implementation need to consider many factors, such as different cultural values and social values, management systems from a cultural adaptation to another, different economic and cultural environment, different learning styles from cultural differences, more than that, these factors must be considered as a system or a whole, so strategic international human resource management is a new level of transnational human resource management, more complex than the traditional human resources management, and its wide application must be able to get a clear competitive advantage, when multinational corporations operate in the world.

The deepening of international labor's division makes the country open to a deeper level, the market barriers is opened layer by layer. Facing with this opportunity, more enterprises are embarking on a path of transnational business. however, opportunities and challenges exist simultaneously, enterprises which have adopted traditional human resource management mostly ended in failure, these experience of failure led some multinational corporations recognize the problems in human resource management, and speed up the strategic human resource management transformation from theory to practice.

3.2 Improving Transnational Human Resource Management Model

Transformation of transnational human resource management model is inevitable development trend of human resource management. Many factors affect the choice of human resource management models, such as host country policies, the host country's management, education and technology level, the nature of their products and production technology features, enterprise organization and enterprise product life cycle, national cultural differences. Currently, Most representative transnational human resource management models has the following four: Nationalism center model, Multi-center model, Regional center model, Global center model.

However, all these four models have some flaws. in nationalism center model, subsidiary managers find it difficult to coordinate the contradiction between the provisions of the headquarters and employees of the host country; in multi-center model, the allocation of global resources is inadequate; in regional center model, communication and coordination are very limited between each region and headquarters; in global center model, There may be a conflict of interest in subsidiaries and corporate headquarters. because of the deepening of international division of labor, multinational corporations must select the most appropriate management model as soon as possible, and gradually formed a global center model-based, nationalist center, multi-center, regional center model combination situation.

4 The Deepening International Labor's Division Increasing the Challenges of Transnational Human Resource Management

4.1 Fiercer Competition for International Talents

Most countries need a huge demand for talent for the deepening of international labor's division and national differences in economic development levels. An urgent need for a lot of talent in developing countries who are slow in the growth of wealth to change their backward situation; urgent need for developing countries who are rapid in the growth of wealth to add a large number of high-level talents to complete their industrialization, modernization and Adjustment of industrial structure; developed countries who have the huge economies of scale also need to add a lot of talent just in maintaining their economic growth levels. In addition, the population of Canada and Australia are scarce, the European countries and Japan, have the problem of aging population, these countries need to add a large number of foreign personnel just to maintain scale of the existing talents.

Multinational business activity involves two or more countries, on the one hand to some extent it complements the missing of business countries, host countries and third countries in the levels and types of talents, on the other hand, it intensifies the competition for talent among the business countries, host countries and third countries, especially the high-level talents such as high-tech talents and senior management personnel. In addition, from corporate strategy point of view, multinational enterprises need to implement a new strategy to expand foreign investment, including expanding the scale of operation, merger brands, developing new products, the establishment of branches, and the key to achieve these objectives is the development and deployment of international talents. In a word, international competition for talent is not avoidable and fiercer.

4.2 More Obvious Cultural Conflict

Human resources management is largely influenced and constraint by a national culture. Enterprises' management activity is a conscious activity which it will be decided by the values, mode of thinking and social customs. Culture impacts virtually on the enterprise's recruitment, promotion, performance evaluation methods and a series of human resource management policies. So every country will form a unique way of human resource management under their own cultural background, making the same policy in different cultural backgrounds may produce different results.

Since the objective existence of cultural differences, when enterprises in different cultural backgrounds do transnational business, it will face the friction and collision from different cultural systems and cultural conflict happens at last. The so-called "cultural conflict" means the processes what culture or cultural elements of different forms are contradictive and mutually exclusive. From the perspective of internal business, the smaller the degree of combination between multinational corporations and local businesses, the greater the gap between cultural traditions, the more problems, cultural conflicts are more intense. the deepening of international labor's division causes many new multinationals emerging, if these new multinationals don't attach importance to those details what are associated with culture, cultural conflict will happen inevitably within multinational corporations.

5 The Measures of Improving the Level of Chinese Enterprises' Transnational Human Resource Management

Compared with developed countries, Chinese enterprises' management level is lower, and enterprises' transnational business has just started. so our enterprises must take full advantage of the opportunities what the deepening of international labor's division brought, and eliminate or avoid the challenges what international labor's division may bring, achieve the international transition of human resource management functions, in order to improve the level of Chinese enterprises' transnational human resource management.

5.1 Recruitment

In recruitment, multinational corporations must fully consider their policies what the headquarters choose the personnel of key positions for, because the choices of different policy determine how the enterprises' specific mechanisms run. If the multinational give priority to consider organization's control, they can use National Center Policy which officers in all the key positions should come from the home country, and the policy is mainly used for the early stages of internationalization. When the headquarters hope to fully mobilize the enthusiasm of their subsidiaries, giving them more autonomy and flexibility, and taking into account the cost of hiring staff and cultural differences, they often take the recruitment policy of Regional Center model or Multi-center model. The recruitment policy of Global Center model is to select the best in the entire organization to serve as key staff positions regardless of their country, so multinational corporations can be formed from an international senior management team. In short, the corporate recruitment policy must adapt to the process of its international strategy, in order to effectively promote the internationalization strategy.

5.2 Talent Selection

Multinational enterprises' talent selection can be divided into two aspects: Firstly, expatriates' selection from the parent company to the subsidiary; secondly, employees' selection from the host country subsidiary to headquarters. The above two methods are essential for multinational enterprises. To avoid and reduce the risks of failure abroad, Human Resources Department in the choice of expatriates should be taken into account the following factors: expertise, cross-cultural adaptability, family, staff personality. In addition, Language ability is an important selection criteria to deploy employees from subsidiary, the candidates should know how to use the work language to communicate and master the language of the country of the parent company.

5.3 The Performance Management of Expatriates

In the system of performance management, identification of target and performance evaluation are two key factors. Goal is the criteria of performance evaluation, its specification and measurability is particularly important. Therefore, Human resources departments should have a rigid target as the main target and use the soft target to support the rigid target, especially in some aspects that can't quantified. Furthermore, considering the complexity of cross-cultural, the assessment of expatriates should be completed by a variety of staff together, including direct supervisor, expatriates themselves and human resources managers, in order to avoid the unfairness which are brought by a class of people.

5.4 Cross-Cultural Training and Incentives

Multinational enterprises' staff training is very important, especially in cross-cultural training. Cross-cultural training is to strengthen the awareness of the response and adaptability of different cultural traditions, promote communication and understanding from different cultural backgrounds, address cross-cultural conflict, and is the most basic the most effective means for improving cross-cultural integration of human resources. Generally speaking, the main content of cross-cultural training should include: national culture and its former corporate culture, cultural sensitivity, adaptability of the training, language training, cross-cultural communication and conflict management training, the local environment simulation. The specific purpose of cross-cultural training are the following: reducing the cultural conflict which the local manager may encounter, so that they can adapt quickly to local environment and play a normal role in the enterprise; promoting local employees to understand the business philosophy and customary of company; maintaining good and stable relationships within the organization; keeping the smooth flow of information within the enterprise and the efficiency of decision-making process; strengthening team spirit and company's cohesion.

What's more, the incentive of employees is an important content of international human resource management. In fact, the training which company provided not only shows the good faith that company wants these employees to serve for them in a long time, but also is a way of incentives for the best employees.

Acknowledgment. This work is supported by the Ministry of Education Humanities and Social Science Programs "Benefit Allocation of Global Value Chains Specialization". (No. 07JC790023).

References

1. Liu, G.-Y.: Multinationals Human Resources Management Model. China's Collective Economy (9) (2004)
2. Gan, Y.-X.: Transnational Human Resource Management. Wuhan University of Technology (November 2002)
3. Hu, H., Zhang, L.-M.: Dynamic value chain and international human resource management. Decision Consulting Newsletter 1 (2007)
4. Liang, J., Li, B.-.Y, Dai, Y.: The process of Chinese enterprises in international human resource management strategy in the. Business Research 19(327) (2005); overall
5. Xia, X.-Y., Wang, Y.: The deepening of international labor's division and national economic security. Huazhong Normal University (Humanities and Social Sciences) 40(6) (November 2001)
6. Shao, F.: Multinational Human Resource Management Review. Economic Forum (17) (September 2009); the total section 465
7. Liao, M.: Cross-cultural context of human resources management. Modern Finance (3) (2008), (280) overall

Analysis of the Role Played by Garment CAD Technology in China's Garment Industry Development

Baoqin Sun

Shandong Silk Textile Vocational College
Zi Bo, Shandong Province, China
baoqin11@163.com

Abstract. As a modern high-tech design and production technology, garment CAD technology has been recognized and applied by a growing number of garment enterprises and applications. This article analyzes the active role played by garment CAD technology from various aspects in the development of China's garment industry, discusses the existing shortcomings and solutions and explain the three-dimensional garment CAD software technology will play a greater role in the future development of China's garment industry.

Keywords: garment CAD, industry, computer, role.

1 Introduction

With the rapid development of computer technology, the computer-aided design is emerging one after another and widely used in industry, commerce, design, entertainment and other fields. With the highest degree of globalization, Garment industry has developed a variety of sophisticated techniques for the collection as a whole. Also the computer-aided design in garment industry has been widely used and played an important role.

Garment CAD, the abbreviated form of Computer Aided Design, is a modern high-tech specialized for costume design with the knowledge of computer graphics, database, network communications and other fields. With the support of computer hardware and software system, the costume designer designs on the screen through human-computer interaction means. It is the main form of modern garment design with seamless integration of computer system functions and costume designer's idea, experience and creativity.

Garment CAD was first developed in the United States during 1970s and now its penetration rate has reached 90% in the United States, Japan and other developed countries. Garment CAD technology in China started late and even with fast growth there is still a big gap between the technologies of foreign countries and China. As the development of China's garment industry and the arrival of branding, more and more China's garment enterprises are applying garment CAD, but our penetration rate is still far below the European and American developed countries'.

J. Luo (Ed.): Soft Computing in Information Communication Technology, AISC 161, pp. 425–430.
springerlink.com   © Springer-Verlag Berlin Heidelberg 2012

2 Structure of Garment CAD System

Garment CAD system includes hardware and software. The hardware consists of computers, input and output devices. The common equipments are digitizers, scanners, plotters, printers etc. Software has covered the three parts of costume design, namely, style design, structure design and process design, and garment CAD software is generally equipped with several modules which are responsible for the completion of their respective specific design tasks. Summarized in the following, there are several areas of functionality.

1, style design module: This module is designed to supply professional software for costume designer to design garment style. 2, the sample design modules: These modules are professional software designed specifically for sample design and production. 3, sample grading module: This module is based on a sample basis of the garment to complete the scale and drawing of a model of grading. 4, the garment nesting module: This module is a system of using computer to make nesting. There are two ways: automatic nesting and human-computer interactive nesting. 5, three-dimensional simulation Fitting System Module: This module is the most advanced computer simulation technology, providing the designer a powerful three-dimensional garment simulation tool with powerful function of stitching and draping to convert the flat pattern into three-dimensional wear effect quickly. The computer can also store large amounts of clothing styles and other information for customers to browse, select and try.

3 The Active Role Played by Garment CAD Technology in China's Garment Industry

Garment industry is a labor-intensive industry. The operation of garment enterprises in all aspects requires considerable human resources; the ability of new technologies will directly affect garment enterprises in the growth and development. Garment CAD technology, not only saves human resources and cost for the enterprises, but also improve product quality to enhance their core competitiveness of the market; at the same time, it plays a significant role in promoting the development of intelligent integration for the traditional garment industry.

3.1 Shorten the Development Cycle of New Products, Improve Design Quality and Enhance Work Efficiency

Development cycle of new products includes new products' design cycle and production cycle. Designers use the garment CAD not only to design new garment styles with functions of style design, pattern color, fabric design etc, but also to bring up the original data changes with clothing styles, colors, fabrics, patterns can be mixed and matched at will quickly and conveniently. This allows designers drawing from the liberation of heavy labor of drawing and full playing to their imagination to improve design quality and speed. Garment production cycle mainly depends on the cycle of the technical preparation, especially for the features of many varieties and small batch production of clothing. Garment CAD system can greatly improve the efficiency of labor with its functions of plate making, plate grading and nesting. According to relevant statistics, with application of garment CAD, product design cycle can be shortened ten times or dozens of times and the production cycle can be reduced by 30% -80%. As a result, the product development cycle has been significantly reduced and labor productivity has been greatly improved, then companies will have energy to make the replacement product to meet the fast changing market, thereby enhancing the company's own vitality and competitiveness, at the same time satisfying the consumption patterns of fast popularity and short period.

3.2 Save Manpower, Material and Financial Resources, Reduce Production Costs and Labor Intensity to Increase Economic Benefits

Garment industry is processing enterprises; the production cost is an important factor in determining production efficiency. Raw material consumption and labor costs account for a sizeable proportion in production costs. The application of CAD system can make the production simple and efficient while saving labor. One staff is enough to complete the work with CAD system more quickly by reducing labor intensity and more accurately which needs a few people originally. Especially for grading and nesting, it can improve material utilization to reduce cost through the human-machine interaction. And that is the feature of CAD system which is used largest and most useful. The computer system can store a large number of styles and nesting model with its characteristics of computer memory capacity. Information can be easily take out from the repository when needed to query the corresponding effect diagram, sample grading map, nesting material and customer files. And all the diagrams can be readily typed out by graph plotter for easy storage and use.

3.3 Improve the Level of Management in Modern Enterprises; Promote Capability of Rapid Response to the Market

Improving the level of modernized management is also one of the outstanding problems for garment enterprises. The improvement depends on the replacement of concept and technical means. Paper pattern is an important technical resource. And it is obvious that the use of CAD system can create paper patterns efficiently with good quality and low cost. It not only improves the means of the management, but also updates concept of the enterprises.

Application of garment CAD technology can improve the working environment, and at the same time, reduce labor intensity and increase the design quality. You can also predict the production data and production plans, lower inventory funds, connect effectively with the international market, transmit data easily, control production costs, reduce job stress and administrative work, build an effective data management and query and respond quickly to the market change in order to increase satisfaction and service quality.

4 Problems and Countermeasures in the Promotion of Garment CAD Technology in China Garment Industry

4.1 The Lack of Talents in the Application of Garment CAD System

Garment CAD is a high technology which needs to operated and managed by some relevant technical staff. However, because some enterprises are not familiar with garment CAD, the purpose of application is not clear and the use of measures is not effective, some of the equipment is useless. The quality of operator in some of enterprises is too low that the function of all aspects of CAD system can not play its due role. Particularly within the enterprises there is a serious lack of dual talents who are very good at both garment and computer. In style design, designers do not use CAD technology to make original design, that is because the design with the CAD system is accomplished by drawing style to the screen by moving the mouse and transforming the screen and generating the graphics, which makes the performance of artistic thinking too complicated. In that case, CAD design system plays only the role of a database.

Countermeasure: As the main channel of training professional talents of garment, Apparel Training Colleges should contribute to the promotion of garment CAD in China by cultivating practical talents of CAD technology. Modular design of the courses should be realized in teaching. That means make sure the various modules of garment CAD will be penetrated into all courses. Teaching methods should be more appropriate to the actual business and simulate actual production. At the same time, in order to make more people learn and use garment CAD, we can increase training opportunities for employees in the community.

4.2 The Application of Garment CAD System Is Not Deep and Broad Enough

Garment CAD in developed countries has applied to style design, production, sales etc, and gradually into the daily business management. At present, in China garment CAD is mainly used in design and production, or even in the plate grading and nesting wastefully. Designers in some enterprises make sample piece by hand first, and then lead it into computer through digitizer to make changes, grading etc. Some companies use the CAD software, but there is only one staff to operate or the software is just applied in only one sector. In fact, the software itself does not play the desired effect.

Countermeasure: enterprises of different types should choose an accurate garment CAD system based on the features of their respective product. For CMT enterprises, they are generally required to accurately process a certain number of garments in a short time according to the sample given by customers. Functions in CAD system like plate making, grading and nesting can accomplish all these work accurately and quickly and shorten the cycle of production. SO CMT enterprises should choose the system of sample design, sample grading and garment nesting. For enterprises who commit the whole process of design, production and sales it is better to choose a full set of CAD system; garment CAD system can give full play to their creativity and information storage advantages in style design, pattern design, and designers are free to use the original design data to improve the design quality and speed.

5   Three-Dimensional Garment CAD Software Technology in the Future Will Be Widely Used in Garment Industry

Presently there are two categories on foreign markets in the application of three-dimensional garment CAD online. One is tailored for accurate: According to measurement of body parameters for the client and their specific requirements for clothing styles, the appropriate plane clothing samples is generates on the basis of accurate measurement and full communication for the costume design. The business achieved by Internet remote control to make it possible that people can enjoy the efficient service from the world's leading designers with their three-dimensional CAD system staying at home. This Web-based apparel design systems is more advanced in the United States, Britain, France, Germany, Japan and Switzerland. One is fitting for visual simulation: get the three-dimensional data collection of customer's body first, make interactive fashion design and last generate the plane sample to product it. Implementation of such business is also using the Internet-based remote operation of e-commerce. Now, three-dimensional garment CAD software in the United States, Japan, Switzerland and other countries has been basically realized three-dimensional garment fitting which can be designed and modified to make functions in the following possible: animation to reflect the comfort of garment, three-dimensional effect simulation of different fabric draping and 360 degree rotation etc.

China is a big country of garment, but the application of advanced technologies in terms of garment started late. Domestic research in this field has made initial progress in achieving imitative three-dimensional CAD design, but the level of technology is still far behind abroad. There is considerable room for development and broad market prospects in China's garment industry. Thus three-dimensional garment CAD software technology will be widely used in the future in China's garment industry.

6   Conclusion

In summary, garment CAD technology has played a very good role in promoting development of China's garment industry. But compared with developed countries, there is still a wide gap. Speeding up the promotion and application of garment CAD technology is the key to push technological updating of garment enterprises and also the means to change the status of China's garment enterprises. Only when we strengthen the promotion and application of garment CAD technology can we enhance their core competitiveness and effectively promote the development of China's garment industry in order to achieve the early realization of the dream to be the strongest garment power.

References

[1] Zhang, W. (ed.): Garment Design and Engineering, vol. 6. Donghua University Press, Shanghai (2003)
[2] Tan, X., et al. (eds.): Garment CAD, vol. 11. China Textile Press, Beijing (2006)

[3] Li, X.: Building of Garment CAD Technology Evaluation System, vol. 27(2). Textile Research (2006)
[4] Chen, X.: Present situation and Countermeasures On the application of garment CAD in domestic enterprises. Scientific & Technical, 04 (2009)
[5] Chen, Y.-H.: How to Improve the application of garment CAD in the domestic garment enterprises. Science and Wealth, 07 (2010)

Cultural Reflection of Experiential Marketing at the Underwear Stores

Aiqing Tang

Shandong University of Technology
Zi Bo, Shandong Province, China
tangaiqing11@163.com

Abstract. Experience elevates emotions and values that are conveyed to consumers by brand culture. With the enhancement of underwear branding in this brand era, it is crucial for an enterprise to enable its consumers to experience the brand culture and understand the brand connotations. This thesis emphasizes the importance of the new experiential marketing mode in the revealment of brand value and the embodiment of consumable brand culture by discussing the applications of experiential marketing at the underwear stores.

Keywords: underwear stores, experiential marketing, brand, culture, application.

1 Introduction

With the brand construction in Chinese underwear enterprises and the booming entry of foreign brands into Chinese market, a modern underwear brand should provide its consumers with not only products of quality but also a set of values and a style of living. In this brand era, attracting customers to experience the brand culture and charm is crucial to its success. Only through integrating experiential marketing with brand promotion can different brands' characteristics and cultural connotations be conveyed, which then changes a brand and its products into the symbols of a certain life style. So how to lead customers to fully experience a brand is the key part in brand construction. As the very frontier to contact with customers, stores are the most important place for brand experience.

Experiential marketing conducted at the stores requires enterprises to understand consumers' minds and accurately collect customers' needed experience types. Although experience is an internal and individual feeling, which cannot be easily observed by others, in fact, people share general characteristics in mental and psychic demands. Based on these human general characteristics and successful enterprise practices, stores can adopt the following experiential marketing strategies, namely, sensory marketing, emotional marketing, life-style marketing and atmosphere marketing, etc.

2 Application of Sensory Experiential Marketing at Costume Stores

Sensory experiential marketing is based on people's sensory experience in senses of sight, hearing, touch and smell, etc. Its main aim is to create sensory experience.

J. Luo (Ed.): Soft Computing in Information Communication Technology, AISC 161, pp. 431–435.
springerlink.com

Through this kind of marketing, products of different companies can be differentiated, consumers can be motivated to buy, and additional values of the products can be raised.

2.1 Creation of Elegant and Comfortable Shopping Environment

The design and decoration of an underwear store will not only impact the immediate profits, but the growth and development of a brand as well. Therefore, for the in-store design, not only the brand characteristics but also the style, ideas and human cultural characters should be displayed to varying degrees so as to show the integration of the brand. The store should be kept clean and tidy, free from dirt, stain or sundries in full view of the customers. All kinds of POPs (Points of Purchase), billboards and labels should be hung in order and cleanly. Corridors should be kept clean and open. Fitting rooms should be clean, tidy and free from miscellaneous stuffs. Goods tally area should be maintained in order and litter free, with different articles put in certain places according to the unified standards. Also, non-related sundries should be kept free within the range of customers' vision. All in all, it is necessary for the store to have a bright, fresh, clean and elegant atmosphere as well as a high-grade and fancy shopping environment.

2.2 Background Music to Create Atmosphere

Background music is an important element that can affect customers in the shopping environment. The arrangement and design of the background music will give the direct embodiment of the brand culture and the brand orientation as sound waves travel. The influence of background music, which is of some importance to the choice of customers' purchase, can be a stimulus or a handicap to the brand sales. Therefore, it is necessary to be cautious when choosing the background music. Generally speaking, the background music's style should be in conformity with the styles of the underwear and the store. The age group and preference of the brand clients also are prime considerations. For instance, more light music should be selected for the brand whose target consumers are the middle-aged over the age of 30, such as melodious and soothing music featuring electronic piano, piano or violin and so on. During Chinese traditional festivals, the auspicious traditional Chinese melodies can be better alternatives. Proper background music reflects the brand image and the overall management level of the company.

2.3 Lighting Design to Add to the Atmosphere

As one of the most significant environmental factors of the underwear stores, lighting plays an important role in serving as a foil to the products and adding to the stores' atmosphere, having a close bearing on customers' purchase. Especially for the personal consumption goods like underwear, most women mainly follow their hearts to choose the products. So the influence of lighting becomes particularly prominent on them. In order to make the most of lighting effects, it is essential to consider the product display way and the decoration design style. Instead of striving merely for the perfect order and unique shape of the lights, the decorators should make sure abundant direct light supply in the place with objects and indirect light supply in the place without. In this way, lighting with varying density will add to the sensations of hierarchy, volume and space and create a more comfortable shopping environment.

2.4 Interior Display

Full and first consideration should be given to how to perfectly demonstrate the beauty of the store as well as the goods when displaying and arranging the goods. This beauty-first merchandise-display method is in fact a visual marketing strategy that rearranges the costumes according to the aesthetic law so as to visually exhibit their beauty as much as possible. Mainly based on customers' emotional thinking characteristics, this kind of display can arouse the customers' shopping emotions and inspire their desire to purchase, which can be realized by properly arranging different colors and commodities, or by adding the sense of rhythm to the store by means of matching, i.e. balance, repetition and response, etc. In this way, the expecting shopping atmosphere will be created, which can quickly impress the customers and lead to regular and joint purchase.

2.5 Store Advertisements

The design of the qualified POPs to promote sales at the stores should be beautiful, vivid, straightforward, attractive, and most importantly able to arouse costumers' purchasing desire.

There are no strict standards for the POP fonts. Normally POP orthography, POP living form and POP variant form are preferred, but other forms can also be utilized. However, the general requirement for the fonts is to be beautiful, lively, vivid, and able to meet the mass need. The POP should be hung straight, without crook or crumple. In general, the POP should be hung about 1.7 meters away from ground, neither too low nor too high (except in exceptional circumstances). When the POPs are hung in a line, it is necessary to keep them neat and order, and at the uniform height as well. As for the colors of the POP, the primary requirement should be in harmony with the store's environment. Because of the uniqueness of underwear stores, warm and cozy colors are widely suggested. But distinguishing characteristics should be introduced in order to make a brand stand out from the competitors.

3 Application of Emotional Experiential Marketing

Emotional marketing refers to the marketing method which suggests that businessmen should hold communion with customers to gain their trust and affection, arouse their desire to purchase, then gradually achieve the aim of expanding market shares.

The elementary condition for the success of emotional marketing is the emotional products or emotional services provided by the manufacturers. The reason why these products and services are called "emotional" is that they contain the designers' emotions, which can successfully arouse the echo in consumers' heart and build a bridge to communicate with them. By adhering to the operating idea of "focusing on the people to create a happy life", and acting in the spirit of "concentrating on consumers" and maintaining a high sense of responsibility to consumers, the well-known Aimer underwear company constantly design and develop new products so as to serve its customers wholeheartedly.

In an era of more and more products with same quality, the value of commodity is embodied more in the emotional services. The key to the emotional marketing's success is to think in the clients' shoes. That means the conventional marketing pattern such as "I manufacture, you buy" must be get rid of. Instead, enterprises should fully consider the clients' needs and get things ready beforehand. This interaction between enterprises and consumers develops the two parties' relationship from simply seller and buyer to a long-term partnership with firm foundations, achieving the common growth of the two parties. As the emotional consumption is becoming the major consumption, the emotional marketing should be given priority in all the marketing strategies.

4   Application of Life-Style Experiential Marketing

With the development of market, the style and focus of marketing have undergone great changes. An enterprise's products are not only the goods to sell or the image of the company, but also a person's taste and pleasure of life, or even his or her lifestyle and attitude towards life. Enterprises provide their consumers with certain lifestyle, and the latter experience it after purchase. For example, a cup of instant coffee only is worth a few yuans, however, such a cup of coffee costs a dozen yuans at least or even more than a hundred yuans in the coffee shops. When the mellowness of the coffee as well as the distinctive lifestyle can be sensed and realized by the consumers, there is still more room to increase the value of this cup of coffee.

Experience increases customers' feelings and valuations of a brand, which makes the products become part of consumers' expecting life rather than the needed merchandise for their value in use. When customers buy the products, they satisfy their own expectations and enjoy the process of buying.

5   Application of Atmosphere Experiential Marketing at the Stores

Compared with other underwear brands, the most distinctive feature of the brands at the stores is the business operating space. The planning of the space and the creation of the atmosphere can make it easier to provide consumers with unique consumption experience. Then distinctive brand image can be built in consumers' mind based on this kind of experience. According to some related researches conducted at the stores, consumers' pro-consumption, in-consumption and post-consumption experience is the key to investigate consumer behavior and brand management, because this pro-consumption, in-consumption and post-consumption experience has become the critical element to increase customer satisfaction as well as customer loyalty. Besides, it should be noted that the style of the store must be compatible with other surrounding stores' and buildings' styles, having appropriate proportions and sizes. But some degree of prominence is indispensable in order to attract passersby' attention at the first sight, which is necessary to meet the needs of harmonious and integrated experience.

Atmosphere experience includes interior decoration, colors, smells, layouts and facilities, etc. at the stores. The store's marginal area should be exploited and made good use. The natural color and texture of materials should be properly used according to actual conditions. It is suggested to create natural and lifelike taste and style, while taking fully into account the soundness and durability of the materials.

As the saying goes, one can only see the colors from afar and patterns from near. Colors can on one side transfer the brand's ideas, and on the other side evoke the psychological effect, which tends to be an important psychological condition of purchase. This visual power exercises an extremely strong influence in brands.

The underwear stores' main customers are female, who are sensitive to smells. Therefore, unique and delicate aroma can impress them and become potential elements for them to return and purchase in future.

Store layout should be clear and uncluttered with products of the same series placed nearby for the convenience of customers. In addition, products of the same colors can also be placed together, and customers can then find the needed products in a glance.

6 Conclusion

Nowadays, there exist a number of misunderstandings at many stores, among which putting experiential marketing one-sided or independent is the common one. The real experiential marketing should be developed based on the senses, emotions, living concepts and atmosphere, etc. which can create a comprehensive experience space for customers. With the enhancement of underwear branding in this brand era, it is crucial for an enterprise to enable its consumers to experience the brand culture and understand the brand connotations. Experiential marketing will become the most effective marketing strategy to highlight the brand value as well as shape the consumption brand culture. While consumers enjoy the happy experience, the probability for them to use more products will increase driven by the sensibility. This kind of relaxing and cheerful consumption atmosphere is conducive to promoting consumers' recognition, helping them experience the brand culture and increasing the brand's market share.

References

1. Liu, T.: The analysis of the enterprise experiential marketing strategy. Small and Medium-Sized Enterprises and Management (25) (2010)
2. Xu, D.: The analysis of the experiential marketing. Journal of Hunan Business College (5) (2009)
3. Zhao, P:. Costume Marketing. China Textile & Apparel Press, Beijing (2007)

The Explanation of Modern Public Art Orientation

Baiyang Jin and Yanxiu Jin

Institute of Art & Fashion, Tianjin Polytechnic University, Tianjin, China
baiyang1960@yahoo.com.cn, jinyanxiou@gmail.com

Abstract. When paying attention to orientation creation of public artists look for its connection with relevant creation language and apply them into the creation of public art work. Meanwhile they involve their work into city view and nature view with their insightful by using the force of culture feature presented by sites like building, street, city landmark ,leisure square, greenbelt, create work about public utilities, sculpture and scenery ,cooperate with environment artists.

Keywords: public art, orientation, power of decision.

1   Starting from the Location

When paying attention to orientation creation of public artists look for its connection with relevant creation language and apply them into the creation of public art work. Meanwhile they involve their work into city view and nature view with their insightful by using the force of culture feature presented by sites like building, street, city landmark ,leisure square, greenbelt, create work about public utilities, sculpture and scenery ,cooperate with environment artists.

The public art embodied in different occasions and environment conveys its morphology force to the audience and then forms a cultural dialogue between the works and the audience as well as guide the way to attract eye of different ambiences. In special ambience of various places will produce works with the trend of aesthetics, art and style in different places. At the same time the design elements of these work are closely connected to the history, culture, economics, science, geographic environment, surrounding sceneries of these places, which fully express the minor issues of the environment and depicts the dignity of living. Therefore, the orientation of the place should eb the first issue to be focused on while conducting the recreation of public art, and the orientation is also different if the place is not the same.

When rambling in public sites and enjoying our surrounding scenery, we can enjoy unique work created by artists describing people's lives in different sites, which can not only cultivate people's mind but can make people think and raise their awareness about regional culture. At the same time, regional culture will influence and help fix public spirit and material, affecting and guiding value orientation of public art.

2   Site to Region

The quick change of new technology and material has broaden the virtuosity of environmental public art creation. Manage and preserve citizen's participation and

J. Luo (Ed.): Soft Computing in Information Communication Technology, AISC 161, pp. 437–441.

judgment seems more important than any time before. We should never forget the protection and establishment of regional cultural features when we put much effort in regional culture. It's undeniable that the feature of city exploitation and regionalization are the narrowing distance of vision, retrenchment of difference, the lowness of identification ability and the shortage of characteristic. Regional culture can surely be inherited if we unearth the origin and its connotation of regional culture.

If the site is a point when artists design their work, the region will be dimension. The regional circumstance sidespin is very important too. The designer must take regional and community circumstance into consideration when design public art work. Region is an essential factor because different region will make regional culture and regional characteristics different. The divide of district and site lie in region which will contain their characteristics, so how to grasp its characteristics is the key to create regional culture deposits and artistic beauty when design regional public art. Social environment is a major characteristic of regional environment, so how to seize the specific social environment and space distinctiveness is a part of regional culture art creation. The interactive and rational combination of regional landscape and cultural landscape is also necessary.

3   Seat by Number

The use and stacking of contemporary materials during the urban development process, the uncontrollable exploitation and unreasonable use of the nature, the outrageous scenes, awkward design makes the contract between the environment and human sharp and not so harmonious. The existence of people becomes embarrassed. From the classical relief to the inundant lions in front of so many gates, the odious "bad cravings" has made the public art go to the wrong way or be dissimilated. The fake antique that is just anxious to achieve quick success and get instant benefits or some symbolic landmarks that is only to achieve political achievement, these are the fault for hierocracy and the contamination of vision. People's heart has been hurt and the appreciation of beauty has been dislocated. The spirit doesn't work hand in hand with the modernization which made the public art come to the wrong road.

Therefore, the designer should think during the creating process, the developing should match the planning, the environment with the nature, the site with the local feature, the personality with the culture and the society with the public.

4   Openness and Public

The most important difference between public art and other art form and styles in previous time doesn't lie in its creation method, material applied, media and expressing senses, but in the word "public". In this level, the word "public" doesn't only mean the publicity of art work site, but also social publicity which reflects public spirit.

As the transforming of the society, the public area and the civil society have been come into being. As the publicity, openness, activeness and meaningless, all these issues made the public art can come to the public exchange area through the method public art, so that to make the art has the intension of publicity; make the art come to a series of social affairs such as the cultural share of social equalism, publicty and democratic participation; then the art has come to a method of entering the social public area. Then to some extent, the cultural art has become a temporary way of bearing the publicity, realizing social euqalism and political democracy as well as an internal requirement of the existence in this temporary society. The public art should be democratic and open. Its privacy and closeness is relative, the receivers of the public art is the public.

5 Spiritual World

We shouldn't consider the creation of public art as designing landmarks or monument. It is to accumulate culture blocks, to collect the social culture of various stages to the area of public spirit in modern public art and the morality in social public area.

There is some difference between public art and other art .That is to pursue public spirit. So what is public spirit? The public spirit in contemporary public art is to represent the common benefit of the public in these times as well as bearing the responsibility of maintaining the public awareness. Therefore, the public art creation shouldn't be considered to be a simple artistic behavior during the design process. It should also be considered as an issue to popularize the modern spirit and understand the superiority of the public art.

While generating, expanding and extending the art works in plane to the public artistic space,the artistist should also encourage the public participate actively and produce corresponding social effect. At the same time, the creation of public art should make the work expression to be in accordance with the corresponding public thinking in accordance with the social life. In that case, it can have the public spirit.

If the public spirit has its sharp public spirit, then as the representative of cultural will, valuable will and national cultural feature, it should be combined with the public will through its public natured artistic model, such as, arcturature, statue, painting and environment design. Then it can have the feature of dignity culture, willing, as well as creating the works with cohesive power, creative spirit.

6 Power of Decision and Public

While conduct the public art design, the relationship between the art and the public spirit should be taken into consideration. The main body of the public art is the public, but as for the ingestion of the public art, it can through the government or persons, but mainly through the government and implanted through the guidance of the government. The public can only appear in the name of tax payers among which, the public is the direct or indirect investors. From this, it is not difficult to see that the nature of the public art is to serve for the public. It is result from the development of the national economy. The artists themselves also belong to the public; the rules of the law also provide the money for the re-creation of the artists.

Besides, the decisive right of the public also in the hand of the public. The public art in the art of the public, it is different from the free creation of the artists, it isn't pure artistic behavior. The creating and design right is in the hand of the art. But whether it can be accepted or not depends on the public. Therefore, the public art in developed countries and regions has a complete system and law to guarantee the objectiveness, fairness, equalism until the best one is selected.

7 Aesthetic Perception

Any art should have the value of beauty; there is no exception for public art. During the design process, the designer should try their best to enlighten their feelings of beauty as well as enhance their aesthetic judgment, increasing the ability of creating aesthetic ambience, purifying the mind and heart, developing their thinking and so on.

But people's value for beauty couldn't be embodied easily. The value of beauty is the one that is transferred from various factors, but it needs the support of other factors. Let's take the art of public facilities for example, it is a public art work of three same form sphere and different skin texture, when we analyze the value of such work, superficially, we can only feel the value form pattern, materials and shape. But actually, it is not so simple, it is through people's experience and feeling and then transferred to be a value that people can feel the beauty. That's the aesthetic level of public art in unrealistic condition. Under virtual condition, people's understanding for beauty is generated periodically which surpassed the feelings of real awareness for model, then it spiritual stage is reached which surpasses the reality. Therefore, there will be texture effect for the following sphere ,but the generation of such beauty is usually extracted from people's understandings and experiences. (Figure 1)

Fig. 1. Seoul coach station

Such kind of value can be understood from the following expressions.

A.   We should have the responsibility and thought during the art creation process. The public art integrates the value orientation and creation of the political dream, thinking behavior and other ones to the art creation process. When the public appreciate and feel such works, the public aesthetic level can be enhances through the works, then aim of enhancing moral cultivation can be reached.

   B.   Firstly, we should extract from our knowledge and experience. We have collected lots of experience and knowledge from life and study. These knowledge and experience provide creation media for us, which have the feature of experienced popularity and broad knowledge. It covers the knowledge of history, local customs, psychology, religion, politics, culture and so on. We should use this knowledge in our creation process to make the beauty of public art be felt indirectly and accept such knowledge under free condition step by step. The scientific systematic education make the richness of public art be closely related to the life of the public as well as provide more content for the art creation.

   C.   The inspiration of spirit and will. By creating a space art atmosphere, public art can inspire people to make progress and turn spirit into actual move.

   D.   The value extension of outside world. By appreciating public art, the public can enjoy the beauty of art works as well as improve their aesthetic ability and realm. Their heart and temperament can be brought to a higher level, which can influence development of public culture psychology and intelligence and thetics storage.

References

1. Xi, Y.S.: Public And Public Art. China Central Academy Of Fine Arts, Beijing
2. Yang, J.B., Xiu, J.Y.: Public Art Desing. People's Fine Arts Publishing House, Beijing
3. Hua, S.Z.: Public Art. ShenZhen program Sculpture Institute, ShenZhen
4. Hei, Z.S.: Arts "Public"And Public Art. Academic Of Arts And desing, Tsinghua university, Beijing

New Knowledge on the Penetrating Point of Public Art Design

Baiyang Jin and Yanxiu Jin

Institute of Art & Fashion, Tianjin Polytechnic University, Tianjin, China
baiyang1960@yahoo.com.cn, jinyanxiou@gmail.com

Abstract. Public art works must be placed in the public space or public domains, so they can have social and public contacts. In doing so, art, artists and the public may have mutual exchanges, communication and influence. Public art is a kind of interactive art, two-way communication art. With the public participation, public art interacts with them and affects public views. Public art pursues art social effect. Public art no longer belongs to artists' personal creation. Instead it needs the participation of the public. So, public art can absorb public discourse right and visual experience. This also manifests the concerns to public values equal communication and shorten the distance between the public and works.

Keywords: starting from five aspects, initiative creativity, language expression, public angle.

1 Multi-perspective Design

In the process of art design, one should take various perspectives and channels to think. In other words, one should learn to change his minds according to the specific situation.

The core of innovative design registers a creative mind. The notion of creation means to transcend the old things, inherit and develop them so as to be innovative. In doing so, it can serve the main trend to be convenient to people's use and their views of beauty both functionally and formally.

In today's plural world, mankind's consumption notions and art appreciation capability are becoming mature. One single form and language of art do not lead the fashion as before. People today are pursuing individual expression and views of beauty, making art more diversified. The single way of expression of art is not confined to traditional crafts, art forms, and ways of expression. Instead, it begins to absorb the best of other kinds of art while maintaining its own features. Design in itself is a creation process. So, it can not be separated from the designers' creative minds. The creativeness stressed here refers to creative notion. The essence of modern art design is embodied in the process of creation designing and practices.

1.1 Starting from Five Aspects

Designing starts from meeting the requirements of the will –be-designed subjects, such as collecting the relevant materials. Then, the specific designing scheme should

J. Luo (Ed.): Soft Computing in Information Communication Technology, AISC 161, pp. 443–449.
springerlink.com   © Springer-Verlag Berlin Heidelberg 2012

be formulated, the organizing personnel determined, the work contents of the research edited. One should view design from the perspective of country, region, history, and culture, political and economic environment. As for the product itself, one should make effort from function, material and technology to add beauty and humanity to it. Besides, one should analyze and assess the collected information so as to determine the reasonable design and vision. In other words, it is to design a major frame and direction and use some innovative methods and imagination to formulate a reasonable designing scheme, which also explores one's way of thinking. One's creativity and imaginativeness count most in this period. One can even be free to think and be bold in explore innovation. The five aspects are as follows,

1) Drafts

They include conception draft, format draft and determined draft. One should sketch them according to the concept, look for the right style according to drawing creation art form rule. Sketching more pictures may realize quantitative- qualitative change, maybe there will be a better idea. One should screen these sketches again after in-depth supplement to have style sketch. Again, after repeatedly screening drafts, one should draw three-dimensional effect sketch which is called determined sketch.

2) Model

According to above sketches, one should make corresponding percentage of model (model making materials is usually used as gypsum, clay, foaming plastics, etc.).

3) Scheme

In the final stage, firstly, the best design scheme is selected from multiple optimal solutions. Secondly, it should be demonstrated in the best way of expression. In the final stage of design, one needs to screen accumulated scheme conceived after evaluation. Then, one should, make the expression sketch and model. The expression sketch can give a person comparative intuitive sense. They include, decomposing sketch and profile expression sketch. The appearance effect sketch is made according to the design requirements and creative rendering, and decomposing sketch and profile expression sketch are usually used together, and they are embodied through accurate data for each component from the perspective of position and shape. Design drawing is the engineering drawings mapped out according to the desired chart and model. It should be accurate general, permanent and duplicable. Specific construction can start according to the drawing. And the design model is also is a kind of precise model shaping. Compared with the previously described ones, it is more meticulous.

4) Designing reviews stage

It belongs to the content of construction and management. No details.

5) Designing management stage

It belongs to the content of construction and management. No details,

1.2   Using a Variety of Materials

Today is a media era. The use of different media delivers to us different art information. Together with different aesthetic feeling of the public to different media being different, if we put media to conversion or grafting, it may produce more kinds of visual language.

The traditional sense of material media is wood, metal and stone, clay materials, etc. These materials used bears history and will continue to play an important role. Therefore, in the creation of public art design, researches on how to use traditional media grafting new ideas, putting the traditional media into modern design language, combining the traditional media with high technology content of modern material in work is an important link of public art design application. How to apply the traditional media successfully to public works of art creation, to complete the reflection and criticism the cultural and social significance of traditional materials ,to combine modern thought with public art, is a creative design research content.

With the development of modern art or traditional art ideas alternately influencing each other, artists' knowledge on material are not simply confined to artistic genre media, but use material properties in the theme of works by original creation with more attention. By giving material new form and spirit connotation, it also brings with it the fresher visual effect. Artistic pioneering art conception and artistic creation material scope is growing. When doing works, one should start from the material properties and familiarity, and select material from the form and the artistic idea expressed so as to broaden one's horizon. Besides, material should also be selected from traditional ones to modern ones and then to the new media images or now finished ones in the information society and so on. Therefore, by borrowing some of it, one can fully exert harmony, idiosyncratic characteristics and tension between and of materials and specific modeling photograph, providing a new starting point for exploring in visual arts continuously used for new materials.

The booming of modern art has caused revolution in the material world. Artists, designers are beginning to realize that material is not only visually can touch, and still can smell. The transformation from tangible materials to the intangible materials, from the objective figuration to subjective abstractions and the pursuit of individual language drives artists to find out the material and technical means adapted to unique personal language. And a diversified pattern has been formed.

Starting from material media development history, the material which serves as a medium to public art can be divided into traditional media, new media and super media. This classification has its relativity. One should analyze the relationship between context meaning and environment according to its as language elements. This relationship is a combination of clarity and fuzziness, certainty and uncertainty, sensibility and rationality.

Material technique is not strictly restricted to the steps and certain techniques. It allows much freedom. Compared with traditional paintings, in addition to considering the same modeling, color which constitute the basic problem, one should pay attention to textural effect of a variety of materials and should generate the inspiration of creation from material of its own characteristics. One should has good control of contrast and adjustment of factors such, softness, density, ups and downs, the size and a variety of special techniques also can be used. And the use of some material is very extensive. One can combine a variety of art form and material properties to get creative inspiration. This is also very important. The comprehensive use of materials is various by different artists in different works. The reasonable comprehensive use of materials can enriched and expanded creation language expression, greatly tapping the artist's potential.

The use of material can not only convey thoughts, but also express form through form. It is very important to select material in creation, form language features also can't be formed without applicable materials media. However, sometimes one can in change material media adapted to art creation or adjust artistic thought to adapt to the media restriction.

1.3 Initiative Creativity

Creativity here means to initiate. One should actively tap his potential energy and express it meticulously in works. Being active is to be at the forefront of creation, meaning thought before action. Because of educational mode, some people especially students tend to look at things numly in stead of using their brains. They are used to habitual thinking or experience, which brings them with no initiative and no possible creative vitality. As a result, their works lack fresh vision shock. The development of the times promotes people's thinking to keep pace with the Times. Otherwise they are unable to create updated works to meet the needs of the times.

According to objective theory, designed art creation has its own rules. That is to summarize the design program in the long-term practice. But this does not mean designers lost creativity. Instead, in doing so is to extend thinking and imaginations through some design methods in order to realise the design process, also a process of creation and innovation. in the creation process, we need different means and forms of materials to reach our imagined goals.

1.4 Public Angle

Public art works must be placed in the public space or public domains, so they can have social and public contacts. In doing so, art, artists and the public may have mutual exchanges, communication and influence. Public art is a kind of interactive art, two-way communication art. With the public participation, public art interacts with them and affects public views. Public art pursues art social effect. Public art no longer belongs to artists' personal creation. Instead it needs the participation of the public. So, public art can absorb public discourse right and   visual experience. This also manifests the concerns to public values equal communication and shorten the distance between the public and works.

One should stands in the public shoes to appreciate public art. Public art faces a kind of brand-new cultural background. It concerns the new social, cultural, historical and other problems. Postmodernism changed people's art angle, making art not independent. People's visual creation opens to the public and artistic operations are also interactive. The way how art created changed to interaction and communication. The sacred art no longer appears important. Art is much closer to the public.

1.5 Language Expression

l hape, color, texture and other factors constitute an important part public of design language research on form here is not the basic the form training. It is a amplified shape. Any form has its appearance referring to the exterior. Actually different outer shape can produce different languages by artist choice and processing. The design model includes external contours and internal structure. Form languages include figurative and abstracted ones. Form language is the image existed in natural state, called figurative language while abstracted form language is a consciously processed one. And there also exists kind of half a figurative and half a abstract design language.

Public art form language is different from language form in graphic works. Public art can not be created without material texture. Different material texture and mechanism affects forms. Although a material is a practical medium of creation, it will become a form of language after art use. Material has another language, that is, intuitive psychological feeling. Processed plastic material gives people the feeling of softness while processed hard metallic materials give people strong and powerful feeling. Materials in design is attached to the artistic creation needs (sometimes material itself has artistic tension). It can use different materials to help us reflect more about external form, providing a bounteous creative design space for our design. Different texture t gives people different feelings. It is also very important as how to apply into design the texture language generated by the mixture of texture and techniques No matter using physical form or to inspire people association, one should always ensure that the complete function of material.

Color is a physiological response. the color language used in the design can give people different feelings. When designing color language, one should grasp the different psychological reactions to color and design them into works and so on. These design language provides certain association space for creating design.

2   Much Language Research

The process of art creation is diversified. In public art creation, all aspects of exploration should be involved, namely from different angles and in various ways. This is to comprehensively examine various factors of internal and external relations, produce works diversiform expression, methods and positive effects of works. One should select the best creative solutions to better promote his works.

In our exploration of various languages, we are used to the existing expression way, actually this kind of habit is understandable. But we cannot stay on existing expression means. Instead, we should find differences of expression language. We should respect existing forms, tolerance differences and coexist with heterogeneous design, thus forming a kind of "harmony with differences" benign relationship. Whether worldwide modernization or multicultural development, they both require diversified art in modern conditions. We can take more open stance in absorbing heterogeneous culture quality elements, thus advancing with time and realize their own cultural change. This is not only a multicultural adjustment, but also an effective ways and inevitable path for the multiculture to adapt to modernization and realize the multicultural independent development.

2.1   Multiple Combination

Element is foundation of design and is an indispensable part of creation. Element is important, but individual element is difficult to constitute works which need element combination and the method of elemental combinations. The different ways present different visual effect.

Below is from several aspects explaining ways of element combinations. considering the varied ways of elemental combination, I just make a summary. If we put these elements together in different ways, the ways will more diversified. Meanwhile elements also have direct and abstract elements.

- the deconstruction: to criticize and develop the existed concept and category.
- disorder: to dismember, scattered and reorganize traditional elements.
- incomplete: to purposely damage elements, emphasizing incomplete state.
- variation: to put several different elements with no coherence together.
- weightlessness: to bend descramble and twist elements
- being unconventional or unorthodox: to break away from conventional elements arrangement, treating abnormal as normal.
- divert: to put new elements in context.
- irony, funny: to sneer at dynamic rather than still
- element deformation: to deform object image.
- prototype decomposition: to decompose and scatter the original elements.
- additional paraphrase: language.
- multiple meanings: various ambiguous semantics.
- ambuguity: an ambiguity of two different kinds of questions.
- metamorphosis: unreasonable.
- scene: various use of scenes
- misreading: the conscious misreading
- sophistry: to use unjustifiable excuses.

2.2 Multiple Visuals

The examining of public art design from various visual angles in fact is constructed on the foundation of innovation. it is to expand creation horizons. With creative as the forerunner, one should seek a special way of expression to make explanation, which can draw people's attention and impress them with the visual image.

Meanwhile, visual communication process is an open system. and it is updated and expanded continually with the impact of new technology. Visual innovation is not only referred to the concept innovation, but also innovation in the use of media and material. The so-called concept of "new" is compared to the historical, traditional and past meaning. Innovation specifically is to transfer information on the subject content, the expression of creative conception. It is not to please the public with claptrap or deliberately mystifying, but congeal beauty and reality so as to serve visual convey.

2.3 Various Means

Art design tools are varied and colorful. They are indispensable for art creation. Different means of design produce different result. Different designs require different methods. In basic teaching and primary design training, we adopt relatively simple pure or very direct means. it facilitates our learning and understanding. If we face practical-oriented people or consumption masses, we should then combine the user's intention and designer's ideas to determine expression means. Art design especially public art design course involve different means, not technology but the idea and thinking. it is to put the feeling into mind and then present them in art forms. Therefore, the designers usually adopt philosophical, symbolic and associative means to make work produce abundant association and sense of beauty. In this way, the work, while conveys information, also gives more aesthetic experience.

Along with the global development of world economy and culture, the trend of modern scientific integration is more apparent. Various disciplines are mutually penetrating, interacting and communicating in a much closer way. As the top-level means artistic design, thinking needs touch various fields. Designers should adopt an omni-directional, three-dimensional and open attitude to seek method. Below are some means:

1) Representation methods: to use signs directly related to object with typical features to straightly illustrate the object. This kind of method is direct, clear, and is easy to understood and memorized rapidly.

2) Symbol means: to adopt graphics, words and symbols, the color ,etc that relate to the meaning of content and use metaphor and descriptive methods to represent abstracted content, such as using cross sickle axe to symbolize worker-peasant alliance, using uprising seedling symbolize children's vitality and growth, etc. Symbolic mark usually adopts already established connection object as a valid proxy.

3) Moral means: to adopt images relevant to the meaning of the subject or full of moral message and using innuendo, hint, signal to demonstrate the content and characteristics of the subject.

4) Imitation and assimilation: it is a kind of method to use image with similar characteristics of the subject to imitate and assimilate the features and meaning of the subject.

5) Visual means: it is a kind of method to adopt concise abstracted graphics, words and symbols with no special meanings and unique form to give people a kind of strong contemporary feeling, visual impact feeling or intimacy so as to cause people to pay attention and hard to forget it. This kind of method does not rely the meaning of the graphics but the visual force of the symbols to perform.

References

1. Xia, B.: Visual Art of Public Space——Statue in New York Street
2. Zhao, J., Liu, J.: Crafts Study on Designing Material. High Education Press
3. Cheng, Y., Li, C., Liang, X.: Training Program on the Creative Mind of Art Design. China Textile Press
4. Bao, S.: Urban Public Art Landscape, China Architecture Industry Press

Analysis on the Competitiveness of Chinese Steel Industry and That of South Korean

Kijo Han and Jun Liu

Dong-eui University, Busan, South Korea

Abstract. As an important industry to supply raw materials, Chinese steel industry remains high growth momentum in GDP duringg this period. In the global steel industry, the rising of Chinese steel industry is thought to be a threat to other countries in the world, especially for the counter part of Korea and Japan. But research on the competence of Sino-South Korean steel industry is not widely and so far there is no concrete study for the whole industry. Therefore, this study uses trade statistics of steel industry of UN COMTRADE to deduce the influence on Chinese steel industry from China's view based on the analysis on the change of characteristics and the competitiveness of China and South Korea.

Keywords: raw materials, the competitiveness, commodities.

1   Introduction

As an important industry to supply raw materials, Chinese steel industry remains high growth momentum in GDP duringg this period. In the global steel industry, the rising of Chinese steel industry is thought to be a threat to other countries in the world, especially for the counter part of Korea and Japan. But research on the competence of Sino-South Korean steel industry is not widely and so far there is no concrete study for the whole industry. Therefore, this study uses trade statistics of steel industry of UN COMTRADE to deduce the influence on Chinese steel industry from China's view based on the analysis on the change of characteristics and the competitiveness of China and South Korea.

The research employs classification materials of trade commodity statistics SITC Version 3 from 5 units to make overall strict analysis on steel industry. In order to grasp the recent competence status between steel industries of China and South Korea, this paper uses the UN COMTRADE China and South Korea ranging from 2000 to 2008 to educe general trade indexes such as trade specialized index, revealed comparative advantage index by analyzing trade structure of Sino-South Korean steel industry. In addition, it will make an analysis on Chinese steel industry's advantages and disadvantages over Korean commodity and contrast between overall competitiveness of the steel industry in the two countries.

2   The Trade Structure of Steel Industry in China, South Korea and Japan

Figure 1 shows the production tendency of crude steel in China, South Korea and Japan. The average annual rate has been in an increasing trend since 1985. China's

J. Luo (Ed.): Soft Computing in Information Communication Technology, AISC 161, pp. 451–459.
springerlink.com   © Springer-Verlag Berlin Heidelberg 2012

occupation of crude steel in the world total production has risen to 37.6% in 2008 from 8.6% in 1990, which indicates that China enjoys the farthest increase rate. In 2008, the gross production of crude steel of these three countries occupies about over 50% of the world total output and within the industry, they can realize industry cored by scale economy. So the competition among them becomes more important for maintaining mutually profitable liberal relationship makes a big difference in stabilizing the global steel industry. (Figure 1)

	1985	1990	1995	2000	2005	2008	2008年 2009
World	71,892	77,049	75,257	84,787	114,669	132,912	[illegible]
China	4,679	6,635	9,536	12,724	35,579	50,031	[illegible]
	6.5	8.6	12.7	15.0	31.0	37.6	[illegible]
South Korea	1,364	2,313	3,677	4,311	4,782	5,365	[illegible]
	1.9	3.0	4.9	5.1	4.2	4.0	4.0 4.0
Japan	10,528	11,034	10,164	10,644	11,247	11,874	[illegible]
	14.6	14.3	13.5	12.6	9.8	8.9	[illegible]

Fig. 1. The production tendency of crude steel in China, South Korea and Japan
Note: The Statistics of IISI

Table 2 is the status quo of the import and export in steel industry in China, South Korea and Japan. Because of its big size and weight, regional steel trade is more outstanding than that of long-distance transportation, which facilitates the import and export trade in the three countries. Table 2 shows the status of 2008 that regional trade (export + import) plays a significant role in the global steel import and export for the three countries, in which China occupies 27.9%, South Korea 51.7% and Japan 44.9%. Besides China' 21.2%, South Korea' 27.9% in regional export, the proportion of whole regional import and export is very high in China, South Korea and Japan.

From the trade balance in Table 2, South Korea is in deficit in steel industry compared with China (-12.871 billion US dollars) and Japan (- 7.84 billion US dollars). The export situation marks the end of China's dependence on Japan in import. But in 2008 the steel industry of South Korea mainly relies on import from China and Japan. China is in surplus to South Korea and in defiict of 6.142 billion to Japan. Japan is in surplus compared to China(5.168 billion US dollars) and South Korea (7.283 billion US dollars) . (Figure 2)

(単位：金額100万 US$, 比重%)

Partner -->		中国		韩国		日本		区域内		对世界
China	export			16,380	(16.1)	5,263	(5.2)	21,643	(21.2)	101,892
	import			5,158	(14.7)	11,405	(32.5)	16,563	(47.2)	35,082
	Total			21,538	(15.7)	16,668	(12.2)	38,207	(27.9)	136,974
South Korea	export	4,476	(14.7)			3,948	(13.0)	8,423	(27.7)	30,397
	import	17,346	(41.2)			11,752	(27.9)	29,098	(69.1)	42,110
	Total	21,822	(30.1)			15,700	(21.7)	37,522	(51.7)	72,506
Japan	export	11,020	(20.8)	11,023	(20.8)			22,042	(41.6)	52,927
	import	5,852	(33.3)	3,739	(21.3)			9,591	(54.6)	17,551
	Total	16,871	(23.9)	14,762	(20.9)			31,633	(44.9)	70,478

Fig. 2. The regional import and export of China, South Korea and Japan
Note: the proportion of China, South Korea, Japan in the world, ranging among the steel products classified by SITC 5-digit
Material: the statistics of UN COMTRADE

3 Analysis on the Results

3.1 China's Competition Tendency to South Korea in Steel Industry

Based on the method of analyzing competitiveness mentioned above, the competitiveness of Chinese steel industry to South Korea will first be analyzed in 2000, 2005 and 2008. In Table 3, it's clear that there are 158 kinds of commodities in steel industry in China's import and export to South Korea in 2000 and the trade volume of export amounts to 856.4 million US dollars. In 2005 the number increases to 4.2132 billion and in 2008 to 16.2876 billion. During the 8 years from 2000 to 2008, the average annual growth rate is about 31%.

Compared to the trade volume of Chinese steel industry exported to South Korea 1.9278 billion US dollars in 2000, it has risen to 5.1101 billion in 2005 and 5.1385 billion in 2008. The average annual growth rate of China's import from South Korea in steel industry is 9.3% from 2000 to 2005, but it slows down from 2005 to 2008 (the average annual growth rate 0.1%). (Figure 3)

		COMMODITY	COMMODITY	COMMODITY	COMMODITY	COMMODITY	total
2000	export	679.3	1.8	10.1	23.3	141.9	856.4
		79.3%	0.2%	1.2%	2.7%	16.6%	100.0%
	import	73.2	0.3	58.2	543.9	1,252.1	1,927.8
		3.8%	0.02%	3.0%	28.2%	65.0%	100.0%
	trade balance	606.1	1.5	-48.1	-520.6	-1,110.2	-1,071.4
	item number	62	6	3	62	45	158
2005	export	2,116.6	234.1	67.9	1,797.6	7.0	4,213.2
		50.2%	5.6%	1.4%	42.7%	0.2%	100.0%
	import	653.2	104.5	376.2	3,463.7	613.7	5,110.1
		11.0%	2.0%	7.3%	67.6%	12.0%	100.0%
	trade balance	1,663.4	129.7	-317.3	-1,666.1	-606.7	-897.0
	item number	63	16	4	43	32	158
2008	export	15,659.9	403.3	26.4	176.2	22.8	16,287.6
		96.1%	2.5%	0.2%	1.1%	0.1%	100.0%
	import	2,709.7	71.9	70.2	1,574.6	712.1	5,138.5
		52.7%	1.4%	1.4%	30.8%	13.9%	100.0%
	trade balance	12,950.2	331.4	-43.8	-1,399.3	-689.3	11,149.2
	item number	84	14	2	32	26	158

Fig. 3. The import and export of commodities in steel industry in China and South Korea Material: the statistics of UN COMTRADE

Such results indicate the same tendency in China's trade balance to South Korea in steel industry. In 2000, China's trade balance is in deficit of 1.714 billion US dollars to South Korea in steel industry and it reduces to 897 million in 2005 but amounts to 11.1492 billion in surplus. So China gains the advantageous status to South Korea in the trade of steel industry in 2008.

First, the commodities with absolute competitive advantage in Chinese steel industry are increasing in number, and so is the trade volume of import and export. Moreover, the commodities of steel industry with absolute competitive advantage in China to South Korea take over the highest import and export proportion. Simply speaking, not only the export of the commodities with absolute competitive advantage is the highest in Chinese steel industry to South Korea in 2008 compared to 2000 but also the import of that. And its trade balance has kept trade surplus and it raises to 1.5534 billion US dollars in 2005 and 12.9502 billion in 2008 from 661 million in 2000.

Second, there are only 6 kinds of the second commodity group with competitive advantage in 2000 and the trade balance exported to South Korea is 1.8 million US dollars holding 0.2% of its total export volume. In 2005, the three numbers are respectively 16, 234.1 million, 5.8% and in 2008, respectively 14, 433 million, 2.5%. In 2000, China's import of the second commodity group with competitive advantage from South Korea is 0.3 million US dollars and only 0.02% of the total imported products of steel from South Korea. In 2005, the import increases to 143 million and 2% of total import, and it holds the smallest proportion among the five commodities. In 2008, the trade is 71.9 million and proportion 1.4% which lead to the increase tendency of trade surplus.

Third, the number of the commodities in the third commodity group with balanced competitiveness is 3 in 2000, 4 in 2005 and 2 in 2008 and the export to South Korea has increased to 58.2 million US dollars in2005 from 10.1 million and reduced to 26.4 million in 2008. The proportion of Chinese steel industry to South Korea is 1.2% in 2000, 1.4% in 2005 and only 0.2% in 2008. From the perspective of import, the number is respectively 58.2 million US dollars IN 2000, which increases to 375.2 million and deduces to 70.2 million. The proportion of the import is 3%, 7.3% and 1.4% and all of them are less than 10%. The trade balance is 48.1 million US dollars in deficit in 2000 and the deficit increases to 317.3 million but reduces to 43.8 million.

Fourth, the number of the commodities of the fourth commodity group with competitive disadvantage is deduced to 43 in 2005 and 32 in 2008 from 52 in 2000. This commodity group of 2000 exported to South Korea is 23.3 million US dollars accounting for 2.7% of South Korea total export. In 2005, the value of export increases to 1.7976 billion, while the proportion amounts to 42.7%. While in 2008 the two figures changes into 175.2 million and 1.1%. Correspondingly, the import of this group from South Korea is 543.9 million in 2000, 3.4547 billion in 2005 and 1.5745 billion. The proportion Chinese steel industry to South Korea of total exports is respectively 28.2% in 2000, 67.6% in 2005 and 30.6% in 2008. The trade balances of the three years are all in deficit: 520.6 million in 2000, 1.6561 billion in 2005 and 1.3993 billion in 2008.

Finally, the number of commodities with absolute competitive disadvantages is 45 in 2000 and reduces to 32 in 2005 and 26 in 2008. The value of export 141.9 million holds 16.6% in 2000 and deduces to 7 million occupying 0.2% in 2005 and grows to 22.8 million taking over 0.1%. The import and its proportion are respectively 1.2521 billion and 65.5% in 2000, 613.7 million and 12% in 2005 and 712.1 million and 13.9%. The fourth commodity group of Chinese steel commodity to South Korea is in comparative disadvantages and the balance of trade is in deficit too, which is respectively 1.1102 billion in 2000, 667 million in 2005 and 689.3 million.

According to the above analysis, it's not hard to get the conclusion that imported goods Chinese steel industry to South Korea mainly locate in those with absolute competitive disadvantage and competitive disadvantage in 2000 and 2005, while most of the exported goods are those with absolute competitive advantage. Because of the bigger volume of import than export, there drops into a serious deficit in the trade balance of steel industry and China is in the bad situation. However, in 2008, such phenomenon reverses to the opposite. The exported goods Chinese steel industry to South Korea mainly fall in the first commodity goods with absolute competitive advantage and imported goods focus on goods of the first and fourth commodity groups. In such way, the export is much bigger than the import, so the trade balance turns into surplus and China wins the upper position.

3.2 The Competitiveness of China and South Korea in the Commodity Groups of Steel Industry

First, Table 4 and 5 show that goods (a+b) with competitive advantage in Chinese steel industry in 2008 are 98 more than those in 2000 and goods with competitive disadvantage in 2008 are 58 less than those in 2000. The number of goods indicates China's comparative superiority to South Korea in 2008. From the value of export and import of steel industry in Table 9 and 10, it's easy to see that goods exported are composed of 99% of commodity group with competitive advantage and 1% of commodity group with competitive disadvantage, while more than 64% goods imported are goods of commodity group with competitive advantage.

	absolute competitive advantage	competitiveness	balance	competitive disadvantage	[illegible]
[illegible]	67121 67141 67149 67151 67152 67154 67155	67139 67131 67133 67153		28221 28239	28229 28231 28232 28233 67122
semi-product	67241 67245 67261 67262 67269 67249	67282			67281 67247
plate material	67351 67431 67452 67533 67534 67538 67541 67542 67553 67300	67531 67537 67551 67552 67571		67352 67413 67414 67421 67441 67442 67443 67444 67511 67522 67532 67535 67543 67554 67555 67556 67561 67574 67573	67353 67411 67412 67422 67432 67451 67512 67521 67536 67562 67572
[illegible]	67611 67629 67646 67681 67682 67683 67685 67687 67688 67701 67709 67829 67610 67620 67630 67640 67810	67625		67639 67647 67648 67684 67686 67821	67612 67633
steel pipe	67911 67912 67913 67914 67917 67931 67942 67944 67951 67952 67953 67954 67955 67959	67915 67916 67932		67939 67941 67943 67949	67933 67956
steel products	69119 69211 69241 69243 69311 69351 69421 69731 69732 69733 69741 69744 69751 69921 69922 69931 69932 69961 69962 69963 69965 69967 69969 74831 74839 81215 69111 69112 69113 69114		69422 74832	81211	69352 69941

Fig. 4. The Classification 1 of Commodities with China's Superiority to South Korea in 2008
Material: statistics of UN COMTRADE

	absolute competitive advantage(a)	competitive advantage	(a)+(b)	balanced competitiveness	competitive disadvantage	absolute competitiveness(e)	(d)+(e)
first hand material	7	4	11	0	2	7	9
semi-product	6	1	7	0	0	2	2
plate material	10	5	15	0	13	17	30
bar-shaped material	17	1	18	0	6	2	8
steel pipe	14	3	17	0	4	2	6
steel products	30	0	30	2	1	2	3
total	84	14	98	2	26	32	58

Fig. 5. The number of commodities China to South Korea in 2008
Note: made based on Table 7

Among the goods with competitive advantage (a+b), goods with the greatest competitiveness include 11 goods of expendable material, 18 goods of stick shaped steel, 17 goods of steel pipe and 30 goods of steel products. The number of reverse panel material and goods of commodity groups with competitive advantage grows to 15 and the number of goods with competitive disadvantage is 30. These figures manifest China's poor competitiveness to South Korea.

But the value of export in Table 9 demonstrates that expendable material (3.9%, 37.6%), plate (58.6%, 17.1%) and steel pipe (4.1%, 41.9%) have a high proportion in export of the commodity group with competitive advantage. Correspondingly, the

	Competitive absolute advantage	Competitive advantage	equilibrium	Competitive disadvantage	Competitiveness absolute disadvantage	Total
first hand material	606,476	151,666	0	545	3,682	762,370
	79.6	19.9	0.0	0.1	0.5	100.0
	3.9	37.6	0.0	0.3	16.1	4.7
semi-product	248,317	8,534	0	0	228	257,378
	96.5	3.4	0.0	0.0	0.1	100.0
	1.6	2.2	0.0	0.0	1.0	1.6
plate material	9,174,085	68,936	0	135,770	9,285	9,388,076
	97.7	0.7	0.0	77.5	0.1	100.0
	58.6	17.1	0.0	77.5	40.7	57.6
bar-shaped material	3,383,918	4,943	0	23,644	0	3,412,505
	99.2	0.1	0.0	0.7	0.0	100.0
	21.6	1.2	0.0	13.5	0.0	21.0
steel pipe	640,946	168,914	0	14,749	4,751	829,360
	77.3	20.4	0.0	1.8	0.6	100.0
	4.1	41.9	0.0	8.4	20.8	5.1
steel products	1,606,112	0	26,410	518	4,893	1,637,932
	98.1	0.0	1.6	0.0	0.3	100.0
	10.3	0.0	100.0	0.3	21.4	10.1
Total	15,659,854	403,293	26,410	175,226	22,838	16,287,620
	96.1	2.5	0.2	1.1	0.1	100.0
	100.0	100.0	100.0	100.0	100.0	100.0

Fig. 6. The situation of competitiveness and export of commodity group China to South Korea in 2008
Material: statistics of UN COMTRADE

proportion of semi-finished products and steel products is lower in export. The plate (57.6%) and steel pipe(21%) also take over a high proportion. However, in Table 7 the plate has a big input (80.3%, 71.9%) among imported commodity groups with competitive disadvantage and it holds a high proportion 71.9% in all the imported plate materials.

	Competitive absolute advantage	Competitive advantage	equilibrium	Competitive disadvantage	Competitiveness absolute disadvantage	Total
first hand material	11,478	1,759	0	151,733	17,066	42,295
	6.3	1.0	0.0	53.4	9.4	100.0
	0.4	2.4	0.0	9.6	2.4	2.4
semi-product	6,223	7,873	0	0	78,305	46,669
	6.7	8.5	0.0	0.0	84.7	100.0
	0.2	10.9	0.0	0.0	11.0	11.0
plate material	1,899,689	17,940	0	1,264,333	511,955	4,067,716
	47.1	0.5	0.0	37.5	18.1	100.0
	59.0	24.9	0.0	80.5	71.9	71.9
bar-shaped material	318,034	1,335	0	67,195	6,335	304,930
	30.9	0.3	0.0	17.1	1.6	100.0
	11.7	1.9	0.0	4.3	0.9	0.9
steel pipe	154,345	43,005	0	56,013	46,673	177,623
	51.3	11.9	0.0	23.6	12.9	100.0
	6.8	59.3	0.0	3.5	6.6	6.6
steel products	889,301	0	70,228	8,083	81,778	470,831
	52.3	0.0	9.9	0.7	7.2	100.0
	21.7	0.0	100.0	0.3	7.3	7.3
Total	885,183	104,528	278,252	3,483,874	813,724	8,110,111
	52.7	1.4	1.4	30.6	13.9	100.0
	100.0	100.0	100.0	100.0	100.0	100.0

Fig. 7. The number of commodities China to South Korea in 2008
Material: statistics of UN COMTRADE

	absolute competition advantage	competitive advantage	balance	competition disadvantage
first hand material	67121 67141 67149 67151 67152 67154 67155	67159 67153		20221
semi-product	67261 67262 67249			
plate material	67351 67431 67333 67334 67541 67300 67542	67581		67431 67443 67444 67511 67532 67524 67550 67556 67551 67412
	67529 67545 67551 67507 67587 67588 67701 67538 67610 67620 67630 67640			67821
steel pipe	67512 67513 67914 67917 67944 67952 67953 67904	67915 67916		67941 67943
steel products	67119 67249 67911 67951 67074 67953 67961 67962 67903 67905 67907 67909 68111		68422	

Fig. 8. The Classification 1 of Commodities with China's Superiority to South Korea in 2008
Material: statistics of UN COMTRADE
Note: the trade balance in this table is over 10 million US dollars in surplus or deficit In Table 8

First, there are 19 products in surplus over 100 million US dollars in the trade of commodities with competitive advantage (including absolute superiority), including 2 of expendable material, 1 semi-finished product, 6 of plate, 2 of steel pipe and 5 products with trade balance over 500 million. Among which, the trade balance of 67300 is the most 5.6 billion and wins the very powerful competitiveness.

Second, there are 2 goods of steel products among the commodities with balanced competitiveness

Third, there are 13 products with competitive disadvantage (including the absolute competitive disadvantage) whose trade balance is in deficit over 30 million US dollars, including 1 of expendable material , 9 of plate, 2 of steel pipe and 1 steel product. In addition, 10 products' trade balance are in deficit over 60 million, including the products of 28221, 67413, 67443, 67511 and 67411. Trade balance of all these goods are in deficit over 100 million, which makes China's competitiveness inferior to that of South Korea.

Similar to the situation mentioned above, plate material of Chinese steel grows to the upper land in 2008 from the lower position in 2000 and takes advantage over products of South Korea.

3.3 The Changing Tendency of Products' Competitiveness from 2000 to 2008

Table 9 shows the changing tendency of products' competitiveness from 2000 to 2008. First, from the overall view, there are more goods with competitive advantage than those with competitive disadvantage in 2008 and products of plate material occupying much in the latter part, and the competitiveness of many products is strengthening. Second, in both 2000 and 2008 products of expendable material, semi-finished products, steel products and steel pipe ones are still in higher competitiveness, while many products of bar-shaped steel, steel pipe and steel are turned into competitive advantage from competitive disadvantage with strong competitiveness. Chinese steel industry is undergoing increasing overall competitiveness in products of plate material, bar-shaped steel, steel pipe and steel in 2008.

2000→2008	first hand	semi-	plate material	bar-shaped	steel pipe
absolute advantage, advantage→ absolute advantage, advantage	67182 67184 67121 67141 67149 67131 67133 67153 67199	67241 67245 67251 67262 67269 67252	67351	67629 67652 67655 67620 67646	67959 67951 67952 67953 67911 67912 67917 67916
absolute advantage, advantage→ absolute advantage, advantage balance→ absolute	67155	67249	67431 67452 67338 67882 67853 67300 67854 67541 67542 67855 67851 67537 67851	67701 67709 67651 67653 67657 67628 67610 67650 67640 67611 67658 67628	67913 67914 67951 67944 67954 67955 67916 67932 67942
absolute balance→ absolute			67571	67810	
absolute disadvantage, disadvantage→ ← absolute disadvantage, disadvantage→	28221 28229 28229 28231 28232 28233	67251 67247	67419 67414 67241 67445 67511 67535 67543 67554 67555 67556 67574 67412 67452 67461 67512 67562 67572 67552 67573 67441 67442 67444 67522 67532 67861 67553 67411 67422 67856	67659 67658 67612 67647 67654 67521 67633	67939 67941 67949 67933 67958
absolute advantage, advantage→ balance→ absolute	67122 67123 67132		67821	67649	
absolute balance→					67943
absolute advantage, advantage→ absolute disadvantage					

Fig. 9. The changing tendency of products' competitiveness from 2000 to 2008
Material: statistics of UN COMTRADE

4   Conclusion

This study uses various trade competitiveness indexes to analyze the competitiveness of Chinese steel industry to that of South Korea in 2008. The results are very clear that Chinese steel industry is in comparatively lower land in products of expendable material, semi-finished products in 2000, but many of these products have turned into goods with competitive advantage and enjoy a strong competitiveness compared to South Korea.

According to such results, Chinese steel industry had better take following strategies in order to catch up with that of Japan and keep the competitive advantage to South Korea. First, adjust the structure of M&A to improve the efficiency of production system. Second, introduce engineering technical innovation and create added value and differentiation of the products. Third, employ the revised M&A structure to promote the appearance of large-scale in domestic steel industry and enhance the height of the after-sales service.

References

1. Im, H.J.: The analysis of the South Korean steel industry's Competitiveness to Japan and China's. Journal of Trade Association 32(1), 263–282 (2007)
2. Kim, G.U., Suh, Y.S.: Competitiveness Study about South Korean, China and Japan's Iron and Steel Industry. International Trade Research 11(1), 1–24 (2006)
3. Su, K.P., et al.: FTA among South Korea, China and Japan: The Manufacturing Sectors' Response Strategies-folk Items as the Analysis Center. Humanities and Social Research Council Collaborative Study Series05-04-02, pp. 266–313. Institute for International Economic Policy, South Korea (2005)
4. Kim, S.Y.: Program of South Korean Steel industry to Improve the International Competitiveness. Journal of Trade Association 25(3), 379–402 (2000)
5. Kim, S.H.: The Comparative Advantages Study of China and South Korea Steel Trade, University of International Business and Economics, Paper of the International Trade Master (2006)
6. Li, H.: Analysis of the Establishment of the FTA among Japan, South Korea and China to the China's Steel Industry. Metallurgical Economics and Management (5), 13–16 (2004)
7. Nam, S.K.: The Effects of South Korean Trade Liberation, China and Japan's Steel Industry by the Means of Gravity Model Analysis. POSRI Practice of Management 4(2), 29–53 (2004)
8. Kazuhisa, O.: The Change of Japan's Trade Structure after a Great Growth in World Trade Volume. Economics Essays 33(1), 96–135 (1996)
9. Pang, D., Huang, R.: Integrated Analysis of the Three Top Steel Nations of Northeast Asia. Northeast Asia Forum (2), 8–14 (2007)
10. Shin, H.G.: Comparative Analysis of Export Competitiveness and Inspirations of South Korea, China and Japan's Iron and Steel. POSRI Practice of Management 4(1), 5–28 (2004)
11. Sohn, S.S., You, S.L.: Enormous Impacts of Korea-Japan FTA on South Korea's Steel Industry. Economic Research 23(2), 71–94 (2005)
12. 韓基早 · 金玲瑾: The Competitiveness Analysis of China's Steel Industry to South Korea and Japan. Journal of the Korean Data Analysis Society 10(1(B)), 379–397 (2008)

13. Wu, D.: China's Steel Industry Development Status and Inspiration Points to South Korea. KIEP, Word Ecnomics 34, 72–81 (2001)
14. Xie, X., Zhang, X.: International Steel Trade Statistical Analysis of New Developments. Journal of Institute of Wuhan Metallurgy Management (3), 16–18 (2003)
15. Zhao, C., Zhao, C., Liu, Z.: International Trade Policy Options of China's steel Industry. University of International Business and Economics (8), 15–20 (2005)
16. IISI Steel Statistics, http://www.worldsteel.org
17. UN COMTRADE, http://comtrade.un.org

Research on the Vulnerability of Coastal Zone System Development

Weixi Zhang, Ruifeng Zhang, and Yefeng Wang

School of Economics
Tianjin Polytechnic University
Tianjin, China
Zhwx_tqg@126.com

Abstract. Economy advanced and populous coastal zones employed large amounts of resources and polluted external environment. Therefore, economic and social system of these zones exhibits vulnerability and sustainable development is profoundly affected. Even the system may breaks down. This article discusses the cause for the vulnerability of coastal zone and establishes its evaluation system on the basis of analysis on coastal zone development. Analytic hierarchy process, method of calculating vulnerability and method of gray prediction are used to evaluate bearing capacity, recuperability, susceptibility and stability exerted by coastal zones in the course of time.

Keywords: Coastal Zone, Vulnerability, Index System, Evaluation System, Fragility Degree, Gray MGM (1, n) Model.

1 Introduction

Located in the intermediate zone between the land and the ocean, coastal zone includes marine water and zones where bordered the land and drowned by seawater. It is the interface of seawater, the land and the atmosphere where is endowed with abundant and multiple resources. It is also the zone with dense population where society and economy is prosperous relatively. Population, resource and environment are issues the entire globe encountered. They are most serious in coastal zones due to the vulnerability of ecological environment. Under the condition that external environment and resources had changed, the prerequisite for the social and economic development of coastal zones lies in evaluating the economic vulnerability, mastering the dynamic characteristics and figuring out existing or potential problems of the coastal zones. Regarding coastal zones as resources is a kind of thought emerged in last few years. It is the furtherance and enhancement of research on geosciences of coastal zones. Increasing of the needs, dependence and degree of cognition of human beings to resources are unprecedented nowadays. For the time being, coastal zones in our country are breeding more than 40% of the entire population where lands take up 15% of the total land area. More than 70% of the cities in our country locate around the belt of coastal zones and the output value of industry and agriculture takes up 55% of the Gross National Product.

J. Luo (Ed.): Soft Computing in Information Communication Technology, AISC 161, pp. 461–469.
springerlink.com
© Springer-Verlag Berlin Heidelberg 2012

2 Manifestations of the Vulnerabilyi of Coastal Zones

Development for economy is indeed the process where people act on natural resources on the basis of mastering social resources which include assets, capital, science and technology. Human beings can always benefit from during the process of developing and utilizing resources. Process for the development of social natural and social economy can not be separated. Utilization and management of resources is the juncture of natural structure and social structure of coastal zone system. Through a series of feedback and negative feedback mechanisms, utilization and management of resources propels the evolution of nature, society and economy of coastal zone system. It is a significant feature of modern coastal zone. It is priority of figuring out what kinds of pressure interferes with and leads to the reversion of the fragile economic system for solving the vulnerability of economic development of coastal zones; in the next place, analyzing what kinds of key factors which lead to the functional and structural degradation of the whole system owing to their incapability for maintaining stability under the occasion that external world is disturbed strongly; further more, analyzing what kinds of states are the entire system and these key factors in at present; and then analyzing responses and responses that may be taken of internal and external system to existing pressure or state. Only in this way can problems be solved.

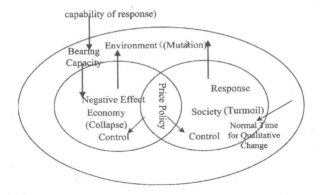

Fig. 1. Picture for Developing Coastal Zone System

3 Establishment of Index System for the Vulnerabilty of Coastal Zones

Subtle balance among seawater, the land and the atmosphere in these dynamic areas enable them become very sensitive to changes of natural conditions. According to a lot researches for the time being, it is a series of environmental changes that exert deep influences over coastal zone in Hebei Province: climate change, sea level rising, coast erosion, saltwater encroachment, loss of wetlands of tidal flats, frequent occurrence of red tide and so forth. The consequence is increasing vulnerability of natural resources in supporting system of socioeconomic system. Sequentially, the vulnerability of development for socioeconomic system is aggravated. Therefore, the selection for indexes should be in accordance with the following principles: a) scientificalness: each index within index system should be capable of reflecting the implications and abilities of economic development vulnerability in coastal zones; b) operability: indexes should be obtained easily and be possessed comparability. Data should be made easy for calculating and method of calculation should be practicable; c) significance: indexes should be capable of reflecting main features and conditions of economic development in coastal zones; d) hierarchical property: index system can be divided into different levels according to the requirements of evaluation and the degree of detail; e) Dynamic property: selections for indexes should better demonstrate and measure future development or development trend for coastal zones.

Establishment of index system: according to the above-mentioned principles and characteristics of coastal zone development, this article establishes an index system at four levels. Its level for ultimate goal is the fragility degree of economic development in coastal zones. The second and the third level is level for standardization. The fourth level is level for indexes. As shown in chart1, this article selects indexes like agricultural acreage according to principles of comprehensiveness and significance.

4   Evaluation Method of the Vunerablity of Coastal Zone Economy

4.1   Evaluation Procedures

It is not necessary to select all the above-mentioned indexes when evaluating the vulnerability of one specific coastal zone. Indexes that can reflect characteristics of the zone should be selected and initialized according to practical situations. Then, these indexes should be weighed in a certain way according to their relative importance degree. Comprehensive fragility degree evaluation is made on the basis of date obtained in practical surveys.

4.2   Standardization of Indexes

Initial data series of all the collected indexes should be standardized in order to eliminate the influences from disparity exists in initial data dimension. Maximum standardization method is adopted for standardizing initial data series in this article:

When index Dij correlates negatively with fragility degree, meaning the larger the amount of Dij the better,

$$p_{ij} = \frac{D_{ij}(0)}{D_j(\max)}$$

In this formula, pij is the standardized value of each index; Dij (0) is the initial value of the number j index in number i area; Dj (max) is the maximum value of index j in all the zones.

When index Dij correlates positively with fragility degree, meaning the smaller the amount of Dij the better.

$$p_{ij} = 1 - \frac{D_{ij}(0)}{D_j(\max)}$$

In this case, its correlation can be integrated.

4.3 Confirmation of Index Weight

Index system for evaluating the vulnerability of coastal zones is a decision-making issue with multiple targets, levels and principles. Analytic hierarchy process (APH method) is adopted in this article to evaluate each evaluation factors. This method is a kind of decision –making analytical method with combination of fixed quality and quantity put forward by American operational research expert A.L.Saaty in the nineteen seventies. This method makes decision-making process by decision makers a modeling and quantified one. Analyses and evaluations should be conducted level by level by experts and decision-makers according to paired comparisons between two listed indexes in the aspect of importance degree on the basis of breaking down complicated problems into several levels and several factors (or indexes). And according to the weight of each index judging by feature vector of the matrix, references for decision-making can be provided.

4.4 Outcome Evaluation on Fragility Degree

All kinds of fragility degree (G_{ij}) can be worked out according to the following formula once different zones in need of evaluation been selected, index system established and weights of each index given.

For instance:

$$G_{1i} = 1 - \frac{\sum_{j=1}^{n} p_{ij} \bullet w_{ij}}{\max \sum_{j=1}^{n} p_j \bullet w_j + \min \sum_{j=1}^{n} p_j \bullet w_j}$$

In this formula, G_{1i} represents retrospective fragility degree in zone i; p_{ij} is the standardized value of each index;

$\max \sum_{j=1}^{n} p_j \bullet w_j$ is the summation of the maximum value of indexes in all the zones and product weights; $\min \sum_{j=1}^{n} p_j \bullet w_j$ is the summation of the minimum value of indexes in all the zones and product weights; n represents number of the evaluated zones.

4.5 Trend Assessment of Fragility Degree

Gray theory can be used to make descriptions with fixed quantity on trend assessment of each index value. Because of the fact that gray theory takes variational situations of index into consideration on the whole, it is better than average increasing rate index of many years. Prediction method used widely can be divided in two categories: one category is explanatory prediction method, which means figuring out every influencing factor being predicted and establishing analysis of regression model; the other category is time series analysis approach, which depends simply on historical observation data being predicted and data pattern. Its order change law can be observed on the basis of sequence analysis. Fragility degree is affected by many factors and is possessed with certain gray features. Therefore, prediction conducted by making use of gray system model accordance with few data in recent period and development trend is one kind of effective method, especially when reliable date of longer series is unavailable. GM(1,1) model can not reflects constraints, interrelationship and coordinated development among multiple variables by making use of single time series data. GM(1,n) model used mainly for describing interrelationship between variables and can not applied to prediction. For the above-mentioned reasons, this article adopted MGM(1,n) model which can take self-adaptability to gray prediction model of multiple related variables into consideration. It is the extension of GM(1,1) model under the condition that variable n has many variables. But it is not simple combination of GM(1,1) model and is different from GM(1,1) model which establishes only first-order differential equation with single variable n. MGM(1,n) model establishes n differential equations with n variables. Parameters in MGM(1,n) model can reflect mutual influence between variables by adopting the method of simultaneous equations.

MGM(1,n) model adopts generate sequence modeling, supposing that there are n time series data with m dimensions to the question. Every serial represents dynamic behavior of one factor variable. A set of data of fragility degree $x_i^{(0)}$ can be worked out based on six sets of data obtained from average values once every five years according to data of the above-mentioned indexes of the eleven coastal counties in Hebei Province. $x_i^{(0)}$ can be obtained through index data set, as shown in the formula,

$$x_i^{(0)} = \left\{ x_i^{(0)}(1), x_i^{(0)}(2), \cdots, x_i^{(0)}(6) \right\}$$.

Gray system modeling is relative successful in conducting short-term prediction. But with the expansion of predictive period, future perturbation or random factors can exert deep influence to the system. In order to predict situation of the system in longer period, gray recurrence dynamic modeling of same dimension can be adopted: adding $\widehat{X}^{(0)}(m+1)$ into the initial series and omitting initial data. Next step can be predicted once the new series of same dimension being used for modeling on the basis of $\widehat{X}^{(0)}(1)$: $\widehat{X}^{(0)} = \left\{ \widehat{X}^{(0)}(k) \middle| k = 2, 3, \cdots, m \text{ and } \widehat{X}^{(0)}(m+1) \right\}$. Predicted values of fragility degree of each item in twenty-five years successively can be worked out: $\widehat{x}_i^{(0)} = \left\{ \widehat{x}_i^{(0)}(1), \widehat{x}_i^{(0)}(2), \cdots, \widehat{x}_i^{(0)}(5) \right\}$.

5   Case Study-Take Coastal Zones in Hebei Province as Example

With coastal zone gross area of eleven thousand and three hundred seventy-nine point eighty eight square kilometers which takes up 32% of total area of Qinhuang Island, Tangshan and Cangzhou, coastline along Hebei Province lasts Four hundred and eighty-seven kilometers. Its land area is three thousand and seven hundred fifty-six point thirty eight square kilometers; its inter-tidal zone area is one thousand and one hundred sixty-seven point nine square kilometers; its shallow sea area is six thousand and four hundred fifty-five. It has one hundred and thirty-two islands with island shoreline being one hundred and ninety-nine kilometers and island area being eight point forty three square kilometers; with straight shoreline, small wave and below three-level stormy waves under general occasions, coastline of Hebei Province is mainly composed of bass land mass, chiltern and muddy coast. It has a semi-humid continental monsoon climate warm temperate zone; Chief ocean industries in Hebei Province include aquaculture, communications and transportation, ship building and mending, raw salt industry, salt chemical engineering, petroleum industry and tourism. Along coast belt of Hebei Province there are three cities: Qinhuang Island, Tangshan and Cangzhou, six counties: Funing, Changli, Leting, Luannan, Tanghai, and Haixing, two county-level cities: Fengnan and Huanghua, three districts: Shanhai Pass district, Haigang district and Beidaihe district and three economic and technical development zones in Qinhuang Island, Jingtang Port and Huanghua Port as well as two county-level farms in Zhongjie and Nanda Port. In this article, results obtained according to the above-mentioned evaluation theory and method is shown in chart2.

Evaluations on current situations: in this evaluation, evaluation results of bearing capacity and recuperability in coastal zones is better than those of susceptibility and stability. It is primarily due to the fact that with relative reasonable economic structure, higher per capita income level, small disparity between the poor and the rich and abundant talent resources, coastal zones are economy developed zones. But with the development of the economy, demand for resources has increased at a quick pace and is bigger than its supply capacity. As a result, there emerged resource exhaustion; at the same time, inappropriate handling of large amounts of rubbish generates by industries and daily lives had given rise to serious environment pollution; Trend assessment: In the future twenty-five years, with the development of the economy, the vulnerability of economic and social system displayed in various zones would decrease. Economic and social system tends to develop stably. With a strong stability, Haigang district would keep its leading position in performance results and cause of formation. Shanhai Pass, Beidaihe and Fengnan enjoy strong stability as well in the process of development; standing from the perspective of susceptibility and stability, Changli, Leting, Luannan and Tanghai are in stable states. But standing from the perspective of fragility degree of bearing capacity and recuperability, Changli and Tanghai is higher than Leting and Lunan; while, the stabilization of susceptibility and stability in Haixing had increased by 0.035, but fragility degree of bearing capacity and recuperability had increased by 0.051.It illustrates that its economy is developed at the expense of environment and resources, which also means that the loss outweighs the gain.

Table 1. Index system of the vulnerability of coastal zones

Level for Objective A	Level for Standardization B_i	Level for Standardization $2C_i$	Level for Index D_j	Sources of Materials
Vulnerability of Coastal Zone Economic Development in Hebei Province	Retrospective Evaluation B1	History Evolution C_1	Way of production and life D_1	Marks Graded by Experts
		Other C_2	Illiteracy RateD_2 Undergraduate Ratio D_3 Per Capital Gross Domestic Product D_4	Obtained Directly from Statistical Data
	Evaluations on Current Situations B2	Bearing Capacity C_3	Per Capita Land Area D_5 Per Capita Agricultural Acreage D_6 Per Capita Volume of Water Resources D_7 Per Capita Water Area D_8 Fishery Resources (Reserves of Representative Specialties in a Certain Region) D_9 Mineral Reserves D_{10} Annual Port Capacity D_{11}	Obtained Directly from Statistical Data
		Recuperability C_4	Vegetation Deterioration Rate D_{12} Economic Losses Caused by Natural Calamities D_{13} Environmental Quality Index D_{14}	Obtained Directly from Statistical Data
		Susceptibility C_5	Dependency of Resources Industry Output Value D_{15} Economic Diversified Coefficients D_{16} High and New Technology Industries Ratio D_{17} Growth Rate of Per Capita Gross Domestic Product D_{18} Economic Structure (Ratio of the Three Industries) D_{19}	Percentage Taken up by Resources Industry Output Value of Total Output Value Obtained Directly from Statistical Data

Table 1. (*continued*)

		Stability	Engel Coefficient D_{20} Average Life Span D_{21} Poverty Gini Coefficient D_{22}	Obtained or Calculated Directly from Statistical Data
B_3 Trend Assessment		Bearing Capacity C_5	The Above-mentioned System Predicts Indexes in Twenty Years (Once Every Five Years) and Draws up Graph $D_{31} \sim D_{34}$	Gray Prediction
		Recuperability C_6		
		Susceptibility C_7		
		Stability C_8		

Table 2. Chart of results of the economic vulnerability of coastal zones in hebei province through statistical calculation

			Shanhai Pass	Haigang District	Beidaihe	Funing	Changli	Leting	Fengnan	Luannan	Tanghai	Haixing	Huanghua
Evaluations on Current Situations			0.764	0.781	0.752	0.612	0.654	0.687	0.740	0.672	0.631	0.548	0.611
Trend Assessment	Bearing Capacity Recup-erability $_8$	05~10	0.620	0.638	0.67	0.571	0590	0.606	0.697	0.590	0.540	0.447	0.501
		11~15	0.618	0.635	0.663	0.560	0.576	0.598	0.692	0.587	0.532	0.438	0.492
		16~20	0.612	0.634	0.654	0.542	0.552	0.592	0.685	0.570	0.526	0.430	0.487
		21~25	0.598	0.632	0.632	0.524	0.534	0.581	0.678	0.561	0.519	0.421	0.472
		26~30	0.594	0.630	0.627	0.503	0.511	0.572	0.671	0.556	0.512	0.405	0.465
	Suscep-tibility Stability	05~10	0.798	0.842	0.810	0.658	0.729	0.762	0.787	0.742	0.723	0.608	0.708
		11~15	0.799	0.851	0.816	0.665	0.734	0.768	0.794	0.748	0.727	0.612	0.715
		16~20	0.821	0.857	0.819	0.672	0.740	0.773	0.798	0.752	0.736	0.621	0.721
		21~25	0.824	0.863	0.824	0.679	0.746	0.781	0.804	0.761	0.745	0.629	0.732
		26~30	0.832	0.869	0.832	0.684	0.756	0.793	0.811	0.774	0.754	0.638	0.746

Acknowledgements. Project of Ministry of Education of the People's Republic of China. (10YJC630389)

References

1. Jiang, T., Wang, Z.: Research on Sustainable Development of Marine Economy along Bohai Rim. Maritime Press, Beijing (2000)
2. Gao, J.: Exploration on Sustainable Development Theory. China Environmental Science Press, Beijing (2001)
3. Liu, S., Guo, T.: Gray System Theory and its Applications. Science Press, Beijing (1999)
4. Yun, C., Jiang, X.: Coastal Sustainable Development and Integrated Management. Maritime Press, Beijing (2002)

5. Liu, Y., Li, X.: Fragile Ecological Environment and Sustainable Development. The Commercial Press, Beijing (2001)
6. Zhao, Y., Zhang, L.: Method Study on Quantitative Evaluation of Fragile Ecological Environment. Scientia Geographica Sinica, 73–80 (May 1998)
7. Ran, S., Jin, J.: Evaluation Theory and Method on Fragile Ecological Zone. Journal of Natural Resources, 117–123 (January 2002)
8. Li, B., Xu, S., Bo, G.: Prediction of City Water Supply Quantity with Gray-Neural Networks Combine Model. China Water & Wastewater 18(2), 66–68

How to Resolve Problems of Strategic Positioning of Higher Institutions—From the Perspective of Corporative Strategic Analysis

Fazong Qiu and Sisi Chen

School of Management Tianjin Polytechnic University Tianjin, China
qiufazong2222@sina.com, angel_css123@sina.com

Abstract. At the present stage, there exist some basic contradictions which still have not been resolved for many years in strategic positioning for the development of higher institutions of China. The way to deal with the problems is to break through case-style thinking method and expand visions of strategic analysis. Since theories and experience of corporative strategic analysis are old and profound, they will definitely provide a salutary lesson to do research on strategic positioning of higher institutions.

Keywords: strategic positioning, discipline category structure, hierarchy.

1 Positioning of "Discipline Category Structure"—Analysis of Contradictions between Specialization and Diversification

Concerning the basic connotation of strategic positioning of higher education, Wu Shulian stated in 2002 that the category of university is constituted of two parts --- "type" and "model". That is to say, higher institutes should be divided according to "discipline structure" and "hierarchy structure". [1] On December, 2006, Tang Weimin put forward three aspects---"hierarchy structure", "discipline structure" and "regional distribution structure" in developing higher education of China, when he analyzed the structure of private higher education of Japan. [2] Meanwhile, Xie Weihe carried out detailed analysis and evaluation in popularization process of higher education of China, also in accordance with the above three aspects--- discipline structure, hierarchy structure and distribution structure. [3] The paper analyzed problems of strategic positioning of higher education from two basic aspects--- "hierarchy structure" and "discipline structure".

Scientific distribution of "discipline structure" is the priority to analyze.

From the scope and size of courses involved, the "discipline structure" of higher education can be generally divided into single-science university, multidisciplinary university and all-around university. Single-science university put emphasis on selection and arrangement of specialization strategy. All-around university belongs to multiplied strategy. And multidisciplinary university is closed to the multiplied strategy. Thus, the discipline structure of higher institutes is attributed to the analysis of contradictions between diversification and specification in essence.

J. Luo (Ed.): Soft Computing in Information Communication Technology, AISC 161, pp. 471–479.
springerlink.com   © Springer-Verlag Berlin Heidelberg 2012

1.1   Corporations: How to Resolve the Contradictive Relationship between Diversification and Specification

Whether to choose diversified or specified management is one of the key points and difficult points in strategic analysis of corporations. Specification emphasizes one product or industry with an eye on deep exploration. While diversification touches upon many products and industries focusing on wide exploration.

Which one is better? Actually, diversified strategy should be carried out with conditions. Firstly, the prerequisite of diversification requires mature diversified management. The original product or industry should be consolidated in the first place. And expanding business can be considered later unless the former industry has lost its reason to exist and space to develop. Secondly, diversified management must be moderate. Not everything is practical just like the bigger the stall puts up, the better. Thirdly, diversification should be based on the order of complexity. According to feasibility, the degree of relativity must be paid special attention. Then, transformation to uncorrelated diversification or cross-business management can be taken into consideration. Fourthly, diversification is closely related with the life circle of product. If an industry stands at the early stage in the life circle of product (investment period, growth period) and possesses great developing potential, we should enter such industry as soon as possible. But if joining at the later stage in the life circle of product (mature period, recession period), there will be few chance to win. Since market is closed to or already saturated, the competition must be intense unconventionally.

In recent years, many eminent corporations have continuously sent signals that they retired at the height of the career. Siemens reluctantly sold the production line of color TV. Philips remised the subsidiary which produces large electronic appliances. Philip Morris auctioned off "non-core business", laid off employees and regrouped. And GM Motor will also sell 15 spare factories in order to improve competitiveness furthermore. [4]

In a word, many famous corporations at home and abroad all have experienced the path of management from specification at first, diversification in the middle, and then diversification with limitation. They do not pursue large-scale production and diversification without any analysis and consideration.

1.2   Higher Institutes: Imbalance of "Discipline Structure" Positioning

Understanding the contradictive relationship between diversification management and specification management of corporations can be a significant inspiration for strategic positioning of "discipline structure" of nearly 2000 higher institutes in China.

Before and after the 20th century, higher education in China has entered a period for expansion and rapid development. According to the strategic positioning of "discipline structure", many higher institutes set their targets to build multi-subject, all-around universities. How to understand and evaluate such a strategic option? The expansion on the scale and quantity of higher education in China plays an important and necessary role in economic and social development to some extent. However, when the principal contradiction in higher education shifted from quantity and scale to quality improvement and structure optimization, the defects of pursuing "all-around" "diversified" university one-sidedly and absolutely become evident increasingly. What we cannot fail to point out demonstrates that there are a large number of universities which not only lack scientific analysis and prediction toward external environment, but also lack support from internal resources and competitiveness during expansion with great blindness. That will definitely make the contradiction between quantity and quality much more acute. As described concretely as follows: large and all-inclusive specialty, relatively high cost, excessive indebtedness, irrational discipline structure, structural difficulty of students' employment, lack of school-running characteristics, low-level competitiveness and questionable quality improvement etc.

Specifically, the misunderstanding is mainly concluded in following aspects:

First, bracket "making bigger" with "making stronger", and equate "expanding scale" to "competing for the best" without analysis. Actually, both "making bigger" and "making stronger", plus "expanding scale" and "competing for the best" have no obvious inner relations. On contrary, "specialization" and "small scale" also can "make stronger". And it's not rare for them to "compete for the best". For instance, California Institute of Technology in U.S. is world famous university acquiring lots of Nobel Prizes. While only 2000 students or so study on campus. Even if in the most eminent university in U.S. --- Stanford University, only about 5000 students are enrolled for further education. Harvard University, with the largest enrollment, merely has 20,000 students in all. [6] That does not hinder them to be the "world- class universities". However, only founded less than two decades, some higher institutes in China have already enrolled 20,000-30,000 students, contained eight to nine of all twelve major subjects and covered over fifty disciplines among the first and second level of disciplines. Of course, due to hysteretic higher education of China in history and a large mount of debt for a long period, it is also necessary to establish more disciplines and enroll more students to satisfy people's needs for higher education to some extent. But there is a long way to go because rationality and scientific research should be considered. Many companies like "Sanzhu Group" and "Giant Group" etc which led to rapid closure because of blind expansion in the initial period of China's reform and opening up attracted our great attention and calm reflection.

Second, violate the general laws of transferring specification to diversification in the educational process to pursue multi-disciplines and diversification. From the perspective of strategic positioning of corporations, the path from specific management at the very beginning, then the correlative diversification, and irrelevant diversification at last, is regarded as a general rule to follow according to relative costs and the theory of comparative advantages. Taking General Electric Company for example, it belongs to a typical company for specific management in its early period. It spent more than two decades only on a light bulb, achieving 90% market share. From then on, it began to undertake another product, neither of which was network nor finance, but electro mechanics and other correlative products, closely abiding by the above principles of correlative diversification. Years later, GE spread towards other industry. [7] So do higher institute. But many of higher institutes in China violate the principle obviously when putting the diversified strategic positioning into practice. They spend too much money and cover too many disciplines regardless of conditions. After large investment in professionals, the prospect still looks bleak because of lack of market and comparative advantages. Of course, both passion and ambition are necessary. But any activity has to obey the general rules and objective conditions, which belongs to the common sense of materialism. Otherwise, any action will lead to passive situation and frustrated result.

Third, mishandle the relation between specification and diversification, and the "main business" appears not prominent. According to the requirement of strategic management, optimizing the allocation of resources towards main business should be the general rule. When the wind of "integrated management" swept the insurance company, China Life Insurance (group) corporation yet adopted to "strengthen main business and diversify others appropriately" as its development strategy. It took life insurance as the core business, which property insurance, pension, and annuity were attached to. Besides, bank account, security, fund and entrust etc were all included in non-core insurance business. Thanks to scientific positioning, China Life Insurance Corporation grows rapidly and has become a truly "world-class". From the year 2003 to 2009, the rank of China Life Insurance in the World 500, namely: 290, 241, 212, 217, 192, 159, and 133. [5] However, in many higher institutes of China, there exists irrational allocation of finance and human resource in strategic positioning and selection. They make average efforts and that results in a situation that we "only see the plain but no peak", which has to be considered as a strategic mistake. In fact, even the world-class university definitely cannot make all its disciplines become the best. A university will exert its influences at home and abroad if it has one or more brilliant subjects, or some characteristic disciplines.

2   "Hierarchy Structure" Positioning—Analysis of Contradictive Relation between High and Low Level

From the perspective of hierarchy of personnel training, "hierarchy structure" in higher institutes can be divided into research type, specific type and skill type in general. Following these principles, many scholars divide higher institutes of China into three basic levels: university of research type, intermediate university (university for education) and university of skill type. Research type is located in the highest level with purpose of cultivating talented people for research. Intermediate university (university for education) focuses on cultivating advanced expertise in the industry. And university of skill type puts emphasis on training skilled personnel as its basic positioning. Theoretically, it can be concluded at the understanding and analysis of the contradictive relation between the high and low level of higher institutes.

2.1   Corporations: How to Straighten Out the Contradictive Relation on the Strategic Positioning of Hierarchy Structure

Hierarchy structure of corporations can be divided into two aspects: positioning on industrial hierarchy and product hierarchy. Industrial hierarchy refers to enterprise choosing knowledge-intensive, capital-intensive and labor-intensive positioning. While, product hierarchy refers that companies choose to position on investment, production and marketing of products on high, middle and low levels. How to choose the positioning of hierarchy is based on system. Hierarchic quality is one of the important characteristics of a system. Everything has to exist as a system.

How to choose and decide products on high and low levels or exploit market on high and low levels? From the perspective of strategic analysis of corporations: demand of the market should be firstly considered. That is to say, consumers' requirement towards products, receiving level of the price and its distribution will be the priority. Second, competitiveness of the corporation has to be analyzed. Many factors including technology, talented personnel, marketing and intangible assets should be considered when comparing with other corporations in the same business. Third, on the basis of comprehensively analyzing the matching of demands in market and capabilities of the corporation, the strategies that are in line with reality can be adopted.

When a grade of product possesses great opportunity for marketing and strong competitiveness compared with other companies in the same business, a proactive strategy should be adopted such as enlarging the finance etc in order to consolidate and expand market share. When the company stays at a disadvantage and lack competitiveness, even if its product has great opportunity for marketing, effective measures have to be taken to reverse the poor situation as soon as possible and strive for a certain market share. When the market of a grade of product has been shrinking or the peak period has passed, even though the company has an industrial advantage for the product, diversification and multi-level management should be expedited to develop, which may secure the whole business. It will be a wise option for the company to withdraw from the grade of product as quickly as possible when the market has shrank and the peak period has passed. Meanwhile, the company does not have any advantages in the business.

Based on the above descriptions, the diagram is shown as follows (generally speaking: SWOT analysis):

2.2　Higher Institutes: Imbalance of "Hierarchy Structure" Positioning

Imbalance of educational structure in higher institutes is reflected not only in discipline structure, but also in hierarchy structure. "In practice, some parts of universities still follow the traditional cultivation model with single object. They put emphasis on cultivating research and academic talents, and blindly compare with elite universities for expansion, which leads to a result that the trained talents all loss touch with society."[6] The following two aspects will illustrate the problem explicitly.

Firstly, the imbalance of undergraduate students compared to the proportion of skilled occupational talents

What kind of proportional relationship should high-level specialists have comparing with skilled personnel? From the view of developed countries, according to research from U.S. Department of Labor, the proportion of high-level specialists among practitioners has not yet changed and stayed at about 20% for a half century. While the proportion of skilled personnel has turned from 20% up to 60%, which fully illustrates the overpowering need for the skilled people during the development of economy and society. Some scholars also find the training structure through quantitative research: "the proportion of skilled personnel among the labor force reaches up to 75%...... in developed countries. Many indications show that partial shortage of skilled personnel has evolved into a general shortage of supply when industrialization developed till now. " [7]

What about the situation in China currently? Li Peilin and Mo Rong mentioned in "2008 Social Blue Book of China": "On one hand, the situation with limited shortage becomes much more prominent and the shortage of skilled personnel increases with large gap. But on the other hand, shortage of high-level labors is much larger. That is to say, the tension in the labor market composed of urban employment from university students above exacerbates the difficulties. " Therefore, it has already been a well-known fact that the cultivation of college students is relatively surplus and high- quality skilled personnel becomes hard-up. According to statistics from Tianjin Human Resources and Social Security Bureau, 5000 university students went back to technical schools for further study from the year 2008 and the first half of 2009. Why? First, students lack teaching practices and are out of touch with the requirements of employers. Take some hot professionals in the university for example, students majored in machinery manufacturing don't know repairing; those who study automation have no chance to operate CNC machine tools but return to technical school. Second, many undergraduate students are cultivated to excess ion. Graduates majored in law, accounting, management and so on could not find suitable work and went back to technical school for skill. [8]

Secondly, the imbalance of the ratio between undergraduate and graduate education

In recent years, the difficulty of undergraduate students' employment began to spread to graduate level. Some deputies to the People's Congress pointed out that "if the popularization of higher education results in difficulty of undergraduate students' employment, the cultivated graduates as the elite will also feel headache, which can not but arouse reflection from higher institutes and education department."[9] Actually, enrollment of graduates is same as that of undergraduates. They both have to be limited and influenced by economic and social structure, and even the structure of higher education. Early in 2004, scholars wrote: "the scale of graduate enrollment is closely related to economy, which is also bound by the constraints of economic development and at the range of national economic development. Otherwise, it will bring heavy burden towards the evolution of economic and social development. Because not only the national development of graduate education, but also graduate students need appropriate economic capacity, potential boundary must be found and reasonable growth model for the scale of graduate enrollment should be established."[10]

The changes of the number of graduate applicants in China for recent ten years can prove this point.

As shown from the above data, the number of graduate applicants in China had increased from the 2000 to 2006. But during 2006 and 2008, increase of graduate enrollment in China tended to be smooth and the gross enrollment began to slide down. After 2008, graduate enrollment resumed again. The contradiction of structural adjustment became more and more prominent along with rapid growth of graduate enrollment in absolute quantity. Higher educational circle in China has called for "transformation from compensatory increase to adaptive increase proactively". Adaptive increase refers to appropriate growth strategy emphasizing on optimizing structure and enhancing quality. In accordance with requirements from Ministry of Education, graduate recruiting units should decrease enrollment of academic graduates by 5% to 10% based on the enrollment last year from the tear 2010. The reducing part will be used to recruit graduates for professional degree until the number of academic and professional graduates is controlled according to the ratio of 1:1. "Strategic transformation from fostering academic talents as the main task to cultivating applicative talents" will be meaning to promote graduate education of China.

3   Basic Ideas of Adjusting Imbalance of Higher Institutes' Strategic Positioning

How could higher institutes position scientifically in connection with high tendency of unilaterally pursuing to be "large and comprehensive", and advocating "academy" existed in higher educational circle of China under the huge background of transforming quantitative growth to quality enhancement and structure optimization in higher education of China?

Table 1. Diagram of tactic analysis on basic corporative strategies

		Inner factor	
		advantage	disadvantage
External factor	opportunity	1、growth strategy	2、reversal strategy
	threat	3、multi-level management strategy	4、defensive strategy

Table 2. The table of the number of graduate applicants in China between the year 2000 and 2011

Year	2000	2001	2002	2003	2004	2005	2006	2007	2008	2009	2010	2011
Number of applicants (thousand)	392	460	624	797	945	1172	1271.2	1282	1200	1246	1400	1512
Compared with the last year(%)	——	17.3	35.7	27.7	18.6	24.0	8.5	0.8	—6.4	3.8	12.4	8

resources: (1) http://yz.chsi.com.cn/z/yzrxks/information network of graduate enrollment in China;
(2) http://edu.qq.com/y/yztstart.shtml/QQ education

According to SWOT strategic analysis method, the following steps should be made from the perspective of strategic positioning of higher institutes:

Firstly, scientific analysis on "external environment" should be done well. Make sure the structure and ratio of skilled, professional, and academic talents needed in the development of economy and society. And then do scientific analysis on the two tendencies of "opportunity" and "threat". The focus is to figure out which period the social needs for each level and each type of the talents experiences. Do they go through the period of strategic opportunity, saturation, or even excessive supply?

Secondly, scientific analysis on "inner environment" should also be done well. The competitiveness of current resources in higher institutes should be analyzed. Compared with other universities, what kind of situation do the current resources (teachers and talents, disciplinary construction, infrastructure, management, and historic foundation) stay in? Do they stay in the upstream, midstream and downstream without developing potential etc?

Thirdly, comprehensive analysis of both "external factors" and "inner factors" should be made to ensure an appropriate strategic tactics. For example, if "external environment" possesses great opportunity and a large gap exists in social needs, meanwhile, a large number of resources and great competitiveness are plentiful in the higher institute; "growth strategy" can be adopted actively. If social needs have been in saturation status basically and the higher institute's competitiveness is just ok; prudence should be considered. If the social needs turn to surplus with poor competitiveness and resources, decisive exit will be a wise choice and "defensive strategy" can be made use of. If a higher institute pursues to be the highest one-sidedly, and continuing to sustain the "riding tigers" situation will cost a large mount of money without hope etc; it should adopt "reversal strategy" timely to reducing the scale or exit directly.

Acknowledgment. Fund Project: the paper is the periodical result of "the research on employment enhanced by entrepreneurship of humanities and social science university students" (TJJX10-1-849) in planning project in philosophy and social science of Tianjin in 2010.

References

1. Wu, S.: Discussion on classification of universities again. Accession of Higher Education of China (Shanghai) (4) (2002)
2. Tang, W.: Structural analysis of private higher education in Japan. Heilongjiang Researches on Higher Education (12) (2006)
3. Xie, W.: Structural analysis of popularization of higher education in China. Educational Science and Publishing House, Beijing (2007)
4. Sun, J.: Improving the focus capacity and growing in temptation (July 11, 2005), http://www.boraid.com/
5. Fortune 500, China Life Insurance ranks 133, http://www.e-chinalife.com/ (July 09, 2009)
6. Zhan, X.: Meditation of talents cultivation goals towards undergraduate students in local colleges. Heilongjiang Education (Research and Accession of Higher Education) (6) (2006)

7. He, Y., Zhang, B.: Theoretical analysis on insufficient supply of skilled talents in China. Education and Vocation Theoretical Edition (1) (2008)
8. Yang, X.: Universities' diploma was challenged by vocational skills. News Tonight, Tianjin (August 11, 2009)
9. Liu, Y., Gao, F.: Talents also encounter a difficult winter Deputy of People's Congress consult on graduate employment (March 13, 2007),
    1.http://www.xinhuanet.com/edu/
10. Chen, S., Wang, J., Yin, L.: Discussion on recent and long-term developing models of recruiting scale of graduate in China. Research On Education Tsinghua University (4) (2004)

Analysis on Tianjin Flavor in Adornment of Kite Wei

Xi Gai and Jicheng Xie

Institute of Art & Fashion Tianjin polytechnic university Tianjin, China
{214869516,1378759338}@qq.com

Abstract. As a representative part of folk culture in Tianjin, Kite Wei is leading the development of intangible culture of Tianjin. It wins an international reputation for its exquisite form and unique decoration style which contains rich cultural accumulation and folk cultural essence. So far how to understand Tianjin culture has become an urgent issue.

Keywords: Tianjin flavor, decorative style, aesthetic meaning.

Tianjin Kite Wei has occupied a significant status in developing local culture as an important part intangible culture heritage of Tianjin. Different from painted sculptures by Ni Ren Zhang and New Year's painting by Yang Liu Qing, besides a good decoration, Kite Wei is more a sporting game for playing and exercising. All kite, regardless of the material, are critical in form and pay more attention to develop local culture in decoration besides considering the harmony with space and environment. The little kite concentrates 600-year Tianjin culture and people learn something while flying kites. Moreover, the decoration of the kite contains generous local culture in aesthetic value and aesthetic idea. So how does Kite Wei improve to be the leader of the intangible culture heritage of Tianjin by its chic decoration and combine together with local folk cultural essence? How does it become such a popular art work and influence the development of other folk art works?

1 The Origin of Tianjin Flavor in Adomment of Kite Wei

As a part of folk art in Tianjin, the decorative style of Kite Wei has always been influenced by the multi-culture of Tianjin. Tianjin, located in the lower reaches of nine rivers, is a city originating from wharf culture and is strongly affected by foreign culture as a concession of nine countries. The kite became the tool transmitting Chinese culture to western countries, inevitably it was impacted by western culture and decorative style while accepted by westerners, especially the composition and color employed western aesthetic ideas. This originates the special history of Tianjin.

Tianjin is a city strongly influenced by shop culture and shop is an earlier operation pattern with local characteristics. Even in ancient time, there was a popular saying in business that "successful business brings a great fortune". And shop is quite important in the traditional operation pattern. The ancient businessmen would like to sell goods on stalls beside the road rather than keep a business in a deep street in order to win the effect of prosperous sales. As a result, Kite Wei was set up very early in the form of shop. Wei Yuantai (1872—1961) first founded the first Chinese kite shop named of " Wei Changqing Kite Shop" after his father's name at the end of Ming Dynasty and the beginning of Qing Dynasty. Because of the expanding requirements, the kite was made in the model of mass and artistic production. So it's different from other kites both in composition and decorative style and its decorative meaning caters to the popular aesthetic need of Tianjin via its marketization.

wKite Wei has been popular for such a long period of time because its decorative style complies with the thinking pattern of folk art and decorative elements express local folk cultural customs. No matter the traditional pattern design or the modern abstract subject, both of them consider how to explore profound the esthetics connotation of folk art, how to inherit and the specific form of folk art, and how to involve more periodical elements and respect traditional culture. Wei Yuantai, the founder of Kite Wei, improves the decorative skills to a high level by employing skills of color change, comparison of cold and warm colors and central symmetry in traditional Chinese detail painting, which is acknowledge as " continuous innovation while learning". It's meaningful that he ingeniously combines the essence of folk art and kite together. In addition, only the close integration of the kite and the thinking pattern of folk art can reflect its tenacious vitality, inspect its development and maintain its vigor in decorative style.

2   The Methods of Tianjin Flavor IN Decorative Composition of Kite Wei

All kinds of kites must take strict composition form in decorative style and provide a whole and happier composition form in the sky no matter how the kite is operated or decorated. The composition devotes particular care to the purpose and image and the purpose always comes first. So that the kite can possess a perfect form with salient image, various veins and harmonious changes. And only a good composition in the sky can make the player and the audience relaxed.

The makers of Kite Wei lay great emphasizes on the composition in the process of decoration. They stress the site of the thematic figure, the anathematic figure in the picture and the relationship between the bottom configuration and the figure. Moreover, they are good at creating the kite by coordinating the part and whole. Wei Yuantai has won a prize in an international exhibition via the exquisite skills and perfect decoration of his Kite of Centipede with 100 knobs and 200 beards. This five-zhang kite can be placed in a small box which shows its high exquisiteness in skill. And the kite is decorated by clearly unified color and delicate strips in detail.

Furthermore, Kite Wei pays great attention to Tianjin flavor in decorative composition which is often decorated with local model drama full of Tianjin flavor, such as Pecking Opera with Tianjin flavor, famous ping opera, and characteristic Hebei Bangzi. The makers employ their regional elements in clothes and facial masks and input these elements of cultures of opera, facial mask and clothes in the picture in proportion. The typical decorative composition with strong Tianjin flavor makes Kite

Wei stand out when they are flying in the sky with Kite Cao in Beijing and Kite Yang in Weifang, Shandong together.

3   The Decorative Style of Kite Wei with Tianjin Flavor

Different from other decorative styles, Kite Wei stresses more on Chinese folk elements, for instance, folk auspicious totem and pictures are always used in the exterior decoration and the makers often turn to the painting skills and ideas of traditional Chinese painting in the painting style.

The use of the bone pen endows the kite lively. Besides, it also emphasizes the popular folk image, such as bat, carp, sand martin, magpie, dragon, phoenix and Kylin and so on. And the decorative styles can be divided into following aspects.

First is the decoration of the picture. Many of the kites made by Kite Wei possess rich decoration. The decorative style can form many visual stimuli for the audience in distance and avoid the little object in the sky. For example, the Kite of Butterfly (see attached picture 1) just uses the facial mask in Jin Opera on the kite, and its decoration gives up the realistic decoration and turns to the separating composition, in which the facial mask locates around the picture to make the kite more beautiful, create an auspicious atmosphere and give people an impressive visual feeling. (Figure 1)

Fig. 1. Kite Wei

Second is the folk tone. Citizens in Tianjin are very hospitable and people with a good knowledge of Tianjin culture are all familiar with the hospitable and positive tone. So Kite Wei tries to make an obvious contrast in color. On one hand, they want to cater to the buyers; on the other hand, kite with obvious contrast can still be seen even in the high sky. So big color lump is one of Kite Wei's characteristics. The kite in attached picture two highlights the ambition and tension of male butterfly by the contrast of yellow and black. And this skill that decorating the kite with color lump looks plain and graceful. Furthermore, the Kite of Sand Martin also makes use of bright colors of black bottom of a pan, blue one and red one. After careful disposition, the color lumps display very intensive decorative effect. (Figure 2)

Fig. 2. Kite Wei

Third is the exaggeration and abstraction. One of the most important features of Kite Wei is the transformation and exaggeration of the material in decoration, including the general abstraction of the color and size and the abstraction of the image. Abstract decoration provides a feeling of simplicity, colorfulness and ruggedness. The kites by Kite Wei aren't totally created by fancy but understanding and reappearance of the object, that is to say, the abstraction doesn't equal to fancy. Both the models of animal and plant are the artistic treatment in natural status and the creative doesn't get rid of the manifestation of the object. The decoration on the upper part of the kite shows all the traits of the object. In order to commemorate the success of Shenzhou V, Kite of Kylin Delivering a Child was made. And this kite exaggerates the proportion of Kylin to express a feeling of dignity and safety. So the exaggeration of the object Kite Wei employed skillfully transmits the spiritual essence and suitable. (Figure3)

Fig. 3. Kite Wei

4   The Decorative Meaning of Kite Wei with Tianjin Flavor

As kites of "Golden balls strengthen the friendship", "Silver balls strengthen the friendship" fly in the sky of Tianjin, the successor of Kite Wei, Wei Yongzhen spreads the kite culture all over the world, and makes kites popular worldwide. Besides edutainment, improving body condition, kites also bring auspicious ornamental meaning to people with enjoyment of mind and passing belief to each other. Every kite has a meaning of suspicion. Flying kites is not only for fun, but also for bless and pursue dreams.

What auspicious word that those totem on Kite Wei stand for? Let's take some example and try to appreciate.

4.1   First Is to Bless

"Good fortune is as long as Eastern Sea flow, longevity is like evergreen in Southern Moutain"; Good fortune and longevity are the everlasting pursuit by Chinese people. Kite Wei also takes fortune as a common theme to use; bat (bianfu) is a theme that is favoured by the most. Because of the homophonic, bat is always a popular role in Chinese traditional ornamentation. Though in the west people think it as evil and dark, ancient mysterious eastern culture regarded it as a unique mascot. In order to seek fortune, people were willing to buy kites decorated with bat picture, which means "obtain both fortune and longevity", "Five fortunes to give longevity", "Five kinds of blessings all came", etc.

Besides of praying for blessing, they also pray for longevity, so many Kites of Wei have symbols like evergreen, spiritual grass, crane for people to pray. The relevant types of kites are "Eight immortals wish for longevity", "Auspicious cloud wish for longevity" etc.

4.2   Second Is to Exorcise Evil Spirits

In Tianjin folk tales, there were many heresy and stories of evil, Tianjin local people like to dress the new born baby with "tiger hat on his head, tiger pinny around the neck, tiger warm sleeve on his hand, tiger bellyband on his belly, and tiger shoes for his feet". A baby is armed with tigers which can avoid the doom. Honestly, many folk cultural legacies are media that carry these signs. So is Kite Wei. In ancient times, many people think kite is spiritual being that can contact with the God and can be used as a way of communication, so is for pray and exorcism. Therefore, Kite is seen as an excellent thing to exorcize evil spirit. In folk tales, there were stories of using kite to evocation and drive away the harm. So Kite Wei was covered with mystery at this aspect, which requires it to be ornamented with exorcism.

Frankly, no natter exorcism or blessing, the ornamental meaning of Kite Wei indicates that Tianjin people's pursuit and wishes for beauty and harmony. Either auspicious symbols or dazzling decorative technique bring about harmonious and glorious praying from people.

5   Conclusion

To overview the Tianjin-featured and ornamental meaning of "Kite Wei", we can understand clearly that, during the long history of its development, "Kite Wei" becomes the treasure in ordinary people's heart with its vivacious entertainment, infinite fun and piety. Wei Kite is treated as a famous cultural totem for its unique decorative style, also work of art that is always glamorous and with longevity. It will definitely come to the whole world together with other non-material culture legacy such as painted sculptures by Ni Ren Zhang and New Year's painting by Yang Liu Qing.

References

1. Pan, L., Tang, J.: Introduction to Chinese Folk Fine Arts. Heilongjiang Fine Arts Publishing House, Harbin (2000)
2. Xu, Y.: History of Kite. Beijing Arts and Crafts Press, Beijing (1992)
3. Li, Y.: The Way of Decoration. Chinese Light Industry Press, Beijing (2004)
4. Yu, P.: My Opinion of the Origin of Kite. Folklore Studies, Beijing (1997)
5. Dong, X.: The Three Famous Places of Kites. Tourism (1995)

Analysis on the Food Packaging
of Honored Brand in Tianjin

Xi Gai and Jicheng Xie

Institute of Art & Fashion Tianjin polytechnic university Tianjin, China
{214869516,1378759338}@qq.com

Abstract. The packaging of honored brand in Tianjin is expressing its unique cultural value besides the historical value through refining the folk culture of Tianjin and transmitting regional history and civilization as a regional cultural mediator. Moreover, the localized packaging has become a carrier of the material and spiritual culture of Tianjin to show its uniqueness in the more and more fiece commodity market economy.

Keywords: Three treasures of Tianjin, honored brand, packaging.

As signature products of honored brand in Tianjin, the three treasures have grown into the representatives, and the food packaging of the honored brand is delivering the local culture through its own characteristics and cultural affective language. The continuously improved packaging has made the products the most popular food of Tianjin citizens and also pushed the development and innovation of the food culture in Tianjin as a brand and mediator to the outside. The famous Guifa Xiang Mahua, delicious steamed bun of Go Believe and tasty Er Duoyan fried cake are the most outstanding honored brand in Tianjin and all of them are attracting people via their particular form. In addition, their packaging is also undergong breakthrough with the food together to satisfy the aesthetic need of Tianjin citizens.

The packaging of honored brand in Tianjin is expressing its unique cultural value besides the historical value through refining the folk culture of Tianjin and transmitting regional history and civilization as a regional cultural mediator. Moreover, the localized packaging has become a carrier of the material and spiritual culture of Tianjin to show its uniqueness in the more and more fiece commodity market economy. Because the competition of packaging is much more shaper in today's market economy and a good packaging can help the corporation develop the market. Nowadays people's consumption concept has turned into cultural consumption model from traditional demand one. So the design concept of the food packaging has become a tendency of the honored brand centered culture.

1   National Design Element in the Packaging Ethos of Honored Brand of Tianjin

The honored brand demands that the modeling, color and picture must reflect the products' superiority and attract the buyers by the perfect external packaging so that

J. Luo (Ed.): Soft Computing in Information Communication Technology, AISC 161, pp. 487–491.
springerlink.com

the products can get a better transmission. Many honored brands pay much attention to the design elements of national and regional style and they usually use national visual figure to express the concepts of packaging, such as traditional auspicious pictures. The connotation of the auspicious totem is used to transfer the corporation values and try to appeal to the consumers. For example, the design aspiration of Zhizun Pianfeng, a new product of Jinjiu Group, comes from traditional Chinese phoenix culture and the modeling design originates the excavation Pianfeng pot. Its noble design manifests China's ancient culture. And the Pianfeng pot reflects its accumulation of culture by the mediator of phoenix, and combines the excavation and the sculpture together by making use of the plasticity and decoration of the pottery. Also the pot adopts the new technology of colored glaze to distinguish the design of the shape from others to realize the perfect combination between national elements and the cultural concept of Jinjiu so that the long history of wine culture in Tianjin can be inherited and embodied. And the Zhizun Pianfeng grows to a representative of northern delicious wine. (Figure 1)

Fig. 1. Picture one the packaging of Pianfeng pot

1.1   The Graphic Characteristics of National Design Element in the Honored Brand

The main carrier of the packaging design of the honored brand in Tianjin is the picture which is the most visualized form meeting with the consumers. The picture is the most important element to arouse buyers' interests. The national elements in the packaging are mainly reflected by a series of visual pictures that are very popular in people's daily life. There are several traits in this aspect. First is the design with national characteristics, such as taotie design, dragon pattern, phoenix pattern, fish pattern, cloud pattern and colored pattern. All these designs are representatives of auspicious omen in Chinese national culture and are thought to seek for blessing. Second comes from local popular animals and plants which are endowed auspicious connotation in the long-term development, so they are always used in the packaging design. For instance, fish, lotus, pine, crane, bat and butterfly are popular almost around the whole country. Many designers show great favor to them among which the fish stands first because Tianjin is a city originating dock culture. Fish can be seen in many crafts including cutting paper, embroidery, tye-dyeing, and the combination of fish and lotus means love. Lotus leaf is a warmly welcomed art form in packaging for it's a homonym of peace meaning a harmonious union. Third is the special symbol. Many symbols are used in the packaging of the honored brand, for example, wan pattern in Buddhism which means a continuous development without stop. The Chinese character wan is also acclaimed in national design elements. In the packaging of the honored brand, it always plays a role of assistant but occupies a significant part in packaging style.

1.2   The Color Characteristics in the Packaging of Honored Brand

The color in the packaging of honored brand in Tianjin enjoys strong local characteristics. Tianjin citizens' unique feelings towards color culture offer denotative meaning to its packaging. First is the utilization of color in tradition packaging. Golden and red are national traditional colors which implicate good luck and joy and can be seen in packaging of many honored brands. People prefer them a lot. For instance, Guifa Xiang employs these two colors in its packaging. Second is the mixing of various colors. While maintaining the traditional colors, many honored brands attach a lot of other colors, including white and blue. The differences between the rich color and light color can highlight the style and taste of the packaging. Guoren Zhang is a good example in this aspect.

The color in the packaging of honored brand is the most sensitive element and the color can grasp consumers' attention and gain broad acceptance and awareness, which is particularly meaningful. The artful combination of generality and individuality of colors has become a classical case in the packaging with national culture and takes over a dominant position in the market.

2   Divers Packaging of Honored Brand

2.1   Guifa Xiang Mahua

In contrast to the bankruptcy of many other domestic honored brands, Guifa Xiang is celebrating another spring in its development and its success provides a successful example for other honored brands. Besides its unique taste, its brand image gives people a profound impression. Guifa Xiang is famous for its various but unified packaging and establishes its own market image in the market. The main reason for its success is its special packaging. Its packaging differs depending on the products. But the whole style still relies on the traditional model, which creates a united brand model by a brave trial of combing modern composition method and traditional construct way. The attached picture two is the packaging of Guifa Xiang. The integration of warm and cold colors looks very comfortable, and the contrast of bright and dark colors feels luxury. (Figure 2)

2.2   The Steamed Bun of Go Believe

The steamed bun of Go Believe is well-known worldwide and its special take-out packaging offers it an opportunity to go overseas as an essential specialty. In later period of Qing Dynasty, the Queen Mother Cixi gave it a high praise that none of the food is more delicious than Go Believe and a long period of taking in can have longevity. After that, it became renowned domestically. In the long-term development as a successful honored brand, its innovation in technology, products, service, marketing and perfect enterprise system, the modern design of packaging plays a marked role. The picture three shows its packaging, whose design elements involve colors of Chinese traditional decoration, coordinated by Chinese folk decorative pattern. The calligraphy conveys its cultural deposits of hundreds of years. The unified but changeable tone in the packaging seems especially effective and is warmly welcomed. It's the unique packaging combing the traditional essence and bold creativity in packaging that propels the development of the honored brand. (Figure 3)

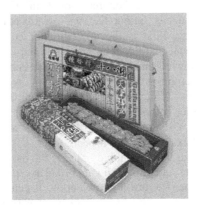

Fig. 2. Picture two the packaging of Guifa Xiang Mahua

Fig. 3. Picture three the packaging of steamed bun of Go Believe

2.3   Er Duoyan Fried Cake

It's not easy for Zeng Shengcheng to enhance the little fried cake to a necessary part in the food culture of Tianjin. Besides its exquisite skills and crafts and delicious taste, the diversified operation mode is a crucial role in maintaining its development and prosperity. The attached picture four is the packaging of the fried cake. Different from the packaging of steamed bun and Mahua, the fried cake reserves more traditional packaging methods. The characteristics of combination of bright and dark colors stand out in the fierce competition. Consumers prefer to the honored brand for its traditional packaging. The red color in its packaging expresses joy and dignity. The traditional packaging form leads to its popularity worldwide. (Figure 3)

Fig. 3. Picture four Er Duoyan fried cake

3   Conclusion

It's easy to see that nationalized packaging design must contain its local cultural characteristics otherwise it won't be accepted and inherited from the packaging form and method of the honored brand in Tianjin. The packaging of honored brand not only needs its independent language, values, thinking pattern, and aesthetics, but also complies with social development. The honored brand can't get rid of the fate of being replaced by new ones unless it can seize social development. The three treasures of Tianjin keep moving forward in their brand and packaging, in which people hold a strong feeling of national identity and sense of times. They will continuous improve their brands and quality as time goes and make progress in unceasing innovation.

References

1. Li, Y.: The Way of Decoration. China Renmin University Publishing, Beijing (1993)
2. Liu, L.: Nationalized Packaging Design. Hubei Arts Press, Wuhan (2004); Li, Y.: The Way of Decoration. Chinese Light Industry Press, Beijing (2004)
3. Guo, S., Tan, M.: The Packaging Design. Hefei University of Technology Publishing House, Hefei (2005)
4. Wang, W.: Three-dimensional Constitution and Packaging Design. Packaging World (2000)
5. Dong, X.: The Three Famous Places of Kites. Tourism (1995)

The Relationship between the Development of Chinese Modern Art Industry and Art Sponsorship

Analysis on Art Biennials of China and South Korea

Yanxiu Jin and Baiyang Jin

School of Art & Clothing Tianjin Polytechnic University Tianjin, China
jinyanxiou@gmail.com, baiyang1960@yahoo.com.cn

Abstract. In modern art, art Biennial is a great power to drive the development of art industry, but also a significant stage to exhibit new achievements of art. This paper took China Shanghai Biennial and South Korea Gwangju Biennial as examples and made a comparison of their developments. Then it tried to probe into the effects of art sponsorship on Chinese art industry, extant problems and some solutions by studying foreign art system and art market. It is concluded from the experience of South Korea Gwangju Biennial and biennials in European countries that a successful biennial must combine expenditure of national cultural industry, corporate sponsorship and market economic operation.

Keywords: Biennial, Art sponsorship, Art Industry.

1 Introduction

Chinese modern art industry is rising to have a place in the international art stage. It is easy to see its prosperous development during its 20-year development from Guangzhou Biennial in 1992, Shanghai Biennial and Beijing Biennial, but another problem also comes to us - the capital, so-called the endowment for the Biennial. Arts sponsorship has a long history overseas and a sound system and the abundant capital brings art infinite liberation and creativity. In turn, art produces much invisible economic value, especially the Biennial, representative of new art, plays an important role in pushing a city's even a country's cultural industry. However, art sponsorship is still a new concept in China. For foreign art market, the emergence and formation of art sponsorship totally depends on capitalist economic system. So it is necessary to do a research on the relationship between the development of Chinese modern art industry and art sponsorship in China's socialist economic system.

This paper took Gwangju Biennial (South Korea) and Shanghai Biennial (China) as examples and made a comparison of their developments. Then it tried to probe into the effects of art sponsorship on Chinese art industry, extant problems and some solutions by studying foreign art system and art market.

2 The Development of China's and South Korean Biennials

In modern art, art Biennial is not only a great power to drive the development of art industry, but also a significant stage to exhibit new achievements of art. Therefore, it

J. Luo (Ed.): Soft Computing in Information Communication Technology, AISC 161, pp. 493–500.
springerlink.com　　　　　　　　　　　　　　　© Springer-Verlag Berlin Heidelberg 2012

is becoming a preferred spokesman of modern art in the international art stage. Moreover, it is the best way to dig into new art and cultivate young artist and conduct international academic communication. The Biennial also indirectly pushes a country's cultural industry and makes a contribution to improve a city's image and cultural environment, tourism, international fame and mass sentiment to bring invisible long-term economic value rather than short-term value.

In early 1990s, in order to adapt to the development of international art industry, China brings the concept of Biennial. Until now, Shanghai Biennial is one of the most influential. After 8-administration efforts, Shanghai Biennial has become the most important modern art exhibition in China and successfully squeezed into the international art exhibitions. But the way to success is not even. Gwangju Biennial ranks No. 3 in the world and Gwangju is the exhibition site of "8.15" political event. It has profound influence on South Korean art industry, this city and even the whole country. Gwangju Biennial not only aims at improving South Korean international competitiveness, but also relates to its national policy of "cultural nation".

Table I describes the development of Gwangju Biennial and Shanghai Biennial [1]-[2]. The statistics reflect following aspects.

First, in the beginning Gwangju Biennial has gained great support of Gwangju municipal government and South Korea's Foreign Ministry and a consortium for the Biennial comes into being which is headed by the mayor of Gwangju in the board. However, the first two Shanghai Biennials were held only by the gallery itself and the third one by special organizing committee, but still no consortiums for it so far.

Second, Shanghai Biennials in 1996 and 1998 mainly focus on art works of modern domestic oil painting and national painting without participation of foreign artists. By contrast, there are 50 countries in the first Gwangju Biennial and the number of participating countries is about 1.5 times over that of Shanghai Biennial.

Third, Gwangju Biennial has special pavilion which provides professional places and exhibit facilities for diverse art works. The pavilion locates in South Korea Gwangju Jungoe Park where the location of Gwangju Gallery and Folklore Museum is. The three museums exhibit the goods art works home and abroad together. In addition, the park lies in the cross section of express way to West Gwangju with convenient transportation and completed infrastructural facilities. Because of its favorable location in the center of Gwangju and courtesy cars during the period of exhibition along with many other buses, the Biennial develops well. In comparison, Shanghai Biennial employs Shanghai Gallery as its exhibition site and so far there is no specialized pavilion for it. Contrast to the rapid development of Shanghai Biennial, the Gallery seems smaller and narrower. Shanghai Gallery was rebuilt in the original polishing hall in 1986 and it has shared no more places in the development of Shanghai. The insufficiency of infrastructure must lead to some limitations to the Biennial [3].

From above comparison, it is clear that the government plays no decisive role in the preparation of Shanghai Biennial. From the beginning, China is lack of systematic mechanism of exhibition and no special consortium for the biennial and it has never included the Biennial into long-term plan of Shanghai cultural industry. The insufficiency of mechanism and lack of experience must result in China's lag in visitor research, exhibition's promotion, fund raising, capital management and investment and image design of gallery and other aspects.

Table 1. The development of Gwangju Biennial and Shanghai Biennial

Name of Biennale	Venue	
The 1st Shanghai	Shanghai Art Museum	29 artists from China
The 1st Gwangju	Jungoe Park	92 artists (50 countries)
The 2nd Shanghai	Shanghai Art Museum	256 works from China
The 2nd Gwangju	Jungoe Park	35 countries
The 3rd Shanghai	Shanghai Art Museum	67 artists (18 countries)
The 3rd Gwangju	Jungoe Park	46 countries
The 4th Shanghai	Shanghai Art Museum	68 artists (20 countries)
The 4th Gwangju	Jungoe Park	325 artists (31 countries)
The 5th Shanghai	Shanghai Art Museum	More than 120 artists
The 5th Gwangju	Jungoe Park	237 artists (41 countries)
The 6th Shanghai	Shanghai Art Museum	94 artists (25 countries)
The 6th Gwangju	Jungoe Park	127 artists (32 countries)
The 7th Shanghai	Shanghai Art Museum	64 artists
The 7th Gwangju	Jungoe Park	127 artists (36 countries)
The 8th Shanghai	Shanghai Art Museum	47 artists (21 countries)
The 8th Gwangju	Jungoe Park	124 artists (31 countrie)

3   The Relationship between the Development of Art Industry and Art Sponsorship

According to above analysis, Shanghai Biennial is lack of specialized operation system of consortium and special fund like that of Gwangju Biennial, so it is easy to see that the biggest crisis for Shanghai Biennial is fund. For example, the failure of Guangzhou Biennial in 1992 is a good example resulting from the breakup of the sponsorship [4].

Fig. 1 shows the statistics of capital from the first Shanghai Biennial to the eighth one [4]. These statistics reflect following results.

Governmental fund occupies over 50% in the entire fund. Moreover, with the establishment of Biennial's cultural brand and its improving international influence which brings Shanghai great indirect economic value, the local government is paying more and more attention to it, which can be reflected in the increasing fund support. And the governmental support keeps a sustained tendency. Look back to the exhibitions and biennials in China in early 1990s, all of them are funded by the nongovernmental enterprises, unfortunately all the sponsors quit after one exhibition and mainly because the organizers lacks an ideal plan to use money and the commercial feedback to the enterprises is low. Such as the Chinese Modern Art Exhibition in 1989 held by China Beijing Gallery, Guangzhou Art Biennial in 1990s designed by some art critics and Chinese Oil Painting Biennial in 1993 organized by the Research Institute of Fine Arts, National Academy of Arts of China [5]. Such results tell us that exhibition funded by enterprises is not consecutive. In contrast, Shanghai Biennial, funded by the government for consecutive 8 times lasting for 14 years, still maintains a good situation. It also shows that the success of biennial has a close relation with governmental cultural policy.

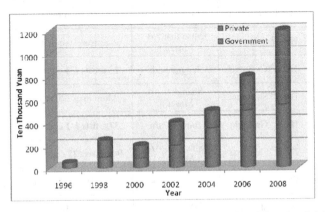

Fig. 1. The statistics of capital from the first Shanghai Biennial to the eighth one.

The first Shanghai Biennial got 500 thousand Yuan from the government, which forms a clear comparison with the 60 million Yuan of the Gwangju Biennial. With the help of sufficient fund, Gwangju Biennial owns a specialized pavilion. However, the 8th Shanghai Biennial has only 20 million Yuan in 2010 after 14-year efforts [4]. Conversely, the 8th Gwangju Biennial has got 200 million of specialized fund.

Chinese art sponsorship includes three types: governmental fund, nongovernmental fund (enterprise and individual) and foundation fund, and the latter two types are also a main resource of the art fund. The statistics in Fig. 1 indicate that nongovernmental fund and foundation fund is not stable compared to the governmental fund from the 2nd to 6th Biennial. There are many factors leading to the instability, including the imperfect mechanism of Chinese art market, low economic feedback to the enterprises and a small amount of foundations in China. Most of the Chinese foundation fund is used to help people in poor or disaster areas and only a small part is input into art. According to the statistical report from the nongovernmental organizations led by the Ministry of Civil Affairs, until the end of 2008, there are about 1600 foundations in China, among which the number of national cultural foundation is only 17, local cultural foundation is over 40 and the private art foundation is countable. Such kind of phenomenon is very different from the situation in Europe and US. 40 thousand foundations exist in US until the end of 2008. It is obvious that China lags behind in pushing the development of art industry. The data of Fig. 1 reveals that the nongovernmental fund and foundation fund are becoming better from the 6th Biennial and both of them has kept a large increase in the 8th one. In addition, the increase rate of two kinds of fund is much bigger than that of the governmental fund in the 6th and 7th Biennial, which indicates the big potential of nongovernmental and foundation fund.

The data of Table I show that the number of participating countries of Gwangju Biennial is about 1.5 times more than that of Shanghai Biennial. If foreign art groups or artists are invited, the big amount of freight charges of the art works, the insurance fee, the artists' transportation fees and the cost of accommodation must be taken into consideration because all of these costs will surely increase the total cost. Without sufficient capital, it is impossible to realize these aims. Therefore, the input of capital, the exhibition's scale (including the increase of the number of the participating countries, the enlargement of the number of the artists, awareness of the propaganda) and the influence (direct and indirect) are proportional. The three elements increase together and keep the complementary cycling relationship. That is to say the increase of the influence will absolutely bring the feedback of capital, the increase of capital will increase the amount of the endowment and sufficient capital can enlarge the exhibition's scale.

The direct influence includes the academic development of modern art, the formation and perfection of a nation's modern art system and art market and a nation's competitiveness in the international stage; the indirect influence includes improving a city's cultural image, driving the development of the cultural industry. South Korea benefits a lot in economic value because of the development of cultural industry. As surveyed, in 2002, the volume of overseas export in cultural industry of South Korea is 15.7 billion and the market scale is 500 million. And the volume of overseas export occupies 1.5% of the whole world [7]. Moreover, the development of the cultural industry will push economic development.

The above analysis indicates that China's biennial can't catch up with the level of biennial in South Korea no matter in internationalization, scale and the quality of exhibition goods. Exhibition needs fund; especially the big event like biennial requires gigantic capital support. From the capital needs of the 8th Shanghai Biennial (Fig. 1), the need is increasing consecutively. With the formation of Shanghai Biennial's cultural brand and increase of international influence, the government turns more attention to it and the fund support increases a lot, but it is not enough. The remainder has to come from the endowment. However, the art sponsorship still sits in the early period which is reflected from the extant problems in the former biennials. The causes for this are all-round, including system, legal system, and national conditions and historical misunderstandings. In a word, thinking about solving the insufficiency of art sponsorship is also change the hard development of Chinese modern art industry.

4   The Difficulties of Art Sponsorship in China

First, China is lack of perfect system of art sponsorship and relative laws. In a mature international culture market, important art exhibition and art performances can't be separated from the endowment of companies and large corporations. The Gwangju

Biennial has about 20 large enterprises as its long-term sponsors. While China's cultural undertakings mainly depended on governmental expenses for culture in the past because at that time the enterprises were owned by the whole people under collectivistic ownership. Various market economic entities have enjoyed enough severity in their operation in the planned socialist market economy since 1990s, but most of them have never realized the importance of strategic charity to the enterprise image and enlargement of the market to make an endowment and they would rather rely on marketing management and big amount of ad investment on TV and some other media for products' promotion [5]. No perfect preferential policies in China about reducing tax on enterprises acting as art patrons should be to blame. Law of Donation for Public Welfare Undertakings in 1999 and Law on Corporate Income Tax in 2008 show that the procedures of donating cultural art program are complicated, the enterprises and patrons enjoy low commutation tax and the amount and scope of the donation are restricted, therefore, corporations have low passion in donating without gaining benefits. Besides corporations and individuals, foundations also play significant role in the development of cultural industry. So far America owns 40 thousand foundations and all of them exist depending on the preferential policies on enterprises' tax and inheritance tax, which entice many rich men to set up foundations to reduce tax. Gwangju Biennial owns abundant fund with the 3rd rank in worldwide biennials. It has a close relationship with its perfect art sponsorship system and strong legal system and both of these systems depend on South Korean developing strategy of cultural industry in the 21st century. In 1994, Ministry Of Culture and Tourism established policy bureau of cultural industry and began to prepare the legal system in cultural industry. At the same time, a lot of comprehensive plan about cultural policies started to stress cultural industry's importance to economic development.

Second, the government holds conservative attitude towards modern art and the art sponsorship.

It is easy to see from South Korean policy of "cultural nation" that art's development and sponsorship require strong supports from the government. In order to guarantee sufficient fund for cultural industry, South Korean government has collected specialized fund through various aspects. For example, the government raised 500 billion Korean won in 2001through budget allocation, investment portfolio, special combination and special fund. Compared with the huge investment, the economic value is more impressive. The market scale of South Korean cultural industry amounted to 15.7 billion dollars in 2002 [7]. Besides, some European countries also discover the fact that art can bring enormous economic value. For instance, Venice Biennale is set up for exploiting marine tourism and blossoming urban economy. Conversely, Chinese government still holds conservative attitude towards modern art and the art sponsorship. In the preparation of Shanghai Biennial, the government almost plays no decisive role. In the light of international conventions, biennial, as a representative of a city's cultural image, should be held by the municipal government and headed by the mayor in the organizing committee in order to deploy all the resources and operate the mechanism effectively and expand the city's international repercussions. Compared with the positive situation of Gwangju Biennial, Shanghai Biennial seems much ordinary. The organizing committee composed of government officials still remains the role of assistance and the chairman of the committee is usually on by an official with low title [8].

Third, China has lack of modern economic management operation mode and stable fund safeguard system.

Biennials in China are often out of scientific and effective economic management operation mode which is embodied by the lack of the research on the market, insufficiency of marketing planning consciousness and short of professional art marketing talents. Most of domestic galleries and art centers only engage in the trade of art works but without mature experience of long-term promotion for art and brand awareness. Although Shanghai Biennial has gained brand effect through unremitting endeavor, it is still unprofessional and immature in long-term promotion for art in the former 8 biennials. All these emerged problems reveal that China is in short of professional marketing management talents in art because of the short period of time of major of art marketing management and small enrollment. It is surveyed that less than 50 students of this major graduate every year and they can't satisfy the overwhelming market demand in gallery, auction house and so on.

5   Conclusion

In the development of Chinese modern art industry in 1990s, the emergence of biennial not only makes up the vacancy of the original mechanism and structure of Chinese art exhibition and makes the most positive contribution to the modern art and art industry. It is concluded from the experience of Gwangju Biennial and biennials in European countries that a successful biennial must combine expenditure of national cultural industry, corporate sponsorship and market economic operation. So some methods are proposed aiming at Chinese modern art and the extant problems in art biennials.

A new pattern that combines art in Chinese market economic system and enterprises together should be explored. The government can set up a long-term effective operation mode of integrating art and enterprises so that corporations can provide long-term fund for art industry and art in turn can feedback the corporation economic values to construct a new multi-win cooperative situation.

The biennial can be set up as a project separately and supported by national special fund. Until now, only Beijing Biennial was totally funded by the government, but it is only one point of the various biennials. Though government funded, it is still weak in pushing the development of Chinese modern art and art industry.

Special fund and consortium for biennial should be established as soon as possible. The special fund and consortium can be an organization responsible for financing to arouse more extensive social power. Gwangju Biennial is a good example in this aspect. The board of directors is in charge of financing and the arts council takes charge of spending. The special fund and consortium free planners of the biennial from the boring task of financing and reduce a lot of burden for them so that they can devote their mind to academics. Current mode that gallery and planners are responsible for financing is not very scientific and gradually can't adapt to modern market economy system. It becomes an urgent problem to be resolved.

All in all, the most important factor for art development of a country or a region is the government's concern and micromanagement, which demands mature laws about modern art's development and art sponsorship system. Such as, the procedures of corporation sponsorship in China are very complex and policies of commutation are not preferential, which can't evoke the enterprises' passion of endowment because the corporation can't always give gratuitous endowment although public sponsorship are occasionally seen, but the economic benefits are their ultimate aim. Therefore, corporation sponsorship for art is often on and off in China and so many art exhibitions failed in 1990s. In order to utterly change such situation, the procedures of corporation sponsorship should be simplified and the government should increase policies of commutation to stimulate enterprises' enthusiasm in endowment. Many achievements have been made abroad and the government can learn from their success.

References

1. http://www.gb.or.kr/
2. http://www.shanghaibiennale.com/
3. http://culture.gxnews.com.cn/staticpages/20101112/newgx4cdd0b29-3399701.shtml
4. Han, Q.: The Prospect and Puzzle of Social Art Sponsorship The Exploration of Funding Issue in Chinese Art Biennial Exhibition. Art Panorama 5, 202–203 (2010)
5. http://artist.artxun.com/11573-yinshuangxi/
6. Lu, J.: The Art Sponsorship and System of Art Foundations in US. Oriental Art 17, 11–13 (2006)
7. Tan, H., Ke, Y.: South Korean Cultural Industry's Inspiration to China. Economic Review 6, 113–115 (2009)
8. Zhu, X.: Shanghai Biennial in Modern Cultural Environment. Central Academy of Fine Arts (2007)

Research on Coordination in Supply Chain for Perishable Goods Based on Quality Risk

Li-na Wang and Li-li Lu

Tangshan Teacher's College, Hebei Tangshan 063000

Abstract. The paper examines the problem of quality control under supply chain coordination for perishable goods based on quality risk. Considering the asymmetric information, the models of principal-agent are formulated based on the share level of the quality risk between supplier and buyer when one of quality evaluation level and quality prevention level is unobservable and when both of them are unobservable. Make the decision of the penalty for defective unites identified during inspection and the external failure's share level between buyer and supplier to optimize the profit of the whole supply chain when there is moral-hazard and the partner is purchasing individual optimization.

Keywords: quality risk, perishable goods, quality prevention, quality evaluation, coordination, quality control.

1 Introduction

Quality is an important feature that enables firms to sustain their competitive advantage and maintain growth levels. Recently, the perishable goods have drawn lots of attention and main research concentrate on the problems of order, stock and cooperation[1-4]. No matter how the situation of the supply chain competition changed, product's cost, quality, price and flexibility are always the focus, and quality become more and more important in market. Some people have been researching on the product's quality, quality control and contract in supply chain under asymmetric information[5-7]. Overseas, Kashi et al.(2005) examine how to make the penalty for defective units through inspection information and external failure information in the case of single moral-hazard and double moral-hazard, to affect the supplier's quality decision[8]. Hwang et al.(2006) examine the buyer's problem of inducing the supplier's quality effort using two arrangements: the appraisal regime and the certification regime[9]. Wei(2001) examine how the contract design solve the converse choice problem effectively[10]. Stanley et al.(2001) research on the effect of quality control on the supply chain's profit[11].

However, the literature on supply chain for perishable goods considering quality risk is a fat lot, the other side is that the quality of perishable goods affects the market competition seriously. Therefore, this paper formulate the coordinate model of supplier and buyer considering the asymmetric information in supply chain coordinate quality control, provide how to identify the penalty for defective units and share in the external failure to optimize the profit of the whole supply chain when there is moral-hazard and the partner is purchasing individual optimization.

J. Luo (Ed.): Soft Computing in Information Communication Technology, AISC 161, pp. 501–509.

2   The Quality Profit Model of Perishable Supply Chain

The paper considered a single-period supply chain with risk-neutral buyer and a risk-neutral supplier. The supplier's quality prevention level, P_s, is the probability that the product performs the desired functions with $0 \le p_s \le 1$, and corresponding quality cost is $C_s(p_s)$. This paper assume $C_s(0) = C'_s(0) = 0$ and $C'_s(1) = \infty$, when $p_s > 0$, $C'_s(p_s) > 0$ and $C''_s(p_s) > 0$. The buyer performs inspection and the inspection technology does not reject a good unit and does not identify all defective units. The evaluation level is p_b, $0 \le p_b \le 1$, and the corresponding quality cost is $C_b(p_b)$. The assume is similar to $C_s(p_s)$. According to the evaluation results, the buyer repairs the defective units and charges a penalty for unit defective W. The repair cost is M, and $M < W$. To keep the model simple we assume the value of the product which have been repaired is equal to the good product. The probability that the product fails when the final consumer uses it, i.e., the external failure is denoted E. We assume that the supplier and the buyer share in the external failure, and the supplier will in charge of αE, with $\alpha \in [0,1]$.

The related parameters are provided below.

TP_b the buyer's profit per year

P the supplier's replenishment rate

$I_{s1}(t)$ the supplier's stock level in manufacture period

$I_{s2}(t)$ supplier's stock level in non-manufacture period

I_{mb} transmissive lot every time

T_b the buyer's replenishment cycle

θ deteriorate rate

TP the supply chain's profit per year

R the buyer's quotient of excessive profit

r the interest rate per year

τ the delay payment interval

$I_b(t)$ the buyer's stock level

C_{bs} the buyer's order cost

n the replenishment time

T the supplier's cycle

T_1 the manufacture time of supplier

T_2 the non-manufacture time of supplier

TP_s the supplier's profit per year

C_{ss} the supplier's set-up cost

C_s the materials cost for unit product

C_{sh} the supplier's hold cost percentage per year

C_{bh} the buyer's hold cost percentage per year

P_m the retail price

C_b the purchase cost

The buyer's profit per year is given by

$$TP_b = P_m d \; -[\frac{C_b I_{mb}}{T_b} exp(-r\tau) + \frac{1}{2} C_b C_{bh} d\,T_b \; +\frac{1}{2} C_b d\,T_b\,\theta$$
$$+\frac{C_{bs}}{T_b}+\frac{(C_b(p_b)-(1-p_s)p_b(W-M)+(1-p_s)(1-p_b)(1-\alpha)E)I_{mb}}{T_b}]$$

$$(1)$$

The supplier's profit per year is given by

$$TP_s = \frac{C_b I_{mb}}{T_b} exp(-r\tau) - (\frac{C_s p T_1}{T} + C_s C_{sh} \bar{I}_s + C_s \bar{I}_s \theta$$
$$+\frac{C_{ss}}{T}+\frac{(C_s(p_s)+(1-p_s)p_b W +(1-p_s)(1-p_b)\alpha E)I_{mb}}{T_b})$$

$$(2)$$

The profit of whole supply chain is given by

$$TP = TP_b + TP_s$$

$$(3)$$

The coordination purpose of the supplier and the buyer is to maximize the supply chain's profit.

$$\text{Maximize } TP(P_m,n,T_2)=TP_b(P_m,T_b)+TP_s(n,P_m,T_1,T,T_2)$$

$$(4)$$

where T_b, T_1 and T is the function of T_2 and then TP is the function of P_m,n and T_2. Solve the formula (4) to get P_m^*,n^* and T_2^* which maximize the profit of the whole supply chain. Derivate formula (3) with respect to P_s and P_b respectively, and set them to zero, we get expressions:

$$C_s'(p_s)=(1-p_b)E+p_b M$$

$$(5)$$

$$C_b'(p_b)=(1-p_s)(E-M)$$

$$(6)$$

formula (5) denote the condition of the supplier's quality prevention level need to satisfy when the supply chain's profit is maximized. Formula (6) denote the condition of the buyer's quality evaluation level need to satisfy when the supply chain's profit is maximized. When the supplier and the buyer have symmetric information of each other, according to formula (5) and (6), we can get the supplier's quality prevention level and the buyer's quality evaluation level when the supply chain's profit is maximized.

3   The Coordinate Quality Control Based Quality Risk

Quality risk denote that when the product provided by supplier have potential defect, the supplier and its customers will suffer loss.

3.1   The Quality Evaluation Information Is Unobservable

When the buyer's quality evaluation information is unobservable and the supplier's quality prevention information is observable, the buyer have advantage information. The buyer is likely to make the practice quality evaluation level smaller than p_b^* to reduce quality cost, or to make the practice quality evaluation level bigger than p_b^* to obtain more penalty. The supplier's problem is provided in the following program.

$$Max\,TP_s = TP_s(p_s) \tag{7}$$

$$s.t. \quad TP_b \geq u_{b0} \tag{8}$$

$$p_b = arg\,\underset{p_b}{max}\,TP_b \tag{9}$$

The supplier must act as contract $p_s = p_s^*$. But the buyer's quality activity is unobservable, the buyer must make the practice quality evaluation level in the purpose of maximizing its own profit. According to literature [15] and formula (9), we get the following expression.

$$p_b = \frac{1}{k_b}[(1 - p_s)(1 - \alpha)E + (1 - p_s)(W - M)] \tag{10}$$

Because $p_s = p_s^*$, the buyer identify its quality evaluation level as follow

$$p_b = \frac{1}{k_b}[(1 - p_s^*)(1 - \alpha)E + (1 - p_s^*)(W - M)] \tag{11}$$

In the other side, the supplier and the buyer make the decision coordinately. Their quality activity should make the whole supply chain's profit optimization. In this case, the buyer's quality evaluation level p_b should satisfy formula (6).

$$p_b = \frac{1}{k_b}(1 - p_s)(E - M) \tag{12}$$

Integrate the purposes of the buyer's profit and the supply chain's profit, we can get the following expression.

$$\alpha = \frac{W}{E} \tag{13}$$

When the penalty for defective units and the external failure's share between supplier and buyer satisfy the relation, the buyer will make the practice quality evaluation as p_b^*.

3.2   The Quality Prevention Information Is Unobservable

In this case, the supplier have advantage information. The problem of buyer is provided by

$$Max\,TP_b = TP_b(p_b) \tag{14}$$

$$s.t. \quad TP_s \geq u_{s0} \tag{15}$$

$$p_s = \arg \max_{p_s} TP_s \tag{16}$$

Derivate formula (16) with respect to p_s and set it to zero, we can get the following expression.

$$C_s'(p_s) = p_b W + (1 - p_b) \alpha E \tag{17}$$

Because the buyer can't observe the supplier's quality activity, the supplier will identify the quality prevention level in the purpose of its own profit's maximization.

$$k_s p_s = p_b W + (1 - p_b) \alpha E \tag{18}$$

In the other side, the supplier make the quality prevention level in the purpose of supply chain's profit.

$$p_s = \frac{1}{k_s}[(1 - p_b)E + p_b M] \tag{19}$$

Integrate the formula (18) and the formula (19), we can get the following expression.

$$\alpha = 1 - \frac{p_b^*(W - M)}{(1 - p_b^*)E} \tag{20}$$

3.3   Quality Evaluation and the Prevention Information Are Both Unobserved

In this case, the supplier and the buyer can't observe each other's quality activity. We assume the supply chain is the dummy principal and the supplier and the buyer are agent. The supply chain's problem is provided in the following program.

$$\max_{p_b, p_s} TP' = TP - TP_2 \tag{21}$$

$$s.t. \quad TP_b \geq u_{b0} \tag{22}$$

$$TP_s \geq u_{s0} \tag{23}$$

$$p_b = \arg \max_{p_b} TP_b \tag{24}$$

$$p_s = \arg \max_{p_s} TP_s \tag{25}$$

In this case, both the supplier and the buyer want to make use of the advantage information. The practice quality activity of supplier and buyer should satisfy formula (11) and formula (18) respectively.

In the other side, the supplier and the buyer make the decision coordinately. The quality activity should maximize the supply chain's profit. The buyer's quality evaluation level and the supplier's quality prevention level should satisfy formula (12) and (19) respectively.

Integrate the purposes of the supply chain's profit and the profit of the supplier and the buyer, according to formula (11) and (18), formula (12) and (19), we can get the following expression.

$$\alpha = 1 - (1 - \frac{M}{E})p_b^{\ *} \tag{26}$$

$$W = E - (E - M)p_b^{\ *} \tag{27}$$

4   Numerical Example

We illustrate these cases with a numerical example using the following parameter values: $a = 3000$, $b = 35$, $C_b = 35$, $C_{bh} = 0.2$, $C_{bs} = 100$, $C_{ss} = 6000$, $p = 3000$, $C_{sh} = 0.2$, $C_s = 20$, $R = 0.5$, $\theta = 0.10$, $r = 0.12$, $W = 20$, $M = 10$, $E = 200$. According to literature [15], assume the supplier's quality prevention function and the buyer's quality evaluation function are $C_s(p_s) = K_s p_s^2/2$ and $C_b(p_b) = K_b p_b^2/2$ respectively, with $K_s = 40$, $K_b = 40$. Table one and table two provide the results.

Table 1. The scenario when quality evaluation level is hidden

W	10	20	30	40	50
α	0.05	0.10	0.15	20	0.25
TP_b	19946	20285	20618	21115	21425
TP_s	5176	4837	4504	4007	3697
TP	25122	25122	25122	25122	25122

The relation of the external failure E , quality defective penalty W and α is provided in figure one. It make out that, when the buyer's quality evaluation information is unobservable, the buyer will charge more penalty for quality defect to supplier, and the supplier will need to in charge of more external failure.

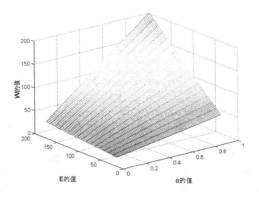

Fig. 1. The relation among E , W and α

Table 2. The scenario when quality prevention level is hidden

W	10	15	20	25	30
α	1	0.798	0.595	0.393	0.191
TP_b	21039	21041	21038	21039	21040
TP_s	4083	4081	4084	4083	4082
TP	25122	25122	25122	25122	25122

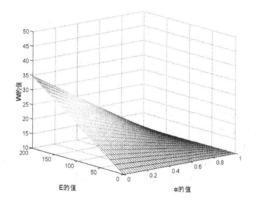

Fig. 2. The relation among E, W and α

In this case, the relation of the external failure E, quality defective penalty W and α is provided in figure two. From the figure, if the external failure is identified, α become smaller as the increase of the penalty for different external failure. From the solution results, we know the repair cost affect the relation of external failure E, quality defective penalty W and α. From figure three, we know when the repair cost become more, the supplier need to pay much penalty to the buyer.

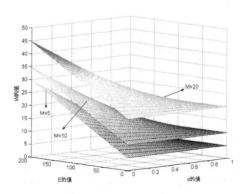

Fig. 3. The relations among E, W and α when M is changed

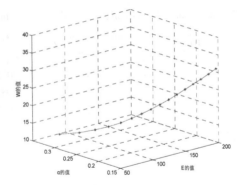

Fig. 4. The relation among E , W and α when both quality evaluation and prevention levels are hidden

In this case, the relation among the external failure, quality defective penalty and α is provided in figure four. If quality activity information of supplier and buyer are both hidden, the external failure's share level and the penalty for quality defect can be identified when external failure is known. In this way, the supplier and the buyer will make the practice decision according to the contract.

5   Concluding Remarks

This paper examine the supply chain coordination for perishable goods considering asymmetric information. Principal's objective function are formulated when one of quality evaluation level and quality prevention level is unobservable and when both of them are unobservable. Three relations among the external failure E , quality defective penalty W and α are provided in different cases. Designing the contract between the supplier and the buyer according to the relations, the members of the supply chain can make the decision to optimize the supply chain while purchasing their own optimization under moral-hazard.

Acknowledgment. The Development Fund Project of Tangshan Teachers' College in the Year of 2010, Project Number: 10C03References.

References

1. Chung, K.-J., Liao, J.-J.: Lot-sizing decisions under trade credit depending on the ordering quantity. Computers operations Research (31), 909–928 (2004)
2. Yang, P.C.: Pricing strategy for deteriorating items using quantity discount when demand is price sensitive. European Journal of Operations Research 157, 389–397 (2004)
3. Skouri, K., Papachristos, S.: Optimal stopping and restaring production times for an EOQ model with deteriorating items and time-dependent partial backlogging. Int. J. Production Economics, 81–82, 525–531 (2003)

4. Anupindi, Bassok: Centralization of stocks: retailers vs. manufacturer. Management Science 45(2), 178–191 (1999)
5. Starbird, S.: Penalties, rewards, and inspection: provisions for quality in supply chain contracts. Journal of the Operational Research Society 52(1), 109–115 (2001)
6. Reyniers, D., Tapiero, C.: The delievery and control of quality in supplier-producer contracts. Management Science 41(1), 1581–1589 (1995)
7. Starbird, S.: The effect of acceptance sampling and risk aversion on the quality delivered by suppliers. Journal of Operational Research Society 45(2), 309–320 (1994)
8. Balachandran, K.R., Radhakrishnan, S.: Quaility implications of warranties in a supply chain. Management Science 51(8), 1266–1277 (2005)
9. Iny, H., Radhakrishnan, S., Su, L.: Vendor certification and appraisal implications for supplier quality. Management Science 52(10), 1472–1482 (2006)
10. Lim, W.S.: Producer-supplier contracts with incomplete information. Management Science 47(5), 709–715 (2001)

Soft Furnishings—Taste of Life Candle

Lei Jiao and Baiyang Jin

Institute of Art & Fashion Tianjin Polytechnic University, Tianjin, China
liangchadanle@163.com,
873494544@qq.com

Abstract. With the pursuit of personal decoration, "soft furnishings" have become a new direction for people to invest. Among the many decorations, the candle is undoubtedly one of the important accessories. However, with the development of technology, its potential is gradually being replaced by high-tech products. Speaking from the origin of this candle to further inquiry by candles, candles evoke a new understanding of life and thus bring an honor to add your color.

Keywords: soft furnishings, candles, life.

1 The Concept and Role of Soft Furnishings

The so-called soft furnishings, after the decoration is finished, make the use of those easy to replace and change the location of the ornaments and furniture, such as curtains, sofa sets, cushions, tablecloths and decorative craft products, decorative wrought iron and so on, for the second indoor furnishings and layout.

As time progresses, people's living standards improve, meanwhile people's ideas and aesthetic standards have changed. Gradually the family pursues personalized decorative style and the rational consumer comfort. So people finally begin to understand that nothing is finished in one step, because too much of the interior is actually a time limit on the interior imagination, restricting the means of decoration and bringing the future trouble to interior repairs. Furthermore, the new decoration concept - "soft furnishings" arises, that is the light fitting and re-decoration in the home decoration. The so-called light fitting, a hard decoration, is no-moving and difficult to change. To ensure perfect function, based on the simplicity of the treatment as far as possible, leaving more room for re-decoration, which is also known as the "soft decoration".

All kinds of decorative objects are arranged in an orderly or disorderly to present a combined artistic phenomenon, so we call the art "soft furnishings". The Soft furnishings, an art style, which is different from indoor decorative arts, is a magnificent style and one-sided pursuit of unbridled ostentation. That emphasizing the scientific, technical and academic is different from the environmental art. It is an artistic effect of the subject inside the building, belonging to the public subject areas. With the rapid development of modern industrial civilization and the development of Del, it develops. As to dilute and soften a sense of apathy the industrial civilization has brought. Soft furnishings creating a warm and harmonious living environment and

J. Luo (Ed.): Soft Computing in Information Communication Technology, AISC 161, pp. 511–516.
springerlink.com © Springer-Verlag Berlin Heidelberg 2012

emotional of comforting people has its own history. Soft furnishings, an extension of the construction and development of visual space, which is relative to the structure of the building itself is put forward. Generally the soft interior decoration can be mobile, that is to say easy replacement of accessories such as curtains, sofas, paintings, crafts, bedding, home, etc. It is the second display and layout of the room. As a removable decoration, soft furnishings break the traditional definition.

2   The Candle Illuminates a New Life

"Candle" comes from the Latin "candere", which means "to become white" or "light flashing".

Lighting a candle, a small act, can give room endless vitality: under the perfect mix of light and shadow, an amazing effect from the dreamy candles brings a trace of nostalgia in the air. In the modern society of Rapid development in science and technology, people still can not ignore the candle to create a comfortable inviting atmosphere. Lighting a candle in the fragrance of the breath of rest and relaxation, bringing a unique sensory experience; a warm table candelabra, when you have tired legs after a day's work, is still the best way to relax.

The trajectory history was removed since it was the candle lighting as a major way. Although it is a cluster of small flame, it symbolizes the hope, truth, life, virtue, wisdom. From the definition of religion, it also symbolizes God's love. The first candle should be traced back to the Roman Empire. Romans at the time roll into small thin volume, all in molten wax and exposure to the sun, so it faded to white. For a long time, this method of manufacturing the candle does not change anything, until today, people are still using the same method.

Over the centuries, due to the high prices, the general public can not afford beeswax candles made, only the wealthy and the priesthood enjoy the chilly, aromatic candle. In the 19th century, citizens used animal fats daily as candles. This candle is made from the rush to justice in animal fat or oil immersed from the system, combustion will produce strong smoke and terrible smell. Candle molds of the invention derived from a Frenchman in the 15th century. Until the mid-19th century, with the stearine technology, candle manufacturers have been permanently altered. The invention, together with the French Jampa's discovering invention of the fold Diego wick in 1825 and paraffin appearing in 1850, declared the end of animal fat candles, so clean, odorless, cheap candle was born. This new, cleaner burning, brighter lighting candles become the main tool. It can be seen that the birth and development of the candle is a very long process.

Today, in advanced and civilized modern society, it no longer plays light and guardian of the image, but it did not disappear in people's vision. Instead, another kind of image accompanies the human. Not only People in special occasions decorate the atmosphere with candles, but also more and more people in daily life are used to light candles in their own homes and let it shine warm home life, in order to add a fun mood. To convey the mood this point, the candle is unique. According to selecting different mood candles, each room will benefit from the light fantastic. Even in the most stereotypical drab room, soft candlelight could improve the atmosphere and provide the bright spot for manufacturers through the features.

3   The Different Kind of Candles, Strange Flavor

Candle reconciles the different styles and designs, making the room fragrant. About the endless shape of the candle, it is sometimes prominent bridge, sometimes honest, sometimes as the Greek columns in the palace. Whether in the magnificent palace or in the corner of the room simple and rustic, it is always to respond to a different attitude around the atmosphere of the home environment for people to bring a little bit warm and adds streaks of color.

3.1   Candle Type

Whether small tea lights or large sculpture, candles of various shapes and sizes are available through the mold shape. People can choose it at pleasure. Complex color patterned candles are suitable for traditional decoration. In the decoration simple, elegant colors of the environment, it is obviously more unexpected. Indoor environment as pristine interior tough Stern design style makes the wide-angle lines simple and make eyes more comfortable. Therefore, slender tapered candles will draw the eye's attention to the wall, or the lower wall setting a floating candle with glass bottles to strengthen the surrounding illumination. A large volume candle or some coarse types of candles can be placed separately, but little small, decorative candles or tea lights can reflective surface to enhance light by mirror.

3.2   Candle Taste

Sense of smell is the most sensitive for human beings, the person can distinguish about 10,000 kinds of different odors on an average. Flavor can be used to adjust mood and create atmosphere, besides, some of the flavor is the best supplement to relax. An aroma taste room will be similar to be home a lot more quickly so that people feel comfortable and pleasant. In recent years, the popularity of home fragrances increased significantly, while a wide variety of spices available in the market evoke any emotions and feelings. The concept of blending incense into the candle is the candle lighting and decoration, in addition to a new added value.

To create a room fragrance, the most simple way is to light a fragrant candle. Moreover, the swaying flame also helps to create quiet and peaceful atmosphere. Aromatic candles emerged in the 19th century. As early as 1893, Britain had created a candle company with blueberry scented candles. However, until recently, the popularity of scented candles really develops. Today, minimalism which prevails at home a large aromatic candles Collection fashion life is "change for the holy family"--- the main elements of this vision. Aromatic candles contain fragrance more and better texture, which is softer than ordinary candles, so incense can be effective for indoor. As with the flame of a candle, the smell can make people feel warm and welcome. A light yellow vanilla candles, you can create a cozy space; and the church contains a lot of beeswax candles which releases into the atmosphere to make the room a sweet honey taste. Scented candles, such as ginger, cloves or cinnamon, can make guests feel comfortable and relaxed, conduciving to conversation and stimulate people's vitality.

3.3   Candle Color

As we know, the colors also affect the mood and feel of the people which are important factors. In interior decoration, people often change their behalf to the people in different rooms of different feelings by changing the colors. Usually, people prefer warm colors, it is appropriate to select the red and orange to create warm and comfortable atmosphere. All the color in the blue, indigo and violet reflect nature, the feeling is cool and quiet. These colors are suitable for modern and strong, or more feminine room. In the modern home environment, you can display different colors in different fashion candles as the crowning touch in the home decoration; and some of the traditional home environment should be in accordance with the original color to match the interior color of the candles. As light yellow, gray and beige colors of these elegant, they look comfort, quiet, introverted. And the pure candle seamless. Simple style of performance should choose white or light yellow candles. This fresh and lively design is designed to meet today's popular home minimalism, deeply loved. Candles should be accompanied by simple but elegant wooden candlestick crude fat, or a simple iron candlestick to maintain the overall style, which is simple and clear.

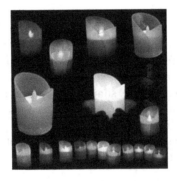

By studying the different colors on different people having emotional feelings, we reach the following summary table:

white	Spotless clean, wash away people
Red	On behalf of warmth and passion, people remain vigilant and enhance energy
yellow	Wisdom lit
green	To bring peace, tranquility and harmony
blue	To relax people
orange	Pleasant mood, inspire creativity

3.4   Candle Temperature

Among rich sense of human organs, candles give people visual and olfactory pleasure and create a warm indoor environment, sent to the warmth of real people's side. Amazing candle dissipate into the atmosphere, this small the flame of a candle. We certainly feel the warmth they bring. In the cold winter, walking into a warm room is a pleasant thing, we immediately feel relaxed and excited. Although the heat generated by the candle itself is not enough to warm the room, but when entering the room, clusters of jump the flame of a candle always make people peaceful and satisfied.

4   Create Candle of Life

In Chinese history, lighting candles has been responsible for the function; however, the romantic Chinese people will not ignore the role of the decorative candle. Traced back to ancient times, candles not only make people appreciate the lingering charm, but also offer people joy. It can be seen from ancient times. In addition to lighting, candles have more other role.

As fashion soft decorative materials, candle affects interior design in the following areas:

4.1   Set Off the Indoor Atmosphere

The atmosphere is that interior environment gives the overall impression. As the ancient Chinese wedding customs, the pairs of red candle warming festive atmosphere can render the same in different occasions with different colors, different shapes and even different flavors of candles, which can create different artistic atmosphere.

4.2   Strengthen the Indoor Environment Style

Different styles of interior designs often correspond to different types of soft decoration materials. Candles, soft furnishings, it is very similar to the lighting, the shape of the corresponding variety of diverse design styles. Candle use and development have a relatively long history, its style can be described as fusion of Chinese and foreign. In other words, whether it is Chinese antique palatial interior design style or design style in Europe and America, candles adapt to different interior design styles with different attitudes.

For example, in 2007, the works receiving the Andrew Martin Interior Design Award is Mr. Zhang Yonghe's design---a cherished idea of the modern Chinese villa projects. He built the indoor environment in the style of this point and stressed the depiction of the light. Scattered throughout the hall, the lights are being immersed in and the source of light, it is a combination consisting of 88 candle chandeliers. The candles formed by the chandelier in the space is much more than a lighting effect, it gives a way to make the body and mind intoxicated, comfortable. And this feeling of warmth and comfort is not for the man to transform the lighting.

5   Conclusion

All in all, as a soft interior decoration materials, the design of the candle reflects the unity of practical and decorative: it not only provides people with both visible light, but also gives people the art of beauty to enjoy. To stress the soft furnishings in contemporary home design is a form of giving life to enrich and beautify the world of people's lives and improve people's lives in important ways.

References

1. Le, B.: Infatuated with manual candles. Shantou university press, Shandou (2007)
2. Xie, B.: Use manual candle decorate your home. Guangxi science and technology press, Guangxi (2008)
3. Jing, F.: Lighting a root small candle. Orient press, Beijing (2010)

Research on the Countermeasures of Tianjin to Join GPA

Yu-yun Zhai[1], Zhuang Zhou[1], Yan-hua Wu[2], and Na-na Chen[3]

[1] Tianjin Polytechnic University, Tianjin, China
[2] Harbin University of Commerce, Harbin, China
[3] Hei Longjian University, Harbin, China
zhaiyuyun@163.com, zhouzhuang@tjpu.edu.cn,
{729810244,454897189}@qq.com

Abstract. China launched formal negotiations to join the GPA in the December 28, 2007; On July 19, 2010 the WTO Government Procurement Committee submitted a revised bid list, the door of China's government procurement market will be open. Tianjin is one of four municipalities in China, the economy is large in scale, rapid development, and the government procurement market potential is also large. In Joining GPA, Tianjin government procurement will face opportunities and challenges. Therefore, this research needs to adapt to the situation and development, provide reference information for Tianjin Joining GPA.

Keywords: Government Procurement Agreement (GPA), Government Procurement Volume, Tianjin Countermeasure.

1 The Meaning of Government Procurement

The "Government Procurement Law" which implemented in our country at January 2003 defined government procurement as "Government procurement refers to all levels of state organs, public institutions and organizations, the use of fiscal funds to develop the centralized procurement law Development within or procurement limits on the goods, works and services activities. "GPA of statistical standards and utility units also include state-owned enterprises, including the procurement of all non-commercial purposes is not limited by financial funds for the purchase. In practice the subject of government procurement should follow the equal competition, transparency, equity and efficiency principles.

2 The Status of Tianjin Municipal Government Procurement

2.1 With Expanding the Scale of Government Procurement, the Procurement of Goods and Engineering Procurement Aret the Body Composition

In 2007 -2009, the Tianjin government procurement shows a rising trend year by year, the Tianjin Municipal Government Procurement in 2007 totaled only 6.786 billion yuan, by 2009 which had risen to 11.645 billion yuan, the scale of procurement billion mark for the first time, three The amount of 4.859 billion yuan during the procurement, growth rate of 71.6%. In 2009, government procurement, the

J. Luo (Ed.): Soft Computing in Information Communication Technology, AISC 161, pp. 517–522.
springerlink.com

procurement of goods 4.1097 trillion yuan, accounting for 36%; engineering procurement 6.106 billion yuan, accounting for 52%, while the procurement of services is only 1.342 billion yuan, accounting for only 12%. goods and government procurement project is the subject of Tianjin.

2.2   Procurement of a Huge Amount of State Organs

In 2009 Tianjin Municipal Government procurement, the state organ procurement stocks 7.284 billion yuan, accounting for 63% of total purchase amount; institution purchases 4.034 billion yuan, accounting for 35% of total purchase amount; and community groups purchase amount is only 324 million yuan, Only 3% of total procurement value, obviously in the Tianjin government procurement state organs and institutions of the procurement account for a large share of almost 98% of all purchases, constitutes the primary unit of government procurement.

Tianjin government procurement funds and capacity mainly used for administrative politics and law, economic development, social security, corporate and educational science. Among them, Chief Political, economic and social security are a subject. For example, in 2009, Tianjin Law for the administrative aspects of procurement funds of up to 3.493 billion yuan, accounting for 30%, of all government procurement funds throughout the year in Tianjin, and the sector in the primacy; procurement funds for economic development to reach 3.115 billion million, second only to the procurement of funds for administrative Law, accounting for 27% of all government procurement funds Tianjin year of procurement funds for social security to 2.309 billion yuan, accounting for 20% of the total weight, only these three departments of government procurement funds used to account for the year in Tianjin, 77% of total procurement funds. The money goes from the point of view, the Chief Political aspects of the procurement funds are mainly used for engineering procurement, accounting for 48.3% of all purchases, accounting for nearly half of the share. Use of the funds remaining in the distribution of goods and services relative equilibrium; economic construction procurement funds are mainly used for procurement of construction works, social security funds and the first two different, its mainly used for procurement of goods. In addition to these three departments, 2009, the Tianjin government procurement for the cultural and educational aspects of science and enterprise procurement funds accounted for 9% and 12%, while for the agricultural finance and other only 2%percentage of procurement funds,. Visiblly with government procurement funds were distributed among the various departments rather uneven. Distribution in the future should be emphasis on the allocation of funds to make use of the funds use more efficient. Overall trend from the point of view, the expenditure for economic construction increased significantly, which is conducive to economic development in Tianjin.

2.3   The Construction of the Tianjin Government Procurement System

In recent years, according to the requirements of the Party Central Committee and State Council, Tianjin continues to strengthen the government procurement system, the construction of the regulating government procurement system referred to the agenda.

First, the detailed examination of the way government procurement of imported products. Research by the member units of Tianjin to join the GPA, Economic and Information Technology Committee and the Municipal Health Bureau, imports were responsible for government procurement of information products, electronic products and medical devices to demonstration, the views put forward industry department, the municipal finance department review approval. Strict audit system for imported products, expanding the number of purchasing domestic products. domestic products improve the procurement budget conscious ideas. Secondly, to support independent innovation, energy saving and promotion of SME development, Tianjin studied and formulated the "Tianjin Municipal Government on the implementation of procurement advice for SME development", from the policy and operational levels will implement the government procurement policy function, making the legal basis for opening the GPA list exceptions.

3   The Influence of Jioning GPA of Tianjin

3.1   The Positive Influence of Joining GPA of Tianjin e

Tianjin as the future international port city、the northern economic center and ecological city, joining GPA is according with Tianjin's economic development strategy. Promoting the internationalization of Tianjin's government procurement, the economic transformation and update, stimulating economic growth, expanding the open policy, making the Tianjin's economy deeper degree integration into the world economy.

Jioning GPA can conducive to regulate the government procurement market. To our city enterprises purchase to the world of high quality cheap products and services. Improving the utilization efficiency of financial funds. It can be saved from other need money to fund the project, such as the state-owned enterprise modification, import advanced equipment, etc.

Tianjin enterprise can obtain more trading opportunities, Join GPA, enterprise can enter GPA contracting open government purchasing market, development of huge international government purchasing market for the enterprise, also export to create more opportunities. Our citi's textiles in world trade competition plays certain advantages, join GPA after can expand the market, to make more profit opportunities.

To promote efficiency, enhance the enterprise the competitive ability.GPA provisions to public bidding mode to identify suppliers, to the enterprise can form a certain amount of stress, prompting them to organize production to the market, and constantly improving technology and strengthening management, improve the production efficiency and product quality, and improve overall domestic enterprises in foreign market competitiveness and resist the ability of the risk, be helpful for them as soon as possible into foreign government purchasing market in the world, the economic competition share.

3.2   Adverse Effect

Add to our country economy's biggest GPA of relevant influence is the impact of the disadvantaged industry, such as automobile industries will be bigger impact. GPA banned counter measures, this to the previous executes local content, technology transfer, production requirements and other provisions of cooperation with related enterprises to participate in bidding industry is very adverse.

Our country government purchasing market influence in the foreign trade balance. Developing countries due to lack of competitiveness of enterprise of difficulty in developed countries government purchasing market in the bid-winning, and its advantage of labor-intensive products and services government purchasing amount relatively few again. Foreign goods and services will depend on its good marketing network and service in China market, while occupied domestic enterprises limited ability to access to foreign markets, from the overall will affect Chinese in government purchasing market trade balance.

Previous depend on the government orders of survival some enterprise (as some software enterprises), is likely to reduce or lose order, increase the number of people unemployed risk, increased economic instability.

4   Countermeasures

4.1   Fiercer Competition for International Talents

In order to attract more developing countries to join, GPA formulated special treatment for many developing countries. For instance, GPA allow developing countries and the negotiators discuss the entity、 product or sernices that do not belong to national treatment principle; developing countries can modify the scope of it according to their needs; developed countries should purchase product and services which related to developing countries export interests, and should provide technical support for developing countries. The latest edition of GPA clause expanded the transition measures for developing countries.

China joined WTO as a developing country; we have to emphasize this position in the negotiations in the GPA, take advantage of the agreement, "special and differential treatment for developing countries", and full enjoy of the rights of a developing member. In the negotiations, these principles should be specific. For example, to promote employment and encourage the development of small and middle enterprises, given the domestic small and middle enterprises suppliers to procurement opportunities, or a special price concessions; provision of goods, works or services in the domestic component; ask other parties for Chinese suppliers about the parties the training of the government procurement system; requires Parties to establish the government procurem.

4.2   Moderate Gradual Opening Up Government Procurement Markett

China's accession to GPA to bear certain obligations, but these obligations can be adjusted through negotiations, such as the only country included in Annex I of the Government Procurement Government entities undertake the procurement before the appropriate areas of open markets; for sub-central government procurement, Member in strict accordance with the principle of reciprocity can be determined in consultation with each other open areas; not implement the MFN principle, the corresponding open field depends only on the bilateral negotiations between the two governments, own have the choice of the countries decided to open areas of comparative advantage. This makes formula for the opening of the implementation of strategies possible.

4.3   Government Should Understand 《Government Procurement Agreement》

At present, although china has not join GPA but as an observer of GPA. In order to fully understand and grasp the implementation of GPA, and other government procurement law and relevant information. China should actively involved in government procurement work. The centre government should search information for the local government, it can reduce the costs and improve information on the accuracy of the information to enable enterprises to seize the opportunity, and common international practices operate.

4.4   Enterprises Should Be Mindly Armed

The items government purchase became widely, from daily office appliance to public project and basic construction, as soon as the company entry the government procurement market, they maybe have enormous development space. And he government procurement market very stable, select and purchase often follow some regulations. Although save cost is very important in government purchase, if some project can contribute very little profit. But the revenue is stable, so it can reduce the business risk. In addition, the government procurement market have authority and enjoy the reputation of the product, so it become easier go in to consumer market, it can reduce propagation cost. Therefore, enterprises should pay attention not only to local and domestic market, but should have the conscious of possession of foreign markets.

4.5   All Enterprises Should Be in Accordance With International Standards in Government Procurement

At present central entity have so many problems in government procurement market competition. For example lack knowledge of how to access to international government procurement market, as a result they often suffer setback in government bid; lack of basic skills which must use in quoted price, put down prices only, even low than cost price, make tenderee have a suspicion the quality; Some enterprises lack the knowledge of biding so that the tender documents do not conform requirements, and so on. All these will affect the branch of the central government in government procurement market competition. After join GPA, the branch of the central government will suffer restrictions in international government procurement market; even they do not opening to the outside world, they must standardize operations when face of competitors from home and abroad, and learn to use GPA's relevant clause to protect their rights and interests.

References

1. Anonymous: E-Procurement goes mainstream. International Journal of Productivity and Performance Management 53(168) (2004)
2. Smith, S.A.: Government procurement in the WTO. Kluner law international, The Hague (2003)
3. Page, H.R. Public Purchasing and marerial Management Mass D.C.Heath & Compamy, 39 (1998)
4. WTO 《Agreement on Government Procurement》
5. Emanuelli, P.: Government Procurement. Lexis Nexis Canada (2005)

Study on Topography Differentiation Characteristics of Spatial Distribution and Dynamic Change of Bamboo Forest

Fengying Guan[1,2], Shaohui Fan[1,2], Jian Liu[3], and Lijuan Miao[3]

[1] International Centre for Bamboo and Rattan
[2] Key Laboratory of Bamboo and Rattan, Beijing 100102, China
{Guanfy,fansh}@icbr.ac.cn
[3] Forestry College of Fujian Agriculture and Forestry University, Fuzhou 350002, China

Abstract. In order to grasp the spatial and dynamic changes of mountain bamboo resources, this study uses the 2007 forestry investigation sub-compartment diagram of Shuncang, Fujian province, two-period TM images, DEM and other data in 1998 and 2007 to extract information of bamboo forest through Remote sensing and Geographic Information technology, create spatial distribution map of bamboo forest and make overlay analysis of it with topographic factors grade map, so as to research the topography differentiation characteristics of spatial distribution and dynamic change of bamboo forest. The results show that the bamboo forest in Shuncang is mainly distributed in the semi-sunny slope and semi-cloudy slope with where are at an elevation of 1200m below with a slope of 25 degrees below, the mountain bamboo forest at the elevation of 1200m below accounts for 99.89% of the total bamboo forest area, the proportion of bamboo forest with a slope of 25 degrees below is90.39%, while the proportion of bamboo forest with a slope of 35 above is less than 3%; the proportion of bamboo forest in semi-sunny slope and semi-cloudy slope is 56.98%; for almost 20 years, the bamboo forest resources in this county are improving rapidly, increase, decrease and unchange, this three kinds of variation types trend to increase firstly and then decrease with the changes of slope direction and aspect, while with the increase of elevation, the types of increase and unchange trend to increase firstly and then decrease, the decrease type is decreasing gradually.

Keywords: Bamboo forest, Remot sensing, Geographic Information, dynamic variation, topographic impact.

1   Introduction

China is one of the distribution center of bamboos in the world, according to the results of seventh National Forest Inventory, the area of bamboo forest is 5.38 million hm2 which accounts for 25% of the total bamboo forest in the world. Our bamboo resources are mainly distributed in the hilly and mountainous areas[1] to the south of Qinling Mountains, Huai River. Fujian is one of the distribution center of bamboos in China with abundant bamboo resources, in recent years, with the activity of bamboo resources market, bamboo resources industry is supported and favored by foresters, the economic benefit of bamboo forest has been improved greatly[2-5]. But because

J. Luo (Ed.): Soft Computing in Information Communication Technology, AISC 161, pp. 523–530.
springerlink.com   © Springer-Verlag Berlin Heidelberg 2012

many places pursue area and scale one-sidedly, emphasize economic benefit, they develop the bamboo pure forest in large areas with high-strength reclamation during the process of development and operation, which aggravates the soil erosion and soil fertility degradation of forest land, causing quality reduction of the ecological benefit of bamboo resources, also has some impacts[6-7] on the stability of regional forest ecosystems at the same time. Therefore, finding out the topography differentiation characteristics of bamboo resources has an important role in developing sustainable development strategy and ecological environment construction for regional bamboo resources. This stufy takes Shuncang county in Fujian province as research area, using ground survey data and two-period TM images for dynamic monitoring of bamboo forest, so as to reveal the topography differentiation characteristics of spatial distribution of bamboo forest.

2   Study Area and Data Selection

Shunchang County is a part of Nanping City, located at the northwest Fujian Province. It is an important forestry county and bamboo base county in northern Fujian and one of the nationally renowned "home of bamboo" with rich bamboo resources and large area of bamboo forest. The geological coordinates of this county are E 117°30′∼118°14′ and N 26°39′∼27°12′. It has subtropical marine monsoon climate, with abundant rainfall, moderate temperature, an average annual precipitation of 1 567.8mm and an average temperature of 18°C above; this county has a typical mountainous and hilly topography, with fertile soil and excellent site conditions that are suitable for the growth of bamboo.

3   Research Methods

3.1   Data Selection Variation

The data adopted in this paper include: 2007 forestry investigation sub-compartment basic map of Shuncang, two-period TM images in 10998 and 2007 with a spatial resolution of 30m; 1:250000 DEM data, 1:10000 topographic map, administrative boundary map and field GPS object coordinate.

3.2   Grading of Topographic Factors

The main attributes that measure the topography differentiation in mountain ecological research are elevation, slope direction and slope aspect, these three attributes also are the dominant factors [8-9] in determining the differentiation of vegetation habitats, such as soil, microclimate, hydrology and other elements. Based on the criteria for national forest inventory investigation, then combine with the actual situation in research area to grade topographic factors (Table 1), extract thematic maps of elevation, slope direction and slope aspect with the use of DEM data.

Table 1. Criteria for the classification of topographic factors

Elevation (meter)	Low :0~400	Lower :400~800	Medium :800~1200	High:1200 above	
Slope direction(degree)	Level slope 0~5	Gentle slope 6~15	Tilted slope 16~25	Heavy slope 26~35	Steep slope:(≥36)
Slope aspect	Sunny slope 135° ~225°	Semi-sunny slop 90° ~135° ; 225° ~270°	Semi-cloudy slope 45° ~90° ;270° ~315°	Cloudy slope 315° ~360° ; 0° ~45°	

3.3 Information Extraction of Bamboo Forest's Distribution

Take the 2007 forestry investigation sub-compartment map as basic layer, carry out superposition, statistical analysis and other spatial operations with elevation, slope direction, slope aspect and other topographic factors thematic maps respectively, getting the spatial distribution (Table 2) of bamboo resources in this county.

Table 2. Bamboo forest area statistics in different topographic factors

	Grade	Low	Lower	Medium	High	
Elevation	Area of bamboo（hm2)	11282.15	28688.33	3820.62	49.9	
	Area ratio of total bamboo area	25.73%	65.44%	8.71%	0.11%	
	Area rate of grade area	10.04%	36.76%	45.65%	18.69%	
	Grade	Level slope	Gentle slope	Tilted slope	Heavy slope	Steep slope
Slope direction	Area of bamboo（hm²）	3167.5	18113.69	18345.5	2944.67	1269.64
	Area ratio of total bamboo area	7.22%	41.32%	41.85%	6.72%	2.90%
	Area rate of grade area	14.63%	21.09%	29.49%	31.67%	17.09%
	Grade	Sunny slope	Semi-sunny slope	Semi-cloudy slope	Cloudy slope	
Slope aspect	Area of bamboo（hm²）	8848.61	12570.82	12411.32	10010.24	
	Area ratio of total bamboo area	20.18%	28.67%	28.31%	22.83%	
	Area rate of grade area	20.81%	24.66%	26.62%	27.44%	

3.4 Interpretation of TM Image

With support from ENVI4.5 software, use the two-period MT images and ground survey data processed with geometry correction, registration, tessellation and cutting to select sixty training samples of bamboo forest and non- bamboo forest each, adopt the QUEST algorithm of RULEGEN module of ENVI for computational analysis, realizing the classification of the remote image and extract bamboo forest information[10-11], the classification accuracy is 83% above which can satisfy the applications needs of dynamic monitoring. A spatial distribution map of Bamboo's changing (Fig1) can be got through overlaying analysis of the two-period remote sensing image thematic maps after classification.

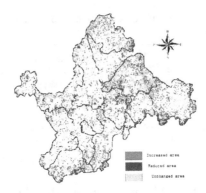

Fig. 1. Spatial change distribution map of bamboo forest in 1988-2007

3.5   Information Extraction of Bamboo Forest's Dynamic Changes

Divide the bamboo forest's dynamic changes into three types: increased bamboo forest, decreased bamboo forest and unchanged bamboo forest, use ENVI software to extract the area and type [11] of changes according to the principle of formula (1).

$$N(i, j) = N(i) \times 10 + N(j) \tag{1}$$

There: $N(i, j)$ — i, j the spatial distribution of Bamboo's dynamic change; $N(i)$ — Bamboo's distribution in year of i ; $N(j)$ —Bamboo's distribution in year of j .

4   Results and Analysis

4.1   Spatial Distribution Pattern of Bamboo Forest

1) Vertical Distribution Characteristics of Bamboo Forest
There are different climatic conditions at different altitude, while different forest vegetation zone simultaneously, there is difference [12] on distribution of one vegetation along altitude. The law that bamboo's distribution area changes (Table 2) following altitude is: lower>low>medium>high. In an area at lower altitude, the distribution area is the largest which accounts for about 0.11% of the total bamboo forest area; the bamboo forest in an area at the altitude of 1200m below accounts for 65.47% of the total area, the proportion of bamboo forest trends to increase first and then decrease along with the increasing of altitude.

2) Slope Direction Differentiation Characteristics of Bamboo's Distribution
The study in Table 2 shows that the distribution areas of bamboo on gentle slope and tilt slope are the largest, they are 18113.69hm^2 and 18345.5hm^2 respectively, they

account for about 83.16% of the total bamboo area. On steep slope, the distribution area of bamboo is the smallest, it is only1269.64hm^2 which accounts for about 3% below of the total area, this shows that bamboo forest is mainly distributed in the gently sloping area, but in terms of the proportion of bamboo area to the land area in this slope, the percentage of steep slope is the largest (31.67%) while the level slope is the smallest (14.63), when the slope is 35 degree below, the percentage will rise gradually with the increasing of slope.

3) Slope Aspect Differentiation Characteristics of Bamboo's Distribution

Slope aspect mainly influences the sunshine received by vegetation and prevailing wind direction, so as to cause significant regional hydrothermal differences further. Bamboo is evergreen shallow rooted plant, it has a high requirements to hydrothermal condition and is hypersensitive, the hydrothermal distribution on earth dominates the geographic distribution of bamboo. Table 2 shows that there is difference of bamboo's distribution on different slope aspect in research area, 56.98% of the bamboo forest are distribute in semi-sunny slope and semi-cloudy slope; the proportions of bamboo forest in sunny slope and cloudy slope are equivalent.

4.2   Dynamic Change Characteristics of Bamboo

With the use of Figue1 and formula (1), it is analysed that the areas of increased, decreased and unchanged of bamboo, and the annual change rate is 3.08%. In order to research the spatiotemporal dynamic change rule of bamboo forest from different angle, overlay Figure 1 with thematic map of topographic factors grade to get the topography differentiation characteristics of bamboo's dynamic change.

1) Vertical Dynamic Changes of Bamboo

It can be seen from Figure 2 that in the three dynamic change types, the distribution areas of increased and unchanged types trends to increase first and then decrease with the increasing of elevation, while the decreased type trends to decrease progressively. The reason analysed is that with 20 years of operation, adjustments of national policy and the development of market economy, the land use condition changes in different degree, the change of decreased type mainly is human interference, with the rise of elevation, human influence will reduce relatively and the climatic conditions are relatively severe also (the solid at high altitude lots is poor), so the changing amplitudes of these three types at high altitude lots are all small. In terms of the increased area, it mainly concentrates on low elevation region and there is few distribution in high elevation region, this conclusion explains the vertical distribution laws of bamboo forest mentioned before. The changes of unchanged type are concerned with the proportion of bamboo forest to the area corresponding slots, the greater the proportion, the larger the unchanged area is, and smaller on the contrary., from the comparison Table 2 we can see that the effect of bamboo's vertical distribution on unchanged type is in accordance with the above-mentioned laws.

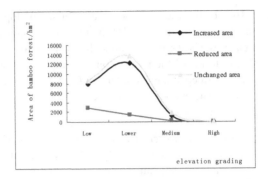

Fig. 2. Distribution variation diagram of Bamboo's changing types following altitude

2) Slope Direction Differentiation of Bamboo's Change

It can be seen from Figure 3 that the curves of three dynamic changes trend to increase first and then decrease. There is large difference in the distribution area of bamboo's change types in every slope direction, the change within the 6-15°of slope is largest; within 0-15°, the area will increase gradually with the increase of slope; within 16-35°, the change decreases gradually; while when the slope direction is 35°above, the changing amplitude is close to zero. This trend accords with biological characteristics, the solid in steep slope slots is poor, so the changing amplitude is small, combined with the forestry statistical yearbook, it is found that there is no insect pest and other natural disasters during study period, therefore, the differentiation characteristics of this change along slope are formed by relative regional human impacts in great slope. In addition, the changing amplitude of slope distribution characteristics for Bamboo's change type is also impacted by the distribution area of bamboo forest in the whole region, the area proportions of the three change types in gentle and tilted slope slots that have great area proportions are all great, this reflects the laws of large base.

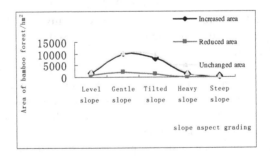

Fig. 3. Distribution variation diagram of Bamboo's changing types following slope

3) Slope Aspect Differentiation of Bamboo's Change

Figure 4 shows that the three dynamic changes of bamboo forest in different slope aspect trend to increase first and then decrease, there is difference in changing amplitudes, the changing amplitudes of unchanged type and increased type are large, while the changing amplitude of decreased type is the smallest, the curve almost is linear.The area-scale of increased type in semi-cloudy slope is the largest, it accounts for 41.63% of the area of this type, next is semi-sunny slope, the area-scale is 30.43%, there is little difference in the proportions of distribution area in sunny slope and cloudy slope. The humidity in semi-cloudy slope is superior to that in semi-sunny slope, this also shows that moisture condition in this region is one of the limiting factors to influence the distribution of bamboo forest, while the low humidity and weak sunlight in sunny slope both affect the distribution of bamboo forest, so both of them have smaller area proportions. For unchanged type, the area proportion of semi-sunny slope is the largest (34.73%), next is semi-cloudy slope (25.95%), compare the slope aspect differentiation characteristics of bamboo above, we can know that changing amplitude is related to the area proportion of bamboo forest in each slope aspect in the whole region.It can be seen from figure that the areas of increased type and unchanged type in every slope aspect all larger than the area of decreased type, this is relative to regional development and natural factors, from the development situation of bamboo industry in Shunchang county, local governments began to place more importance on bamboo industry since 1996, and increased support to promote the rapid growth of bamboo areas.

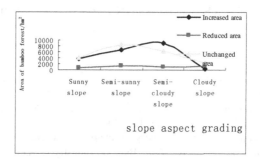

Fig. 4. Distribution variation diagram of Bamboo's changing types following aspect

5   Conclusion

The study on topography differentiation characteristics of bamboo distribution and dynamic change shows that: (1) the bamboo forest in Shunchang county is mainly distributed in the semi-sunny slope and semi-cloudy slope with a altitude of 1200 below and slope under 25 degree. The mountain bamboo forest at the altitude of 1200 below accounts for 99.89% of the total bamboo area, the distribution proportion of regional bamboo forest with the slope under 25 degree is 90.39%, while in the mountain with a slope of 35 degree above, the distribution proportion of bamboo forest is less than 3%; the proportion of bamboo forest in semi-sunny slope and semi-cloudy slope is 56.98%. (2) For about 20 years, the dynamic changes of increased bamboo, decreased bamboo and unchanged bamboo trend to increase first and then decrease along the changes of slope direction and aspect, while with the increase of elevation, the increased type and unchanged type trend to increase first and then decrease, and the decreased type is decreasing gradually; increased bamboo forest is mainly distributed in the mountains of semi-sunny slope and semi-cloudy slope at the altitude of 400-800 meters with a slope under 25 degree, the change law of unchanged bamboo type is the same as increased type's; this indicates that bamboo forest has good adaptability to these topographic conditions; the decrease of bamboo forest in low elevation and gently sloping mountains is the most, and there is no significant difference in aspects. This kind of change law reflects the effects of natural factors and human driving force on the distribution of bamboo forest.

The spatial distribution of bamboo forest is mainly reflected in geography which is the comprehensive effect of bamboo's biological characteristics and site condition, it reflects the adaptation of bamboo species to site condition and is the significant basis for bamboo forest tending guidance, therefore, it is significant to study bamboo and the topography differentiation characteristics of it's dynamic change. For dynamic change, this paper does not research what kind of landscape types have replaced the reduced bamboo and which kind of landscape types have been converted to the added part, this will be discussed further later.

Acknowledgements. Subsidization from State Forestry Bureau 948 subject "Introduction of Dynamic Remote Sensing Monitor Technology for Bamboo Resources" and the Special Funds for Basic Scientific Research Business on International Center of Bamboo and Rattan (No.:1632010012、06/07-A06).

References

[1]   Jang, Z.: Bamboo and Rattan in the World. Liaoning Science and Technology Press, Liaoning (2002)

[2]   Chen, X.: Development Status on Bamboo Industry in Nanping Prefecture, Fujian Province. World Bamboo and Rattan 4(2), 30–32 (2006)

[3]   Zhang, H.: Fujian Bamboo Industry, Embark on A New Journey—Review and Perspective of "The Tenth Five-Year" Bamboo Industry Development in Fujian. China Forestry Industry, 36–37 (2007)

[4]   Chen, L.-X.: The Advice on Quicken The Development of Bamboo Industry More in Fujian Province. Wuyi Science Journal 22, 284–287 (2006)

[5]   Chen, C., Tang, X.: Bases and Prospects of Bamboo Development in Fujian. Journal of Bamboo Research 21(3), 13–20 (2002)

[6]   Li, Z., Mao, Y., Xie, Z.: Study on the Productivity Maintenance of Bamboo and Broadleave Tree Mixed Stands. Journal of Bamboo Research 22(1), 32–37 (2003)

[7]   Xie, J.: Establish Ecological Bamboo Forest to Improve Comprehensive Benefit. China Forestry Science and Technology 17(suppl.), 61–63 (2003)

[8]   Ostendorf, B., Reynolds, J.F.: A model of arctic tundra vegetation derived from topographic gradients. Landscape Ecology 13(3), 187–201 (1998)

[9]   Pearson, S.M., Turner, M.G., Drake, J.B.: Landscape Change and Habitat Availability in the Southern Appalachian High-lands and Olympic Peninsula. Ecological Applications 9(4), 1288–1304 (1999)

[10] Yu, K., Liu, J., Xu, Z.: Study on Bamboo Resources Thematic Information Extraction in the South of China. Remote Sensing Technology and Application 24(4), 449–455 (2009)

[11] Li, G.: Study on Remotely Sensed Information Extraction of Chief Species in Southern Mountain Forest. Fujian Agriculture and Forestry University (2009)

The Study on Management and Development of the Value Chain Construction of State-Owned Forest Resources Operational Organization

Jie Wang[1,2], Yude Geng[1,*], and Kai Pan[2,*]

[1] College of Economics and Management, Northeast Forestry University
[2] Northeast Agricultural University Harbin, China
wangjie-0825@163.com

Abstract. With China's forestry reform, state of forest resources management theory and practice are faced with a historic turning point, as the core of economic growth, promote the optimized allocation of state-owned forest resources, improve operational efficiency and optimize the organization's management structure will be the focus of forest resource management development. By building the value chain of State-owned forest management organization, the value of an enterprise's major business would be identified , prompting the state-owned forest resources dynamic organization, coordination and management, improving operational organizations in the forestry industry competitiveness, and promoting a rapid, sustainable forestry economy development.

Keywords: State-owned forest resources, operational organization, value chain.

1 Introduction

The theory of Value Chain in forestry production and management of enterprises received wide attention in practice and play an important role, companies in related industries have a competitive advantage increasingly depends on the value of the value chain activities for various degree of integration and coordination. For a long time, people pay more attention on the feasibility of classification of management theory instead of after-classified management of small business, with the deepening reform of forest ownership and relocation of people in forestry, forest resource management organizations optimize the value of activities and maintain a sustainable competitive advantage, and the organization of forest resources management performance evaluation will be put on the agenda, this article from a management point of view on China's state-owned forest resources research organizations, in order to build internal and external value chain organization, prompt the state-owned forest resources dynamic business organization, coordination and management, improve business organizations in the forestry industry's competitive advantages, and promote sustainable forestry rapid economic development.

---

* Corresponding author.

2 The Meaning of State-Owned Forest Resources

The concept of forest resource management is divided into narrow and broad. Narrow forest resource management mainly refers to the forest resources and forest resource management. From the management of forest resources management targets, the broad one includes not only forest resources and forest resources, but also covers the forest animal and plant resources, tourism resources; from the management of the business perspective, it includes not only forest resources planning and management of data and survey design, but also the use of forest resources management activities, such as decision-making, organization, coordination and oversight. This article is from the management perspective, positioning the state-owned forest resources management in the broader context, especially how to create a commercial operation of resources, better integrate into the market economy, the increase in efficiency in the use of forest resources, maximize their economic, social, and ecological benefits, a reasonable allocation of resources and use.

For a Long period of time, because the burden of state-owned forest management organizations, ecological, economic and social benefits, operating efficiency is low, there has been state-owned forestry enterprises and workers doom and gloom. With the repositioning of Forestry in China, state-owned forest resources management theory and practice are faced with a historic turning point, as the core of economic growth, promote the optimized allocation of state-owned forest resources, improve operational efficiency and optimize the organization's management structure will be the develop focus of the state-owned forest resource management.

On the basis of theory and practice in the state-owned forest resources classification management, according to the differences between supply types and business process provided by production management, the state-owned forest resources management organization is divided into state-owned zoology forest management organization and state-owned effective operation organizations. State-owned zoology forest management organization is engaged in the cultivation and ecological public welfare forest resources management. They are the enterprises relying on forest resources of forest tourism and various animal of cultivation, breeding business enterprises. State-owned effective operation organization is an economical one engaged in the business of cultivating and forest products, including lumber cultivating industry, wood, wood processing industry, lumbering industry man-made board, furniture, such as pulp manufacturing, and various Lin by-product acquisition processing, chemical products processing, etc

3 The Value Chain Theory and Its Application in the State-Owned Forest Resources Organizations

Harvard Business School Professor Michael Porter puts forward the concept of the value chain in his book

Competitive Advantage, in which states that every business is the collection activities covering the design, production, marketing, purchasing and other items, all of which can be represented by the value chain. As a research tool the value chain have been widely used in many businesses, industries, with increasing emphasis on business management and business strategy to enhance competitive advantage, value chain theory in production management obtains more attention in practice and play an important role. It is necessary to adapt to the changes in forestry enterprises by connecting with the value chain thinking and forest management, Only the each step of value chain is harmonious and unified, thus can guarantee the value activities optimization of forest resources management organization, create and maintain competitive advantage.

Value chain theory which describes the value chain, but as an industry organization, the enterprise value chain is a part of the value chain, so companies should not only be the main value of the identification and analysis of the business, but also recognize the companies industry value chain and competitive position, identify competitors, competitive strategy and the strategy to enhance the organization and advantages of the business shape is very important. The external value chain of state-owned forest resources is a namely industry value chain, that is shown in Figure 1, the internal value chain shown in Figure 2.

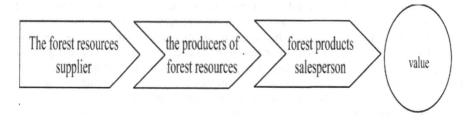

Fig. 1. The production value chain of state-owned forest resources

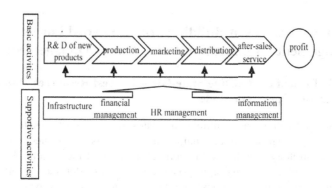

Fig. 2. Value chain of state-owned forestry resources organizations

State-owned forest resources management organizations in the industry value chain, mainly are the resources suppliers, That is to say, with the forestry classified management reform, the property rights and management right of forestry resources is separated, However, its extended value chain might develop further. State-owned ecological forests of business organization can provide ecological public welfare forest social and ecological service, also can become forest resources of producers and sellers of forest products, through utilizing the resources superiority development forests of plants and animals that planting and breeding and the development of ecological tourism economy. The state-owned commercial forest operation can get rid of the shackles of the past and concentrate on engaging in the cultivation and management of commercial forests, due to the special nature of forest resources, it can go directly to the resources of production and marketing chain, to promote forest industries market-oriented role to the greatest extent.

Construction of the value chain of value activities need to be identified, assess the additional value of each step to assess the added value and competitive advantage of linkages. Constitute the main activities of enterprise value, including basic activities and support activities into two categories, shown in Figure 2. Basic activity of which is the formation of the value of research and development and new product design, raw material procurement, production, sales, service, etc., in which production processes can be further divided according to the value of process activities, such as chipping, polishing, drying, sanding, sizing, pressing, trimming, and inspection activities. Support activities to achieve the protection of values, including: infrastructure building, financial management, human resources management and information management. Many activities in the enterprise, not every aspect of the activities to create value, the value created by business, from some of the key activities that may constitute a core competence, competitive advantage for enterprises. Therefore, different key aspects of business or organization is different, the state-owned forest resources for the organization, management of state forests and state-owned commercial forestry organizations, organized its business goals, value chains is a big difference in construction and composition, the value should be targeted to identify activities and key activities, resources and capacity to identify what constitutes organizational core competencies.

4   Countermeasures of Building Stated-Owned Forest Management Organization Value Chain Management

4.1   The Reform of the Existing State-Owned Forest Resources Management System and Operation Mode

This is the protection of building a state-owned forest resources operation management organizations value chain, but also it is the protection of building organizational strategy to enhance the competitiveness of the system. The traditional production process of forestry enterprises did not consider factors other than the impact on enterprise competitiveness, and no exposure to the enterprise value chain in the forestry industry, forestry enterprises truly understand the competitive advantage and competitive position. On the basis of classification management in forestry, the practice of the forestry reform, and understand the development of forestry industry and business problems, from a strategic perspective to improve the vitality and competitiveness of the organization, the reform of state-owned forest resources and a clear management structure and operation mode. Different business objectives to implement the state-owned forest resources in different economic policies, the implementation of state-owned forests resources, although the cause of the management system, but to change the role of government at all levels to manage the market economy as the main forest, improve management efficiency; state-owned commercial forest resource management rely mainly on market mechanism to regulate, give full play to both the efficiency of resource allocation.

4.2   Clearly State the Overall Goal of Forest Management Organizations, and the Sub-Objectives of the Value Chain

Setting the overall goal of enterprise needs to comply with the progress of business and social development, on the basis of that, set the enterprise value chain on all aspects of the target. State Forest Resource Management organization is to achieve the overall objectives of eco-efficiency as the main objective of the value chain of sub-goals which must be coordinated with the overall goals and objectives of the various sub-sectors have to be coordinated synergy to achieve organizational goals. State-owned commercial forest resource management organizations, the overall goal is to improve organizational competitive advantage, then the value chain will build the core competitiveness of outstanding short-term goals and long-term goals should also maintain the same orientation. Only the overall objectives and targets, short-term goals and long-term goal is the same value chain to make the advantages of fully reflected.

4.3   Strengthening the Value Chain Collaboration and Dynamic Adjustment of the Division of Labor

Construction of state-owned forest resources management organization to the original value chain will be restructuring the business process re-examine the organizational value structure, analyze and determine which business or part of the business is critical, and comparing the competitors what competitive advantages it have, which can be organization's core competencies to form the source, from the perspective of maximizing the value of re-integration of business processes to achieve efficiency and value of the entire process optimization. The interface between the various aspects of poor, to strengthen control, collaboration and communication, improve the management of the overall function and improve organizational performance. Organizations such as the state-owned forest resources to achieve a dynamic process management, business processes and management processes of unified planning, design and coordination, and always maintain the value chain of dynamic adaptability.

4.4   To Use the Technology to Ensure the Continuing Escalation of Value Chain Management and Organizational Performance Improvement

Management of state-owned forest resources should be market oriented, to achieve ecological, social and economic benefits, but also to ensure the effectiveness of resource supply. Relying on technological progress to increase the technological content and added value products as the core, through the full utilization of forest resources, so that businesses or organizations can achieve industrialization, to explore the potential use of forest resources and the appreciation of space, the optimal allocation of limited resources and maximum value, then the state-owned forest resources can improve organizational performance, achieve organizational business objectives and achieve the organization's strategic and sustainable development.

4.5   To Strengthen Information Management of Internal and External Links of Value Chain

Information feedback shall be made during the process of building and managing the value chain in order to link the goals and objectives of the adjustment and correction. Information management is critical for organizations to gain competitive advantage from product design, production, pricing, competitor analysis, business goals and competitive position, all aspects need organize their own markets, industries, information on various types of data and intelligence have a clear understanding and knowledge, the data platform for the establishment of state forest resource management organizations to effectively manage internal and external value chain, more rational, and some organizations have begun to use information technology data management throughout the organization, managers can share information and make timely decisions, not only shorten the decision time also increased organizational efficiency, and play an important role in optimizing the value chain.

References

1. He-tengfa, He-tengfa: Forest Resources Management. China Forestry Publishing House, Beijing (2007)
2. Wan, Z.-F., Zhang, Q.: Based on Tenure Reform of State-owned forest Resources Management. Forest Economic Problems (June 2009)
3. Geng-Yude, Wan, Z.-F.: Heilongjiang Province State Forest Region and Optimization of Industrial Structure Adjustment. Forestry Science (June 2006)
4. Lee building: Based on Enterprises value-chain achievements inspection system research. Changchun University of Technology (November 2007)

Author Index